Vitamin E

Chemistry and Nutritional Benefits

Food Chemistry, Function and Analysis

Series editors:
Gary Williamson, *University of Leeds, UK*
Alejandro G. Marangoni, *University of Guelph, Canada*
Juliet A. Gerrard, *University of Auckland, New Zealand*

Titles in the series:

1: Food Biosensors
2: Sensing Techniques for Food Safety and Quality Control
3: Edible Oil Structuring: Concepts, Methods and Applications
4: Food Irradiation Technologies: Concepts, Applications and Outcomes
5: Non-extractable Polyphenols and Carotenoids: Importance in Human Nutrition and Health
6: Cereal Grain-based Functional Foods: Carbohydrate and Phytochemical Components
7: Steviol Glycosides: Cultivation, Processing, Analysis and Applications in Food
8: Legumes: Nutritional Quality, Processing and Potential Health Benefits
9: Tomato Chemistry, Industrial Processing and Product Development
10: Food Contact Materials Analysis: Mass Spectrometry Techniques
11: Vitamin E: Chemistry and Nutritional Benefits

How to obtain future titles on publication:
A standing order plan is available for this series. A standing order will bring delivery of each new volume immediately on publication.

For further information please contact:
Book Sales Department, Royal Society of Chemistry, Thomas Graham House, Science Park, Milton Road, Cambridge, CB4 0WF, UK
Telephone: +44 (0)1223 420066, Fax: +44 (0)1223 420247,
Email: booksales@rsc.org
Visit our website at www.rsc.org/books

Vitamin E
Chemistry and Nutritional Benefits

Edited by

Etsuo Niki
University of Tokyo, Japan
Email: etsuo_niki@yahoo.co.jp

ROYAL SOCIETY OF CHEMISTRY

Food Chemistry, Function and Analysis No. 11

Print ISBN: 978-1-78801-240-9
PDF ISBN: 978-1-78801-621-6
EPUB ISBN: 978-1-78801-733-6
Print ISSN: 2398-0656
Electronic ISSN: 2398-0664

A catalogue record for this book is available from the British Library

The Royal Society of Chemistry is a charity, registered in England and Wales, Number 207890, and a company incorporated in England by Royal Charter (Registered No. RC000524), registered office: Burlington House, Piccadilly, London W1J 0BA, UK, Telephone: +44 (0) 20 7437 8656.

Visit our website at www.rsc.org/books

Printed in the United Kingdom by CPI Group (UK) Ltd, Croydon, CR0 4YY, UK

Preface

Vitamin E was discovered in 1922 by Evans and Bishop as an essential micronutrient for reproduction in rats. Rats fed a diet low in vitamin E had reduced fertility and a high rate of fetal resorption, but supplementation with a lipophilic fraction of lettuce or wheat germ oil resulted in a successful implantation of the embryo, implying an existence of a hitherto unrecognized dietary factor essential for reproduction. The active substance was isolated in 1936 by Evans and was given the name "tocopherol" meaning "to carry a pregnancy" from the Greek words "tokos, birth" and "pherein, to carry". The ending "-ol" comes from its status as a chemical alcohol. It was named vitamin E according to a consecutive alphabetical order preceded by the discovery of vitamins A–D.

Vitamin E is the generic name of all tocopherols and tocotrienols. Vitamin E encompasses a group of eight isomeric molecules, which are characterized by a 6-hydroxychroman ring structure and a side chain. The tocopherols possess a 4′,8′,12′-trimethyltridecyl phytyl side chain, while the tocotrienols differ by the presence of double bonds at the 3′,7′ and 11′ positions of the side chain. The α-, β-, γ-, and δ-isomers of tocopherol and tocotrienols differ in the number and position of the methyl substituents attached to the chromanol ring.

In 1980, Machlin stated in the preface of his book *Vitamin E: A Comprehensive Treatise* that "there is more myth and controversy concerning vitamin E than any other single nutrient." After almost forty years, the situation is still the same, despite the accumulation of much novel scientific information. This reflects that there are many unresolved issues with this vitamin and its metabolites in the fields of chemistry, biochemistry, biology, nutrition, and medicine, which may be ascribed in part to the inherent complexities of the role and action of vitamin E *in vivo* as well as *in vitro*.

Food Chemistry, Function and Analysis No.11
Vitamin E: Chemistry and Nutritional Benefits
Edited by Etsuo Niki

Published by the Royal Society of Chemistry, www.rsc.org

This book is intended to give an overview of the state-of-the-art chemistry of the properties, actions, functions and nutritional benefits of vitamin E and its metabolites by summarizing what has been demonstrated and what has not, which is essential for understanding the chemical, biological, nutritional and medical effects of vitamin E. Information on the chemistry of the role and action of vitamin E based on sound and solid scientific evidence is essential for understanding the advantages and limits of vitamin E and also for its application in the promotion and maintenance of health and prevention of diseases. This is a timely issue before the centennial anniversary of the discovery of vitamin E.

I would like to express my sincere thanks to all the authors for their contributions and their patience in reacting to editorial demands and to the staff of the Editorial Office of the Royal Society of Chemistry, who encouraged and assisted the publication of this book.

Etsuo Niki

Tokyo, Japan

Contents

Food Chemistry, Function and Analysis No.11
Vitamin E: Chemistry and Nutritional Benefits
Edited by Etsuo Niki

Published by the Royal Society of Chemistry, www.rsc.org

CHAPTER 1

Vitamin E: Structure, Properties and Functions

ETSUO NIKI*[a] AND KOUICHI ABE[b]

[a]University of Tokyo, Research Center for Advanced Science and Technology (RCAST), Komaba, Tokyo 153-8904, Japan; [b]SSCI Laboratories, Faculty of Pharmacology, Musashino University, Nishi-Tokyo-Shi, Tokyo 202-0023, Japan
*E-mail: etsuo-niki@emer.rcast.u-tokyo.ac.jp

1.1 Introduction

In 1922, Evans and Bishop demonstrated the existence of a hitherto unrecognized dietary factor essential for normal reproduction in the rat.[1] It was accepted at that time that the most striking function of vitamin E was to provide a normal gestation in a pregnant rat to prevent the resorption of the embryos which invariably occurred in its absence.[2] This unknown dietary factor X was found to be present in green lettuce, dried alfalfa leaves, wheat, and oats. Evans isolated the factor X from wheat germ oil, provided the chemical formula $C_{29}H_{50}O_2$ and proposed the name α-tocopherol in 1936.[3] The structural formula for α-tocopherol was provided by Fernholz in 1938.[4] Tocotrienols were discovered much later than tocopherol and named in the early 1960s.[5,6]

Olcott found that the lipid fractions of vegetable oils contained antioxidants against the oxidative deterioration of lard.[7] Since then, it has been unequivocally demonstrated that vitamin E acts as an essential antioxidant

Food Chemistry, Function and Analysis No.11
Vitamin E: Chemistry and Nutritional Benefits
Edited by Etsuo Niki

Published by the Royal Society of Chemistry, www.rsc.org

in vivo as well as *in vitro* and plays an important role in the prevention of detrimental oxidative damage of biological molecules.[8–11] More recently, the non-antioxidant functions of vitamin E, including cellular signaling, gene regulation, membrane processes, and nerve functions, have also received much attention.[12–14] However, many issues are still controversial and remain to be elucidated. Sound information based on solid chemical evidence is essential for understanding the role of vitamin E *in vivo* as well as *in vitro*.

1.2 Homologues: Nomenclature and Structure

Vitamin E is a plant-derived, lipid-soluble substance whose molecular structure is comprised of a chromanol ring with a side chain located at the C2 position. Vitamin E refers to a group of eight different compounds: α-, β-, γ-, and δ-tocopherols and the corresponding four tocotrienols. The four tocopherols have a saturated phytyl side chain, while tocotrienols have an unsaturated isoprenyl side chain containing three double bonds at C3′, C7′, and C11′. The double bonds of tocotrienols' side chains at C3′ and C7′ have a *trans*-configuration. The α-, β-, γ-, and δ-forms differ with respect to the number and position of methyl groups on the chromanol ring. The α-forms of tocopherol and tocotrienol have three methyl groups at the C5, C7, and C8-positions of the chromanol ring, while the β- and γ-forms have two and the δ-forms have one methyl group as illustrated in Figure 1.1.

In addition to tocopherols and tocotrienols, tocomonoenols and tocodienols containing a single and two double bond unsaturation, respectively,

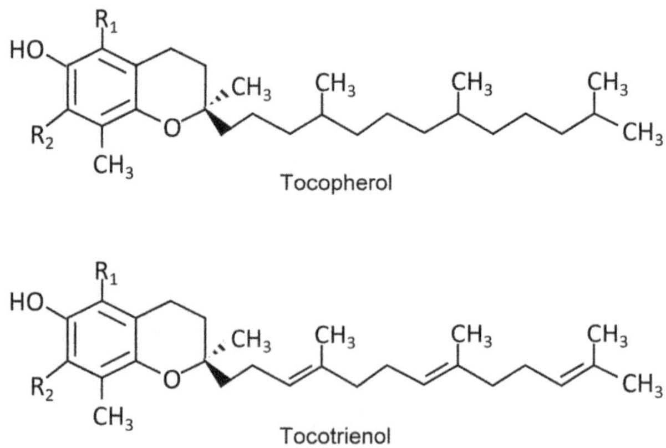

	α	β	γ	δ
R_1	CH_3	CH_3	H	H
R_2	CH_3	H	CH_3	H

Figure 1.1 Chemical structure of vitamin E homologues.

have also been found in nature. For example, a tocomonoenol with a single double bond at carbon 11′, 2,5,7,8-tetramethy1-2-(4′,8′,12′-trimethyltrideca-11′-enyl)-6-chromanol, was isolated from palm and rice bran oils.[15] Since then, several groups have detected tocomonoenols in plants and plant foods, such as α-tocomonoenol in palm oil,[16–22] pumpkin seed oil (*Cucurbita pepo* L.),[23] and sunflower oil (*Helianthus annuus*),[24] γ-tocomonoenol in pumpkin seed oil,[22,23] δ-tocomonoenol in kiwi (*Actinidia chinensis*),[25] and β-, γ-, and δ-tocomonoenol in the leaves of *Kalanchoe daigremontiana* and *Phaseolus coccineus*.[26] A tocomonoenol with an unsaturation at the isoprenoid-chain terminus was also found in the tissues of salmon.[27] Furthermore, tocodienols with two double bonds at carbon 7′ and 11′ were identified in palm oil.[16,21]

Tocopherols contain three chiral carbons, one at C2 in the chromanol ring and two in the side chain at C4′ and C8′. Naturally occurring α-tocopherol contains chiral carbons in the *R*-conformation, 2*R*, 4′*R*, and 8′*R*-α-tocopherol. α-Tocotrienol has one chiral center at C2 in the chromanol ring and natural tocotrienols occur as the *R*-isoform. On the other hand, the chemical synthesis of α-tocopherol produces an equimolar mixture of eight different stereoisomers: *RRR*, *SRR*, *RSR*, *RRS*, *RSS*, *SSR*, *SRS*, and *SSS*. The synthetic α-tocopherol is called all-rac-α-tocopherol. An equimolar mixture of RRR-α-tocopherol and SRR-α-tocopherol is called 2-ambo-α-tocopherol. The IUPAC names of *RRR*-α-tocopherol and *RRR*-α-tocotrienol are (2*R*)-2,5,7,8-tetramethyl-2-[(4*R*,8*R*)-(4,8,12-trimethyltridecyl)]chroman-6-ol and (2*R*)-2,5,7,8-tetramethyl-2-[(3*E*,7*E*)-4,8,12-trimethyltrideca-3,7,11-trienyl]-3,4-dihydrochromen-6-ol, respectively.

Ester forms of tocopherol and tocotrienols, including acetate, nicotinate, succinate, and phosphate, have been prepared and their action and potential applications have been studied. Vitamin E is easily oxidized when subjected to heat, light, and alkaline conditions, but esters are less susceptible to oxidation and therefore more appropriate for food, cosmetic, and pharmaceutical applications compared to the free form. Polyethylene glycol conjugates of tocopherols and tocotrienols have the ability to form miscible micelles in water due to amphiphilic properties and enhance bioavailability in animals and humans *via* improving their water solubility and absorption.[28] It was reported that *RRR*-α-tocopheryl polyethylene glycol 1000 succinate acted as a safe and effective form of vitamin E for reversing or preventing vitamin E deficiency during chronic childhood cholestasis.[29]

1.3 Physicochemical Properties

The tocopherols are viscous oils at room temperature, insoluble in water but soluble in ethanol and aprotic solvents. Vitamin E is a slightly yellow to amber, nearly odorless, clear, viscous oil, which darkens on exposure to air or light by oxidation. The physicochemical properties of α-tocopherol, the most abundant and active form of vitamin E in humans, are summarized in

Table 1.1 Physicochemical properties of α-tocopherol.

IUPAC name	(2*R*)-2,5,7,8-Tetramethyl-2-[(4*R*,8*R*)-4,8,12-trimethyltridecyl]-3,4-dihydrochromen-6-ol
Molecular formula	$C_{29}H_{50}O_2$
Molecular weight	430.7 g mol^{-1}
Physical description	Slightly yellow to amber, nearly odorless, clear, viscous oil
Melting point	3 °C
Boiling point	235 °C
Solubility	Insoluble in water (1.9 × 10^{-6} mg L^{-1} at 25 °C), soluble in ethanol
Density	0.950 g cm^{-3} at 25 °C
Partition coefficient	log P = 12.2
Stability	Unstable to UV light, alkaline, and oxidation
Dissociation constant	pK_a = 10.8
UV absorption maximum	292 nm in ethanol
Fluorescence	Excitation 290–295 nm, emission 320–335 nm
BDE (O–H)[a]	77.1 kcal mol^{-1}

[a]Bond dissociation energy.[31]

Table 1.1.[30] The melting point of *RRR*-α-tocopherol is 3 °C. The optical rotations of tocopherols are very small and depend on the nature of the solvent. The ultraviolet absorption spectra of tocopherols and tocotrienols in ethanol show an absorption maximum at 292–298 nm, while the infrared spectra show OH (2.8 ± 3.0 μm) and CH (3.4 ± 3.5 μm) stretching and a characteristic band at 8.6 μm. α-Tocopherol is fluorescent with an emission maximum about 325 nm in a hydrophobic solution. The bond dissociation energy of α-tocopherol's O–H bond is 77.1 kcal mol^{-1}.[31] The pK_a values for α-, β-, γ-, and δ-tocopherol in a micellar solution were reported as 13.1, 12.8, 12.7, and 12.6 respectively.[32]

The partition coefficient of α-tocopherol is 12.2, while that for ascorbic acid is −1.85,[30] showing their lipophilic and hydrophilic characters, respectively.

1.4 Sources

Numerous foods provide vitamin E. Seeds and fruits are among the best sources of vitamin E. Green leafy vegetables are also important sources. Only plants and photosynthetic organisms are able to synthesize vitamin E.[33,34] The synthesis is facilitated by enzymes to produce stereospecific tocopherols: *RRR*-tocopherols.[35] Tocopherols and tocotrienols can be extracted and purified or concentrated from vegetable oils and other higher plant materials. Dietary supplements are another source of vitamin E. Most vitamin E supplements contain substantially higher amounts of vitamin E than foods. Vitamin E in dietary supplements and fortified foods is often esterified to prolong its shelf life while protecting its antioxidant properties.

Table 1.2 The contents of tocopherols (T) and tocotrienols (T3) in representative edible oils. Adapted from ref. 36–41.[a]

Oil	αT	βT	γT	δT	αT3	βT3	γT3	δT3
Corn	**18–25**	1	**44–75**	2–3	1	0	1–2	<1
Olive	**11–17**	<1	1	<1	<1	<1	<1	<1
Palm	**6–42**	<1	<1	<1	**5–26**	<1	**11–36**	3–8
Rapeseed	**18–24**	<1	**37–51**	<1	0	0	0	0
Rice bran	**1–15**	0–2	0–8	0–2	**0–13**	0–2	**1–23**	0–2
Soybean	9–12	1	**61–69**	**23–26**	0	0	0	0
Sunflower	**32–59**	0–2	1–4	<1	<1	0	0	0
Wheat germ	**151–192**	**31–65**	0–52	<1	2–3	0–8	0–1	<1

[a]In mg per 100 g oil. Major components are shown in bold.

The contents and composition of tocopherols and tocotrienols in natural oils vary markedly between the plant species and even within the same species. Tocopherols are widely distributed in higher plants, whereas tocotrienols occur only in some non-photosynthetic tissues. Some examples of the contents in natural edible oils adopted from several reports[36–41] are shown in Table 1.2. α-Tocopherol is the major tocopherol in palm, olive, and sunflower oils, whereas the contents of γ-tocopherol in some edible oils such as corn, rapeseed, and soybean oils are higher than α-tocopherol.

The major sources of tocotrienols are palm, rice, and annatto, the ratio of tocopherol–tocotrienol in each being 25 : 75, 50 : 50, and 0.1 : 99.9, respectively.[42] Palm oil contains considerable amounts of α-, γ-, and δ-tocotrienols. γ-Tocotrienol is the major vitamin E isoform in the oil extracted from rice bran[39,43] while β-tocopherol is one of the major isoforms found in wheat germ oil.[36,37] Interestingly, the lipid fraction of annatto (*Bixa orellana* L.) seeds contained only tocotrienols, mainly δ-tocotrienol, but no tocopherols.[44]

1.5 Chemical Synthesis

Due to wide applications in the pharmaceutical, food, and cosmetic industries, about 35 000 tons of vitamin E is manufactured per year worldwide.[45] All-racemic-α-tocopherol, all-rac-α-tocopherol, an equimolar mixture of all eight stereoisomers, is industrially the most important product and is prepared either by total synthesis or originates from natural sources. It is mainly applied as an acetate ester. The industrial synthesis consists of three major parts: the preparation of 2,3,5-trimethylhydroquinone, the production of the side chain component, and the condensation reaction.[45]

Pure *RRR*-α-tocopherol is chemically prepared by semi-synthesis. Vegetable oils refined on a large scale and isolated by a combination of several separation methods contain mixtures of *RRR*-tocopherols, which are converted to *RRR*-α-tocopherol by permethylation reactions. The efforts and strategies to synthesize tocopherols and tocotrienols have been described in detail.[45]

1.6 Analysis

Natural products contain mixtures of isomeric tocopherols and tocotrienols together with minor amounts of tocomonoenols and tocodienols. They are metabolized and oxidized to produce many kinds of compounds. The biological samples contain complicated mixtures of vitamin E isoforms and their metabolites and oxidation products. The identification and quantitation of vitamin E isoforms and their products are important for elucidating their role and functions. Sample preparation is the most time-consuming and important step in analysis, because it is the main source of errors. Many analytical methods including colorimetric, fluorometric, gas chromatographic (GC), high performance liquid chromatographic (HPLC),[46] GC-MS,[47] and HPLC-MS[47,48] have been reported (see also Chapter 2). Concerning public documents for the quantitative determination method of vitamin E in active pharmaceutical ingredients, United States and European Pharmacopeia adopt the GC method, while Japanese Pharmacopeia adopts the HPLC method. Each of these methods has its merits and demerits. At the present time, HPLC methods are the most widely applied.

HPLC conditions for vitamin E analysis have been summarized in several reviews.[46,47,49,50] Vitamin E and products are obtained from natural products, food matrices, and biological samples by direct extraction, solid-phase extraction, and saponification. Care should be taken to avoid artifactual oxidation and decomposition during preparation of samples and analysis. The use of electrochemical (amperometric), fluorometric (Ex. 290–295 nm, Em. 320–330 nm), or MS detection is much more sensitive and specific than the conventional UV detection. Tocol, 2-methyl-2-(4,8,12-trimethyltridecyl)-6-chromanol, may be the best compound to use as an internal standard because its characteristics are similar to vitamin E. Deuterium-labelled α-tocopherol acetate is commonly used as a standard in MS analysis. In most vitamin E analyses, atmospheric pressure chemical ionization (APCI) is used in the positive ionization mode. It may be noted that supercritical-fluid chromatography using CO_2 and ethanol as a mobile phase, NH_2 as a stationary phase, and mass spectrometry for the detection and quantification of vitamin E homologues has been proposed as a "greener" analytical strategy.[51,52]

1.7 Functions and Applications

Vitamin E is a physiologically essential micronutrient and has been applied in various fields including medicine, pharmaceutics, cosmetics, and foods. Vitamin E is assumed to play an important role in the promotion of health and prevention and/or treatment of some diseases and disorders. The daily recommended intake is 15 mg (22.4 IU, International Unit) for adults [https://ods.od.nih.gov/factsheets/VitaminE-HealthProfessional/].

Multiple functions of vitamin E have been demonstrated or proposed, including (1) antioxidant by scavenging free radicals, especially peroxyl radicals, and singlet oxygen, (2) membrane stabilization by forming complexes with destabilizing molecules so as to prevent disturbance of the amphipathic balance within the structure, (3) physiological regulator of enzyme activity, cellular signaling, cell proliferation, and gene expression, which is not directly related to antioxidant action, (4) inhibition of platelet coagulation, (5) prevention of diseases including neurological disorders, cardiovascular diseases, age-related eye and skin damage, and infertility, and (6) biocompatible modifier of biomaterials and medical devices, for example in high molecular weight polyethylene used in hip and knee implants. It has been argued that tocotrienols possess additional positive health effects beyond those of tocopherols including, for example, induction of immune responses and lowering of serum cholesterol levels.[53–55]

Among these functions, the role of vitamin E as an antioxidant against free radical-mediated lipid peroxidation has been unequivocally demonstrated and it appears that this is the most important physiological function of this vitamin. Vitamin E also inhibits air oxidation of foods, oils, and industrial materials to extend their lifetime. The physiological significance of other functions that are not related to antioxidant function must be established in a future study.

Vitamin E is widely used as a dietary supplement, by itself or together with other micronutrients such as vitamin C, to promote health and to reduce the risk of or prevent diseases that are assumed to arise from detrimental oxidative modification of biological molecules. Vitamin E is used to fortify some foods and beverages.

Deficiency of vitamin E is uncommon as typical diets appear to provide sufficient amounts, although malnutrition and genetic disorders may result in vitamin E deficiency. Premature babies of very low birth weight might be deficient in vitamin E. Further, people with fat-malabsorption disorders[56] and inherited disorders in which the liver's α-tocopherol transfer protein (α-TTP) is defective or absent[57] or there are decreased levels of selenoproteins[58] are more likely to become vitamin E deficient and require high doses of supplemental vitamin E.

The potential role of vitamin E against periodontal disease,[59] nonalcoholic steatohepatitis,[60] and sarcopenia[61] has received attention. Insufficient micronutrient intake (vitamins and phosphorus) was significantly associated with severity of periodontal disease.[62] A systematic review showed significant improvement of periodontal parameters by vitamin E and lycopene.[63] Vitamin E may mitigate age-associated skeletal dysfunction and enhance muscle regeneration, thus attenuating sarcopenia.[64]

Vitamin E is also used for biomedical materials. α-Tocopherol is blended to a dialysis membrane for clinical management of chronic hemodialysis patients to improve stability and functions.[65] Vitamin E-blended ultra-high molecular weight polyethylene (UHMWPE) has been developed as a material for use in total knee and hip replacements.[66] α-Tocopherol is accepted to improve oxidation resistance while maintaining wear resistance and fatigue strength.

The poor water solubility of vitamin E has greatly limited its application. Investigations to improve both oral and topical encapsulation and formulations of vitamin E have been widely carried out and various delivery systems for vitamin E have been developed to overcome this problem and to enhance solubility, permeation, and bioavailability. Liposomes,[67] nano-emulsions, and lipid nanoparticles[68] are some examples.[69] The tocopheryl ester of polyethylene glycol and tocopheryl phosphate are water soluble. Polyethylene conjugates of tocotrienols have also been studied.[28]

Dietary polyphenols such as quercetin are converted to glucoside and/or sulfate derivatives, or their *O*-methyl derivatives, during absorption into the body, and presents as these conjugated metabolites in circulating blood. Formation of the metabolites of tocopherol such as glucoside, sulfate, and *O*-methyl derivatives *in vivo* has not been reported but α-tocopherol glucoside was synthesized from 2-hydroxymethyl-2,5,7,8-tetramethylchroman-6-ol and maltose by transglycosylation with α-glucosidase from *Saccharomyces* species.[70] The glycosylated product, 2-(α-D-glucopyranosyl)methyl-2,5,7,8-tetramethylchroman-6-ol, is water soluble and acts as a radical scavenging antioxidant.[70,71] δ-Tocopherol glucoside was prepared and its effects during its topical application on skin were studied.[72]

One important question is whether or not each vitamin E isoform exerts its own specific function which is not shared with other isoforms. It has been argued that tocotrienol exerts some functions that tocopherol does not, and that tocopherol interferes with such tocotrienol functions.[42] These important issues must be clarified in future studies.

1.8 Stability

Vitamin E is stable at ambient temperature, but it is readily oxidized at high temperature, under light or in an alkaline medium. Frying is one of the most common methods of food preparation and during the frying process, usually between 160–190 °C, vitamin E is oxidized to give complicated mixtures of products including tocored (5,6-tocopheryldione) and dimers that are assumed to be formed by a recombination of tocopheroxyl radicals. α-Tocopherol is oxidized to α-tocopheroxyl radical by transition metal ions such as ferric ions (Fe^{3+}) and cupric ions (Cu^{2+}).

References

1. H. M. Evans and K. S. Bishop, *Science*, 1922, **56**, 650.
2. E. Niki and M. G. Traber, *Ann. Nutr. Metab.*, 2012, **61**, 207.
3. H. M. Evans, O. H. Emerson and G. A. Emerson, *J. Biol. Chem.*, 1936, **113**, 319.
4. E. Fernholz, *J. Am. Chem. Soc.*, 1938, **60**, 700.
5. J. Bunyan, D. McHale, J. Green and S. Marcinkiewicz, *Br. J. Nutr.*, 1961, **15**, 253.
6. J. F. Pennock, F. W. Hemming and J. D. Kerr, *Biochem. Biophys. Res. Commun.*, 1964, **30**, 542.
7. H. S. Olcott and O. H. Emerson, *J. Am. Chem. Soc.*, 1937, **59**, 1008.

8. L. Packer, *Am. J. Clin. Nutr.*, 1991, **53**, 1050S.
9. W. A. Pryor, *Free Radical Biol. Med.*, 2000, **28**, 141.
10. M. G. Traber and J. Atkinson, *Free Radical Biol. Med.*, 2007, **43**, 4.
11. E. Niki, *Free Radical Biol. Med.*, 2014, **66**, 3.
12. A. Azzi, *Free Radical Biol. Med.*, 2007, **43**, 16.
13. R. Brigelius-Flohé and F. Galli, *Mol. Nutr. Food Res.*, 2010, **54**, 583.
14. F. Galli, A. Azzi and M. Birringer, *et al.*, *Free Radical Biol. Med.*, 2017, **102**, 16.
15. A. Matsumoto, S. Takahashi, K. Nakano and S. Kijima, *J. Jpn. Oil Chem. Soc.*, 1995, **44**, 593.
16. C. Mariani and G. Bellan, *Riv. Ital. Sostanze Grasse*, 1996, **73**, 533.
17. S. Strohschein, C. Rentel, T. Lacker, E. Bayer and K. Albert, *Anal. Chem.*, 1999, **71**, 1780.
18. M. H. Ng, Y. M. Choo, A. N. Ma, C. H. Chuah and M. A. Hashim, *Lipids*, 2004, **39**, 1031.
19. C. W. Puah, Y. M. Choo, A. N. Ma and C. H. Chuah, *Am. J. Appl. Sci.*, 2007, **4**, 374.
20. M. Mozzon, D. Pacetti, N. G. Frega and P. Lucci, *J. Am. Oil Chem. Soc.*, 2015, **92**, 717.
21. P. T. Gee, C. Y. Liew, M. C. Thong and M. C. Gay, *Food Chem.*, 2016, **196**, 367.
22. A. Irías-Mata, W. Stuetz, N. Sus, *et al.*, *J. Agric. Food Chem.*, 2017, **65**, 7476.
23. B. Butinar, M. Bučar-Miklavčič, C. Mariani and P. Raspor, *Food Chem.*, 2011, **128**, 505.
24. S. Hammann, M. Englert, M. Müller and W. Vetter, *Anal. Bioanal. Chem.*, 2015, **407**, 9019.
25. A. Fiorentino, C. Mastellone, B. D'Abrosca, S. Pacifico, M. Scognamiglio, G. Cefarelli, R. Caputo and P. Monaco, *Food Chem.*, 2009, **115**, 187.
26. J. Kruk, A. Pisarski and R. J. Szymanska, *Plant Physiol.*, 2011, **168**, 2021.
27. Y. Yamamoto, N. Maita, A. Fujisawa, J. Takashima, Y. Ishii and W. C. Dunlap, *J. Nat. Prod.*, 1999, **62**, 1685.
28. A. Abu-Fayyad, F. Behery and A. A. Sallam, *et al.*, *Eur. J. Pharm. Biopharm.*, 2015, **96**, 185.
29. R. J. Sokol, N. Butler-Simon and C. Conner, *et al.*, *Gastroenterology*, 1993, **104**, 1727.
30. Pubchem, https://pubchem.ncbi.nlm.nih.gov/compound/alpha-Tocopherol.
31. M. Lucarini and G. F. Pedulli, *Chem. Soc. Rev.*, 2010, **39**, 2106.
32. K. Mukai, A. Tokunaga, S. Itoh, Y. Kanesaki, A. Ouchi, K. Ohara, S. Nagaoka and K. Abe, *Biofactors*, 2008, **32**, 49.
33. S. Munné-Bosch, *Vitam. Horm.*, 2007, **76**, 375.
34. L. Mène-Saffrané, *Antioxidants*, 2017, **7**(1), 2.
35. D. DellaPenna, *J. Plant Physiol.*, 2005, **162**, 729.
36. E. L. Syväoja, V. Pironen, P. Varo, P. Koivistoinen and K. Salminen, *J. Am. Oil Chem. Soc.*, 1986, **63**, 328.
37. H. Schwartz, V. Ollilainen, V. Piironen and A. M. Lampi, *J. Food Compos. Anal.*, 2008, **21**, 152.
38. J. Frank, X. M. Chin, C. Schrader, G. P. Eckert and G. Rimbach, *Ageing Res. Rev.*, 2012, **11**, 163.

39. P. Goufo and H. Trindade, *Food Sci. Nutr.*, 2014, **2**, 75.
40. Y. F. Wong, A. Makahleh, B. Saad, M. N. Ibrahim, A. A. Rahim and N. Brosse, *Talanta*, 2014, **130**, 299.
41. F. Shahidi and A. C. de Camargo, *Int. J. Mol. Sci.*, 2016, **17**, pii: E1745.
42. A. M. Trias and B. Tan, in *Tocotrienols*, ed. B. Tan, R. R. Watson and V. R. Preedy, CRC Press, Boca Raton, 2013, ch. 5, p. 61.
43. P. Sookwong, K. Nakagawa, K. Murata and T. Miyazawa, *J. Agric. Food Chem.*, 2007, **55**, 461.
44. N. Frega, M. Mozzou and F. Bocci, *J. Am. Oil Chem. Soc.*, 1998, **75**, 1723.
45. T. Netscher, *Vitam. Horm.*, 2007, **76**, 155.
46. K. Abe and A. Matsumoto, in *Vitamin E - Its Usefulness in Health and Curing Diseases*, ed. M. Mino, H. Nakamura, A. T. Diplock and H. Kayden, Japan Scientific Societies Press, Tokyo, 1993, pp. 13–19.
47. E. Bartosinska, M. Buszewska-Forajta and D. Siluk, *J. Pharm. Biomed. Anal.*, 2016, **127**, 156.
48. W. M. Stöggl, C. W. Huck, H. Scherz, M. Popp and G. K. Bonn, *Chromatographia*, 2001, **54**, 179.
49. B. Cervinkova, L. K. Kremova, D. Solichova, B. Melichar and P. Solich, *Anal. Bioanal. Chem.*, 2016, **408**, 2407.
50. J. Y. Fu, T. T. Htar, L. De Silva, D. M. Tan and L. H. Chuah, *Molecules*, 2017, **22**(2), 233.
51. T. Yarita, A. Nomura, K. Abe and Y. Takeshita, *J. Chromatogr.*, 1994, **679**, 329.
52. M. Mejean, A. Brunelle and D. Touboul, *Anal. Bioanal. Chem.*, 2015, **407**, 5133.
53. K. Nesaretnam, W. W. Yew and M. B. Wahid, *Eur. J. Lipid Sci. Technol.*, 2007, **109**, 445.
54. C. K. Sen, S. Khanna and S. Roy, *Mol. Aspects Med.*, 2007, **28**, 692.
55. *Tocotrienols*, ed. B. Tan, R. R. Watson and V. R. Preedy, CRC Press, Boca Raton, 2013.
56. K. V. Kowdley, J. B. Mason, S. N. Meydani, S. Cornwall and R. J. Grand, *Gastroenterology*, 1992, **102**, 2139.
57. K. Ouahchi, M. Arita and H. Kayden, *et al.*, *Nat. Genet.*, 1995, **9**, 141.
58. Y. Saito, M. Shichiri and T. Hamajima, *et al.*, *J. Lipid Res.*, 2015, **56**, 2172.
59. G. Zong, A. E. Scott, H. R. Griffiths, P. L. Zock, T. Dietrich and R. S. Newson, *J. Nutr.*, 2015, **145**, 893.
60. A. J. Sanyal, N. Chalasani and K. V. Kowdley, *et al.*, *N. Engl. J. Med.*, 2010, **362**, 1675.
61. E. Chung, H. Mo and S. Wang, *et al.*, *Nutr. Res.*, 2018, **49**, 23.
62. P. P. Luo, H. S. Xu and S. P. Wu, *Aust. Dent. J.*, 2018, **63**, 193.
63. F. W. Muniz, S. B. Nogueira, F. L. Mendes, C. K. Rösing, M. M. Moreira, G. M. de Andrade and R. S. Carvalho, *Arch. Oral Biol.*, 2015, **60**, 1203.
64. A. Ble, A. Cherubini and S. Volpato, *et al.*, *J. Gerontrol. A, Bio. Sci. Med. Sci.*, 2006, **61**, 278.
65. G. D'Arrigo, R. Baggetta, G. Tripepi, F. Galli and D. Bolignano, *Blood Purif.*, 2017, **43**, 101.

66. A. Turner, Y. Okubo and S. Teramura, *et al.*, *J. Mech. Behav. Biomed. Mater.*, 2014, **31**, 21.
67. S. Koudelka, P. Turanek-Knotigova and J. Masek, *et al.*, *J. Controlled Release*, 2015, **207**, 59.
68. S. Hama and K. Kogure, *Biol. Pharm. Bull.*, 2014, **37**, 196.
69. V. Saez, I. D. L. Souza and C. R. E. Mansur, *Int. J. Cosmet. Sci.*, 2018, **40**, 103–116.
70. H. Murase, R. Yamauchi, K. Kato, T. Kunieda and J. Terao, *Lipids*, 1997, **32**, 73.
71. H. Murase, J. H. Moon, R. Yamauchi, K. Kato, T. Kunieda, T. Yoshikawa and J. Terao, *Free Radical Biol. Med.*, 1998, **24**, 217.
72. A. Mavon, V. Raufast and D. Redoules, *J. Controlled Release*, 2004, **100**, 221.

CHAPTER 2

Tocotrienols: From Bench to Bedside

JU-YEN FU*[a], GEETHA MANIAM[a,b], FU-SHUN WONG[a,c], DORYN MEAM-YEE TAN[a,d], PUVANESWARI MEGANATHAN[a,c] AND LAY-HONG CHUAH[d,e]

[a]Nutrition Unit, Product Development and Advisory Services, Malaysian Palm Oil Board, 6 Persiaran Institusi, Bandar Baru Bangi, 43000 Kajang, Selangor, Malaysia; [b]Department of Pharmaceutical Chemistry, School of Pharmacy, International Medical University, 126, Jalan Jalil Perkasa 19, Bukit Jalil, 57000 Kuala Lumpur, Malaysia; [c]Department of Pharmacology, Faculty of Medicine, University of Malaya, 50603 Kuala Lumpur, Malaysia; [d]School of Pharmacy, Monash University Malaysia, Jalan Lagoon Selatan, 47500 Bandar Sunway, Selangor, Malaysia; [e]Advanced Engineering Platform, Monash University Malaysia, Jalan Lagoon Selatan, 47500 Bandar Sunway, Selangor, Malaysia

*E-mail: fujuyen@mpob.gov.my, fujuyen@gmail.com

2.1 Introduction

Vitamin E is a family of compounds consisting of two groups, namely tocopherol and tocotrienol. It was traditionally known as a fertility compound since its discovery in 1936, with the name tocopherol referring to "fertility" or "pregnancy" in Greek. As it is not synthesized in the human body, it plays the role of an essential nutrient needed in the human diet. The major sources of vitamin E are vegetable oils. Among the vegetable oils consumed globally,

Food Chemistry, Function and Analysis No.11
Vitamin E: Chemistry and Nutritional Benefits
Edited by Etsuo Niki

Published by the Royal Society of Chemistry, www.rsc.org

corn, soy bean and safflower oil have the highest amounts of tocopherols, while palm and rice bran oils have the highest amounts of tocotrienols.[1] Most research on vitamin E in the past 80 years has been focused on tocopherol due to its abundance.

The definition of vitamin E as an essential micronutrient is based on its role as an antioxidant and neuroprotective agent.[2] Nevertheless, the only form of vitamin E recognized as a physiologically active vitamer was alpha-tocopherol. As alpha-tocopherol binds effectively to alpha-tocopherol transfer protein (α-TTP), they get secreted by the liver into systemic circulation through incorporation into very low-density lipoprotein.[3] Thus, deficiency in vitamin E is often reported in individuals with suppressed α-TTP and they are recognized *via* neurological symptoms such as peripheral neuropathy and cerebellar ataxia.[2] On the contrary, other isoforms of tocopherols and tocotrienols are preferentially metabolized in the liver due to poor binding to α-TTP.[4] Owing to the lower bioavailability of tocotrienols, their role as a nutrient is often underestimated.

According to the Institute of Medicine (IOM), the recommended daily allowance (RDA) of vitamin E is 15 mg per day for both men and women. Based on age and gender, the RDA for children ranges from 6 mg per day to 15 mg per day.[5] On the other hand, the European Food and Drug Authority (EFSA) proposed to use Adequate Intakes (AIs) as a reference point for the estimation of an adequate vitamin E intake. The AI values were constructed based on dietary surveys done in nine countries within the European Union. The recommended AIs are 13 mg per day and 11 mg per day for men and women, respectively, and ranged from 6 to 13 mg per day for children based on age and gender. Currently, there is no RDA proposed for tocotrienols.

However, various distinctive functions were reported in tocotrienol studies, some of which were not observed with tocopherols. Over the past decade, tocotrienol has emerged as the unsung hero of vitamin E, showing physiological roles such as anti-inflammatory, lipid-lowering, neuroprotection and anti-cancer properties.[6] A current update on the nutritional benefits of tocotrienols will be covered in Section 2.6.

2.2 Physical and Chemical Properties

The yellow viscous oily vitamin E belongs to a family of fat-soluble vitamins. Vitamin E consists of alpha, beta, gamma and delta isoforms of tocopherols and tocotrienols. Both tocopherol and tocotrienol homologues share the same chromanol ring but differ in the number and position of methyl groups attached to the ring. The saturation of the isoprenoid side chain differentiates tocotrienols from tocopherols. Side chains in the tocotrienol side chain have three carbon–carbon double bonds, while tocopherols are saturated (Figure 2.1).

As shown in Figure 2.1, tocopherol has three chiral centres at C2, C4′ and C8′, which produce eight different stereoisomers (*RRR, RRS, RSR, RSS, SSS,*

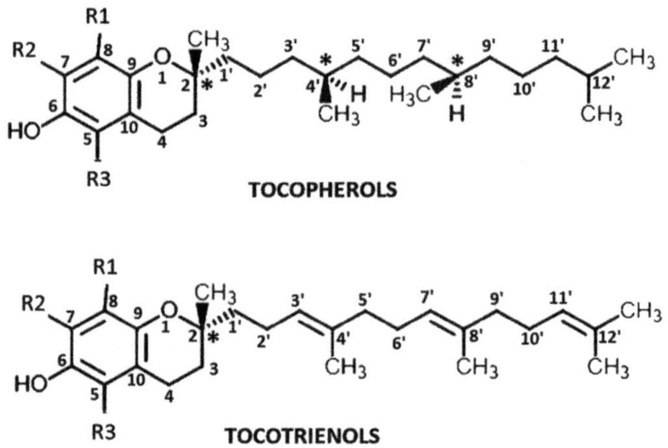

	R1	R2	R3
ALPHA	CH_3	CH_3	CH_3
BETA	CH_3	H	CH_3
GAMMA	CH_3	CH_3	H
DELTA	CH_3	H	H

Figure 2.1 Chemical structures of tocopherol and tocotrienol with homologues named according to the degree of methylation at R1 and R2.

SSR, SRS, SRR). The only naturally occurring alpha-tocopherol is *RRR*-alpha-tocopherol.[2] In the synthetic form, all the stereoisomers are present in equal amounts along with its esterified forms which include all-rac-alpha-tocopheryl acetate and *RRR*-alpha-tocopheryl acetate. Natural *RRR*-alpha-tocopherol is the most biologically active isomer due to its effective interaction with α-TTP. The 2*R*-stereoisomers have a higher binding affinity to α-TTP compared to the 2*S*-enantiomers. Therefore, natural and synthetic tocopherols are not bioequivalent. The International Unit (IU) or α-tocopherol equivalent (α-TE) are commonly used to distinguish the bioactivity of vitamin E based on a rat resorption–gestation assay. One IU of all-rac-alpha-tocopherol is equivalent to 1 mg of all-rac-alpha-tocopherol or its esters or 0.5 mg of *RRR*-alpha-tocopherol.[2]

In contrast to tocopherols, due to the presence of double bonds, tocotrienol has only one chiral stereocentre, which is 2*R*, and thus each isoform has only two stereoisomers, which are dextrorotatory (D) and levorotatory (L) isomers.[3,7] However, in natural sources, tocotrienols exist as D-tocotrienols.

On the other hand, the hydroxyl group attached to the C6 of both tocopherols and tocotrienols can readily donate hydrogen and be oxidized. In reactions with free radicals, the vitamin E homologues neutralize the radicals by donating hydrogen, which can be attributed to their antioxidant property. Free radical production is initiated by metabolic activities in a biological system and can propagate through a chain reaction, unless there is a terminator to deactivate the reactive species. Therefore, antioxidants play an important role in dietary nutrients. Tocotrienol was reported to be a better antioxidant than tocopherol as the unsaturated side chain of tocotrienols takes an "arch" conformation over the chromanol ring.[3] This conformation provides a platform for a better interaction with radicals.

Tocotrienol has a log P of 8.2.[8] In general, a log P greater than 0 is lipid soluble and a log P less than 0 is water soluble. Therefore, this lipophilic tocotrienol is miscible in alcohol, organic solvents and vegetable oils but not water. Besides, tocotrienol has the physical property of being a fluorophore that emits light at a wavelength of 330 nm after excitation at 290 nm.[9] The fluorescence property of tocotrienol is driven by the presence of conjugated and planar rings, non-rotatable conjugation and the electron donating hydroxyl group in the tocotrienol structure. This property has been widely utilized as a means of quantification for tocotrienols in analytical methods.

2.3 Analysis Method

Due to the presence of homologues with similar chemical structures, chromatography is the most commonly used method to analyse vitamin E. Numerous methods were reported to accurately quantify vitamin E depending on the analyte of interest and sample matrix. These include liquid chromatography (LC), mass spectrometry (MS) and gas chromatography (GC).

High performance liquid chromatography (HPLC) is frequently used due to its accessibility and versatility. In an oil matrix, Cunha *et al.* evaluated three detection systems (fluorescence, UV and evaporative light scattering detector) on the separation efficiency and sensitivity of a normal phase HPLC method to quantify tocols.[9] Up to eight homologues were identified within a run time of approximately 25 minutes. Tocotrienols in food matrices such as barley, cereal and coffee beans were reported by Tsochatzis *et al.*, Panfili *et al.*, and Alves *et al.*[10–12] Due to the complexity of food matrices, extra steps in sample preparation were needed, such as solvent extraction and saponification. The study by Tsochatzis *et al.* used direct solvent extraction to separate milled barley with acetonitrile using reversed-phase HPLC coupled with a fluorescence detector.[10] In cereal samples, Panfili *et al.* compared extraction methods with saponification and solvent extraction in a normal phase HPLC system.[11] Solvent extraction is the most common method by which the solvents should be efficient enough to penetrate the water phase to reach the tocotrienols and to liberate them from lipophilic membranes.[11,13] An example of a chromatogram of eluted vitamin E homologues is presented in Figure 2.2. Most methods in an oil or food matrix

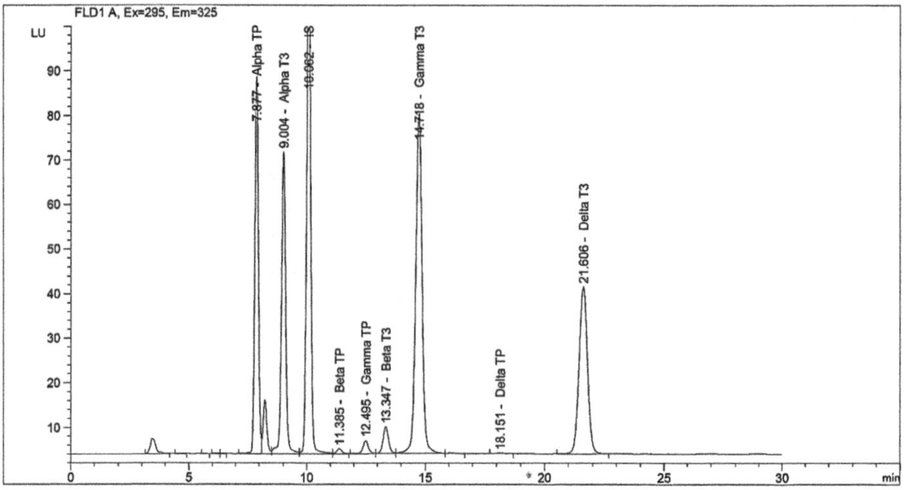

Figure 2.2 Chromatogram of a tocotrienol-rich fraction from palm oil. System: NP-HPLC-FLD, fluorescence detector: Ex = 295, Em = 325, column: Luna 5μ Silica 100A ODS Hypersil column.

were able to identify up to eight homologues of vitamin E. However, some chromatograms may have some additional peaks that represent compounds other than the standard isomers, such as plastochromanol-8, which was identified in vegetable oils[14] and tocomonoenol, which was identified in palm oil.[15] A recently reported method demonstrated the use of supercritical fluid chromatography coupled with ultra-performance convergence chromatography (UPC^2) to quantify vitamin E in a tocotrienol-rich fraction. The authors identified a new vitamin E component (α-tocodienol), eluted between α-tocomonoenol and α-tocotrienol, which constitutes 0.2% of the tocotrienol-rich fraction.[16]

For biological samples from *in vitro* or animal studies, a few methods were reported using reversed-phase HPLC. Schüep *et al.* spiked tocols into homogenized liver samples and plasma whereby six analytes were obtained by reversed-phase HPLC.[17] While the run time was three minutes faster than the normal phase, peak separation between β- and γ-tocols was not possible. In order to improve the sensitivity, an LC-MS method was developed and validated to measure the level of γ-tocotrienol in rat plasma.[18] MS alone is unable to isolate samples with a high number of compounds as it can only differentiate compounds by their mass-to-charge ratio. Hence, MS has not been widely used for the analysis of vitamin E due to the difficulty involved in the ionization of non-polar molecules. Therefore, a combination of MS and LC provides rapid, cost-effective and quantitative measurements of organic molecules for an enormous variety of applications.[19,20] According to Liang *et al.*, a validated LC-MS method was demonstrated to be accurate and precise over the concentration range of 10–1000 ng mL^{-1}, 10 times lower than that reported using HPLC.[18] The use of MS as a detection method was extended to gas chromatography. A study by Zhang and co-workers developed an accurate GC-MS method to quantify the analytes in vegetable oils with full separation of the eight homologues while removing the interference of the oil matrix within 14 minutes.[21] Recently, sensitive HPLC coupled with a coulochem electrode array system (CEAS) was developed by Lee *et al.*[22] Due to the sensitivity of the CEAS detector, a well-resolved separation of analytes was observed with minimal interfering peaks compared to that of the MS and the fluorescence detector.

Moving from bench to bedside, clinical studies with tocotrienols necessitate an accurate method to quantify tocotrienols in human plasma and tissue samples. Quantification of tocotrienols in human plasma was reported by Yap *et al.* using a reversed-phase HPLC method with a fluorescence detector.[23] Che *et al.* and Meganathan *et al.*, on the other hand, developed a normal-phase HPLC method with a fluorescence detector to quantify plasma levels of tocotrienols in human subjects after acute consumption.[24,25] Methods using the normal and reversed phase were both validated. One of the major obstacles in detecting vitamin E in human plasma is the interference peaks from endogenous compounds. Therefore, in the published methods, two to four analytes, such as α-, γ- and δ-tocotrienol, were being studied instead of eight vitamin E homologues. For human tissue, up to three papers reported measurements of vitamin E levels in breast, adipose, skin, liver and brain tissues.[26–29] In human breast adipose tissues, Bartosińska *et al.* developed a HPLC-MS/MS system that was able to quantify tocotrienol levels as low as 10 ng g^{-1} tissue.[27] These methods are useful for clinical studies to determine pharmacokinetics, tissue distribution and compliance. The HPLC systems to quantify vitamin E are summarized in Table 2.1.

2.4 Bioavailability

2.4.1 Animals

The blood levels of tocopherols appeared to be higher than tocotrienols after oral administration, possibly due to their affinity to α-TTP as specific transporters.[35] For tocotrienols, Yap *et al.* have studied the plasma concentration of T3 following oral, intraperitoneal and intramuscular administration.[36] Their findings revealed incomplete oral absorption, with α-T3 having the highest absolute bioavailability of 27.7%, followed by 9.1% and 8.5% for γ- and δ-T3, respectively. The difference in bioavailability of T3 homologues was attributed to the lipophilicity of the molecules, distinguished by the number of methyl groups in the chromanol ring. As shown in Figure 2.1, α-T3 has three methyl groups, γ-T3 has two and δ-T3 has only one. A similar trend was shown with the absorption of T3 into a rat's lymphatic circulation.[37] Surprisingly, intestinal absorption of γ-tocotrienol was found to be faster than α-tocopherol.[35] With an *in situ* rat intestinal perfusion model, Abuasal and his team discovered that intestinal uptake of γ-T3 is concentration

Table 2.1 Examples of HPLC systems reported to quantify vitamin E.

Columns	Mobile phase	Detection	Sample matrix	Analytes	Reference
Alltima SI 5u, 5 μm, 250 mm*4.6 mm	4–5%, 1,4-dioxane in *n*-hexane, 1.5–2 mL min^{-1}	FLD: λ_{Ex} = 294 nm; λ_{Em} = 326 nm	Oat extract, palm oil	α-, β-, γ-, δ-tocotrienol α-, β-, γ-, δ-tocopherol	30
ZorbaxSil, 5 μm, 250 mm*4.6 mm	2.5% tetrahydrofuran and 0.25% methanol in hexane, 1 mL min^{-1}	FLD: λ_{Ex} = 298 nm; λ_{Em} = 325 nm	Palm oil	α-, γ-, δ-tocotrienol α-tocopherol	31
Econosil Si, 10 μm, 250 mm*10 mm	Hexane/tetrahydrofuran (gradient 0–15% tetrahydrofuran)	FLD: λ_{Ex} = 290 nm; λ_{Em} = 330 nm	Soybean oil, wheat bran	α-, β-, γ-, δ-tocotrienol α-, β-, γ-, δ-tocopherol	32
Supelcosil LC-SI, 3 μm, 75 mm*3 mm	2% 1,4-dioxane in *n*-hexane, 0.7 mL min^{-1}	FLD: λ_{Ex} = 290 nm; λ_{Em} = 330 nm	Coffee beans	α-, β-toocopherol	12
Pheomenex silica, 5 μm, 250 mm*4.6 mm	4% tetrahydrofuran in hexane, 1 mL min^{-1}	FLD: λ_{Ex} = 290 nm; λ_{Em} = 330 nm	Seeds and seed oil	α-, β-, γ-, δ-tocotrienol α-, β-, γ-, δ-tocopherol	33
LiChrosorb Si 60, 5 μm, 250 mm*3.2 mm	8% 1,4-dioxane and 0.2% isopropanol in *n*-hexane, 1 mL min^{-1}	FLD: λ_{Ex} = 290 nm; λ_{Em} = 330 nm	Food and tissue	α-, β-, γ-tocotrienol α-, β-, γ-, δ-tocopherol	34
KromasilPhenomenex Si, 5 μm, 250 mm*4.6 mm	1.8% ethyl acetate and 0.9% acetic acid in *n*-hexane, 1.6 mL min^{-1}	FLD: λ_{Ex} = 290 nm; λ_{Em} = 330 nm	Cereals	α-, β-, γ-, δ-tocotrienol α-, β-, γ-, δ-tocopherol	11
Luna 5u Silica 100A, 5 μm, 250 mm*460 mm, 26 °C	2% 1,4-dioxane and 0.5% propanol in *n*-hexane, 1 mL min^{-1}	FLD: λ_{Ex} = 295 nm; λ_{Em} = 325 nm	Human plasma	α-, β-, γ-, δ-tocotrienol	24
Metaphase CrestpakC18S, 5 μm, 250 mm*4.6 mm	0.5% (v/v) distilled water in methanol	FLD: λ_{Ex} = 296 nm; λ_{Em} = 330 nm	Human plasma	α-, γ-, δ-tocotrienol	23

dependent and saturable, and the process is governed by Niemann-Pick C1-like 1(NPC1L1).[38]

For tissue distribution, α- and γ-T3 were detected in adipose tissues of rats after 24 hours.[39] Maximal T3 levels in serum, liver, mesenteric lymph node, spleen and lungs were observed eight hours post-administration. However, no T3 was detected in blood clots, brain, thymus, kidneys, testes, vice-testes and muscles. When given to mice in the form of an emulsion through subcutaneous injection, γ-T3 predominantly accumulated in adipose and heart tissues.[40]

Alternative approaches have been designed to improve the bioavailability of tocotrienols. Most of these strategies were achieved *via* formulation of tocotrienols in drug delivery systems.

- Prodrug[41]
- Chemical modification of isoprenoid side chain (Tocoflexol)[42]
- Self-emulsifying drug delivery system[43,44]
- Cyclodextrin[45]
- Nano-emulsion[46–48]
- Nanocarriers[38,49–51]

2.4.2 Humans

As a fat-soluble vitamin, the influence of food on the bioavailability of tocotrienols at 300 mg per day was investigated in eight healthy male adults under fasted and fed conditions.[52] At least a two-fold increment in plasma concentration of each tocotrienol homologue was observed in fed subjects compared to those who consumed tocotrienols in the fasted state. Food intake stimulates secretion of bile salts, which in turn assist transportation of T3 across gut mucosa for intestinal absorption by emulsifying the lipophilic T3. Furthermore, when investigating the transport mechanisms, Fairus *et al.* reported that the absorption rate of tocotrienol homologues in healthy volunteers followed the order: α-T3 > γ-T3 > δ-T3.[53] The researchers postulated that tocotrienols are transferred into plasma through lipoprotein exchanges as they were detected in triacylglycerol-rich particles, LDLs and HDLs.

The physiological effects of T3 supplementation have been investigated on diverse populations including patients with history of white matter lesions, liver disease, cancer patients, hypercholesterolemic and healthy individuals. Most studies have shown that increasing T3 dosage correlates with increased plasma concentration. Nevertheless, inconsistencies have been reported with reference to plasma T3 concentration in different age groups. Chin and co-workers reported that Tri®E supplementation for six months has shown a significant increase of plasma T3 levels in older individuals aged 50 years and above as compared with younger individuals.[54] However, another study by Heng *et al.* showed an increased plasma T3 concentration in younger individuals aged 32 ± 2 years.[55] Volunteers in both studies were healthy individuals supplemented with Tri®E for six months. The discrepancy in the plasma concentrations of young and older individuals may be attributed to several reasons. Differences in the formulation and dosing regimen of tocotrienol supplementation may give rise to these varying results. Moreover, the analytical method to analyse T3 concentration, as reported earlier in this chapter, also notably contributes to this scenario.[56]

For tissue distribution, an investigation on Malaysian patients with breast lumps revealed that daily consumption of a palm oil diet led to accumulation of T3 in breast adipose tissues.[57] Compared to malignant lumps, higher T3 tissue concentrations were detected in benign breast lumps. In another study conducted by Patel *et al.*, α-, β- and γ-T3 were detected in the adipose, brain, heart and liver tissues of patients undergoing surgical procedures after oral supplementation of 400 mg T3 for 12 weeks.[58] Among the isomers, γ-T3 was found to be the highest in the adipose, heart and liver tissues.

2.5 Safety and Tolerance

Patient safety in clinical trials is given the utmost importance while designing and conducting clinical studies. Although drugs discovered in laboratories are primarily subjected to pre-clinical safety assessments, they may not reflect their actual effect or interactions in humans. Similarly, T3 has been subjected to various safety assessments at both pre-clinical levels and clinical trials. Based on the findings, several health authorities have suggested a tolerable safe dosage in humans. The Institute of Medicine (IOM) has recommended the upper level of vitamin E as 200–1000 mg per day (IOM, 2000). The European Food Safety Authority (EFSA) reported that the no observed adverse effect level (NOAEL) in rats fed with 0.19% TRF in a sub-chronic toxicity study is 120 mg kg^{-1} body weight per day for male rats and 130 mg kg^{-1} body weight per day for female rats. Meanwhile, a chronic toxicity study reported NOAEL as 0.4% palm TRF in the diet (or 303 mg kg^{-1} body weight per day for males, and 472 mg kg^{-1} body weight per day for female rats).[59] Besides, an intake of up to 1000 mg of tocotrienol per day would result in a daily intake of 16.7 mg T3 per kg body weight per day for a 60 kg person and would be seven times below the NOAEL of the rat study (EFSA, 2008). In addition, T3 was granted generally regarded as safe (GRAS) status by the FDA in the year 2000 based on its safety and tolerability.[60]

To date, more than 40 clinical trials have been published with various dosages of T3 available in different formulations in diversified study populations.[61,62] Tocotrienol supplementation was shown to be safe and tolerable in these trials. However, one clinical study reported drug-related diarrhoea at 3200 mg vitamin E delta tocotrienol (VEDT) in pre-operative pancreatic cancer patients.[63] The safety data from both animal and human studies are summarized in Tables 2.2 and 2.3.

Table 2.2 Safety data from animal studies using mixed tocotrienols.

Vitamin E composition	Dosage	Duration	Route	Species	Results	References
Palm oil extract (80% tocotrienols)	2.5 mg kg^{-1} BW/d	30 days	Oral	Rat	No abnormalities	64
Vitamin E from rice bran oil (80% tocotrienols, 20% α-tocopherol)	5000 mg kg^{-1}	14 days	Oral	Rat	No abnormalities	65
78.6% tocotrienols, 21.4% α-tocopherol	120 to 130 mg kg^{-1} BW/d or 0.19% in diet	13 weeks	Feed	Rat	No abnormalities	66
Tocotrienols	1000 mg d^{-1}	5 weeks	Feed	Chick	No abnormalities	67
78.7% tocotrienols, 21.3% α-tocopherol	303 mg kg^{-1} BW/d for male rats, 472 mg kg^{-1} BW/d for female rats	52 weeks	Feed	Rat	No abnormalities	59

Table 2.3 Safety data from human studies using mixed tocotrienols.

Population	Number of subjects	Dosage of tocotrienols	Duration	References
Healthy	22	60 mg d^{-1}	6 months	68
Healthy	64	160 mg d^{-1}	6 months	54
Healthy	8	300 mg	Single dose	69
Healthy	36	80–320 mg d^{-1}	8 weeks	43
Healthy	11	800 mg	Single dose	25
Healthy female	16	150 mg d^{-1}	6 months	70
Hypercholesterolemic	44	240 mg d^{-1}	20 weeks	71
Hypercholesterolemic	25	200 mg d^{-1}	4 weeks	72
Hypercholesterolemic	90	15–200 mg d^{-1}	105 days	73
Hypercholesterolemic	32	300 mg d^{-1}	6 months	74
Mildly hypercholesterolemic	40	240 mg d^{-1}	4 weeks	75
Hypercholesterolemic patients with non-alcoholic fatty liver disease	64	400 mg d^{-1}	1 year	76
Patients with carotid atherosclerosis	50	224–336 mg d^{-1}	18 months	77
Women with early stage estrogen positive breast cancer	240	400 mg d^{-1}	5 years	78
Surgical patients	18	400 mg d^{-1}	Minimum 4 weeks	79
Patients with white matter lesions	88	400 mg d^{-1}	2 years	80
Metabolic syndrome	32	400 mg d^{-1}	2 months	62,81

2.6 Nutritional Benefits

According to the IOM and EFSA, the nutritional role of vitamin E is defined as being part of the antioxidant defense system, involved in scavenging free radicals as well as preventing the propagation of free radical reactions. As an essential nutrient for humans, a deficiency in vitamin E will lead to several conditions, including neurological symptoms, impaired absorption of fats and vitamins, cholestatic liver diseases, severe malnutrition, and cystic fibrosis.[2] While vitamin E consists of both tocopherols and tocotrienols, a large pool of literature was found investigating the effects of tocopherols, with little emphasis on the equally, if not more important T3. This section discusses the various nutritional benefits that are associated with tocotrienol supplementation.

2.6.1 Antioxidant

Similar to α-tocopherol, tocotrienols are potent antioxidants, owing to the radical scavenging ability of the hydroxyl group. Studies have been initiated since 1992 to exploit the antioxidant properties of tocotrienols. In a study by Serbinova *et al.*, a palm oil vitamin E mixture (consists of both α-tocopherol and α-T3) was shown to exhibit greater protection for the heart against oxidative stress than α-tocopherol alone.[82] In the following year, α-T3 was found to be a more potent antioxidant than α-tocopherol when they are encapsulated in phosphatidylcholine liposomes. This suggests that α-T3 requires a membrane environment to exhibit greater antioxidant effects.[83] In a human study, the antioxidant effect of T3 was demonstrated in reducing maximal platelet peroxidation, suggesting its role in preventing, and possibly reversing, the natural course of carotid atherosclerosis.[77] On the other hand, Weber *et al.* showed that topical α-T3 supplementation (5% w/vol) for seven days significantly alleviated lipid peroxidation induced by benzoyl peroxide (BPO). A decrease in malondialdehyde (MDA), a marker for oxidative stress, was observed with α-T3 supplementation.[84]

As most studies showed that T3 displays greater antioxidant activities, some studies revealed that T3 and tocopherol are equally effective. Yoshida *et al.* performed a comparative study on the two, and found that both compounds possess similar reactivities toward radicals, and equivalent antioxidant effects against lipid peroxidation in solution and liposomal membranes.[85] Another study using 0.1% Ricetrienol®, containing a mixture of vitamin E, showed that the mixture could protect against oxidative damage in diabetic rats. The raise in plasma MDA was successfully suppressed with one week's supplementation of Ricetrienol®.[86]

Following evidence from pre-clinical studies, a randomized, placebo-controlled human study was conducted to evaluate the antioxidant effects of T3. Healthy volunteers were supplemented with 80 mg, 160 mg or 320 mg of T3-rich vitamin E daily for two months. At the end of the study, the group that received the 320 mg dose presented a significant improvement in total antioxidant status (TAS) by 9.2%, measured by the suppression of radical formation by the supplement. At lower doses, even though the overall plasma concentrations of all T3 isomers were elevated, TAS did not show significant changes.[87] Despite the lack of evidence in healthy subjects, T3 has been shown to increase plasma TAS in hypertensive rats,[88] suggesting that T3 may be more beneficial for hypertensive patients.

2.6.2 Anti-aging

Aging can be categorized as chronological aging and accelerated aging. Chronological aging happens naturally with time. As cellular metabolic and biochemical reactions take place in our body, reactive oxygen species (ROS) are continuously formed. High level of these radicals are detrimental to our body as they can damage sensitive biological structures such as DNA, lipids and proteins.[89] Damaged DNA that remains in the body may cause genetic changes, disrupt cellular homeostasis and lead to age-related biological disorders.[89] On the other hand, exposure to environmental abuse such as UV radiation can cause accelerated aging. This accounts for a wide range of biological effects, such as physiological processes and changes to the skin.[90] Exposure to UV light can cause inflammation and free radical generation, subsequently leading to coarse and deep wrinkles, mottled pigmentation, sallowness and telangiectasia.[91]

In a preliminary study, tocotrienols prevented DNA damage induced by gamma rays.[92] To further investigate this phenomenon, several *in vivo* studies in mice were carried out. Mice pre-treated with γ-T3 before exposure to gamma irradiation were found to be protected from the radiation, survived with peripheral blood cell recovery and increased granulocyte-colony stimulating factor (G-CSF),[93] recovered hematopoietic stem cells (HSCs)[94] and progenitor cells (HPCs).[95] Similar results were observed in skin studies, where tocotrienols supressed DNA damage caused by UV, along with reduced skin inflammation and depigmentation.[91]

The anti-aging effects of tocotrienols were also investigated in human volunteers where a daily dose of 160 mg tocotrienol supplements or a placebo were given to the subjects for six months. Upon analyzing the blood samples, the group that received T3 supplements had a significant reduction in DNA damage at the end of the study. The frequency of sister chromatid exchange (SCE) was also found to be reduced, especially in subjects over the age of 50 years old. As SCE is a measure of double strand break (DSB) repair processes, the lower frequency of SCE means less DSB repair is taking place, hence an indication of less damage occurring that needs repair. Analysis of urine samples shows a reduction in 8-hydroxy-2′-deoxyguanosine (8-OHdG) excretion. This shows that T3 supplementation reduces DNA damage, hence a reduction in DNA repair.[96]

2.6.3 Neuroprotection

About 20 years ago, tocotrienols were first reported to modulate glutamate-induced neurodegeneration. In rat HT4 hippocampal neuronal cells treated with glutamate, α-tocotrienols were able to block glutamate-induced cell death even at a nanomolar concentration range. It was later found that α-tocotrienols exerted their neuroprotection effects *via* several mechanisms: by suppressing the glutamate-induced early activation of c-Src kinase,[97] hindering the access of arachidonic acid to the catalytic site of 12-lipoxygenase (12-LOX),[98] as well as attenuating phosphorylation of cPLA2.[99] Further to *in vitro* studies, animal models of stroke using rats and canines confirmed the neuroprotective effect of tocotrienols.[100,101] Oral supplementation of α-tocotrienols in the rat models resulted in an increased level of α-tocotrienols in the brains, thereby protecting the rats against stroke-induced injury.[100] In canine models, tocotrienol-enriched supplementation significantly reduced the ischemic stroke-induced lesion volume, prevented the loss of white matter fiber tract connectivity after stroke, and improved the cerebrovascular collateral circulation to the ischemic middle cerebral artery territory.[101] These pre-clinical studies provided strong evidence of the neuroprotection effects of tocotrienols.

Moving forward, two clinical trials were conducted to further investigate the efficacy of tocotrienols in neuroprotection.[80,102] The first clinical trial was conducted on volunteers with cardiovascular risk factors and white matter lesions (WML). The volunteers were randomized to receive either 200 mg mixed tocotrienols or a matching placebo twice a day for two years. The WML volume of the placebo group was found to have increased after two years, whereas that of the tocotrienol-supplemented group remained unchanged. The results indicated that T3 has a protective effect on brain white matter, reducing the progression of WML, protecting the brain against neuropathological changes and tissue damage.[80] Another recently completed clinical trial reported preliminary improvements in lancinating pain in a sub-group of diabetic patients supplemented with 400 mg mixed T3 a day for one year. Nevertheless, an overall improvement in neuropathic symptoms was not seen. Therefore, further investigation is warranted to look into the preliminary results observed in this study.[102]

2.6.4 Anti-inflammation

Inflammation is part of the complex biological response of body tissues to irritants and can be characterized by redness and swelling. Chronic inflammation is associated with asthma, cardiovascular diseases, autoimmune disorders, and cancers, to name a few.[103] Tocotrienols have been shown to suppress inflammation *via* inhibition of certain proteins and enzymes, such as COX-2, NFkB, STAT3, inducible nitric oxide synthase (iNOS),[104–106] proving their role in anti-inflammation as well as prevention of diseases arising from chronic inflammation.

In a study in diabetic rats, T3 in combination with insulin reversed neuropathic pain by modulating oxidative-nitrosative stress, inflammatory cytokine release and caspase-3, whereby administration of insulin alone only resulted in anti-hyperglycaemia and partial reversal of the pain response. Plasma pro-inflammatory cytokine tumour necrosis factor (TNF)-α, transforming growth factor (TGF)-β and interleukin (IL)-1β levels were decreased in a dose-dependent manner upon exposure to T3. The study also showed that T3 was able to restore lipid peroxidation, non-protein thiols, superoxide dismutase and catalase activities in the sciatic nerves of diabetic rats.[107] Tocotrienols were also proven to be superior to α-tocopherol in suppressing pro-inflammatory cytokines and protecting rat bones against osteoporosis.[108] A study in mice showed that γ-T3 modulated the inflammation process by decreasing plasma levels of pro-inflammatory cytokines, decreasing macrophage infiltration, suppressing MAP kinase and inhibiting NFKB activation.[109] The anti-inflammatory activities of T3 were further evidenced by Wong *et al.* who showed that α-, γ-, and δ-T3 reduced inflammatory cell infiltration to the hearts of rats fed with a high-carbohydrate, high-fat diet.[110]

Among the chronic diseases related to inflammation, cancer has been extensively investigated for tocotrienols.[105] Down regulations of NFkB were reported in human breast and lung cancer cells.[111] In addition, suppression of c-reactive protein and tumour necrosis factor α were reported in arthritis rat models fed with γ-tocotrienol.[112] Active studies in chronic diseases proved tocotrienols as a potential nutraceutical agent, especially for clinical application.

2.7 Research Gap

A recognized analytical method for tocotrienols is still lacking despite years of research in laboratory analysis. On the contrary, the American Oil Chemist Society (AOCS) has published an official method to analyze α-tocopherol, which has been widely applied in most certified laboratories.[113] Being a natural compound, the composition of tocotrienols often varies between sources or even batches of production. Issues arise when the research community does not have a reference method to quantify tocotrienols, and results were based on methods developed in individual laboratories. The root cause for this major stumbling block is the lack of certified reference materials for tocotrienols. The isolation of individual tocotrienol homologues is a complex and expensive procedure, with a high variation in purity. Fortunately, certified reference materials have been made available in recent years. Although they come with a high cost, method validation can be carried out to accurately quantify tocotrienols. Work focusing on method validation to address all analytes of interest warrants an enormous effort and would be highly appreciated.

Referring to the EFSA, AI values were derived from epidemiology data on blood levels of α-tocopherol. Data from a large population helps to confirm the association between α-tocopherol levels and nutritional functions. For example, a primary vitamin E deficiency associated with neurological symptoms was reported when serum α-tocopherol levels fell below 2.3 μmol L^{-1} (EFSA). Nevertheless, such data is not available for tocotrienols. Preliminary efforts were published by Mangialasche *et al.* where a lower risk of cognitive impairment was seen in subjects with higher serum levels of γ-tocopherol, β-tocotrienol and total tocotrienols.[114] The limitation of this study is the small sample size (n = 140) from a limited geographical background. Therefore, a joint effort from international institutions is needed to address these shortcomings.

In the latest revision of the IOM report in 2000, only α-tocopherol was recognized as the active form of vitamin E. Despite having similar nutritional functions or rather beyond the function of vitamin E, the role of tocotrienols has been underestimated. Looking at the statistics, the number of tocotrienol research papers was much higher from the year 2000 onwards. While the EFSA revised the report on vitamin E in 2015, most publications on tocotrienol clinical studies were still in the pipeline back then. With a strategic research direction, a concerted effort to address global nutrition and food authorities with credible research data is imperative.

2.8 Conclusion

In conclusion, a large amount of research findings on tocotrienols have been translated from laboratory to human studies. Collective data concluded that tocotrienols behave similarly to tocopherols in terms of chemical structures, analytical methods and safety profiles but exhibit distinctive characteristics in terms of bioavailability and nutritional values. Based on their health benefits, the potential of tocotrienols is beyond vitamin E, as they demonstrate the properties of nutraceuticals with the added advantage of low toxicity. In this chapter, we have identified a few challenges to be addressed. Studies in these areas will bridge the gap in the translation of tocotrienols from bench to bedside.

References

1. K. Nesaretnam, *Cancer Lett.*, 2008, **269**, 388.
2. EFSA Panel on Dietetic Products, Nutrition and Allergies, Scientific opinion on dietary reference values for vitamin E as alpha-tocopherol, *EFSA J.*, 2015, 1.
3. H. Y. Peh, W. S. Tan, W. Liao and W. S. Wong, *Pharmacol. Ther.*, 2016, **162**, 152.
4. R. Brigelius-Flohe, F. J. Kelly, J. T. Salonen, J. Neuzil, J. M. Zingg and A. Azzi, *Am. J. Clin. Nutr.*, 2002, **76**, 703.
5. Food and Nutrition Board, Institute of Medicine, *Dietary Reference Intakes: Vitamin C, Vitamin E, Selenium, and Carotenoids*, National Academy Press, Washington, DC, 2000.

6. B. B. Aggarwal, C. Sundaram, S. Prasad and R. Kannappan, *Biochem. Pharmacol.*, 2010, **80**, 1613.
7. J. Y. Fu, T. T. Htar, L. De Silva, D. M. Tan and L. H. Chuah, *Molecules*, 2017, **22**, 233.
8. N.C.f.B. Information, PubChem Compound Database, CID=9929901.
9. S. C. Cunha, J. S. Amaral, J. O. Fernandes and M. B. Oliveira, *J. Agric. Food Chem.*, 2006, **54**, 3351.
10. E. D. Tsochatzis, K. Bladenopoulos and M. Papageorgiou, *J. Sci. Food Agric.*, 2012, **92**, 1732.
11. G. Panfili, A. Fratianni and M. Irano, *J. Agric. Food Chem.*, 2003, **51**, 3940.
12. R. C. Alves, S. Casal and M. B. P. P. Oliveira, *Food Sci. Technol. Int.*, 2009, **15**, 57.
13. M. Ryynänen, A. M. Lampi, P. Salo-Väänänen, V. Ollilainen and V. Piironen, *J. Food Compos. Anal.*, 2004, **17**, 749.
14. H. Schwartz, V. Ollilainen, V. Piironen and A.-M. Lampi, *J. Food Compos. Anal.*, 2008, **21**, 152.
15. M. H. Ng, Y. M. Choo, A. N. Ma, C. H. Chuah and M. A. Hashim, *Lipids*, 2004, **39**, 1031.
16. P. T. Gee, C. Y. Liew, M. C. Thong and M. C. L. Gay, *Food Chem.*, 2016, **196**, 367.
17. W. Schüep and R. Rettenmaier, *Methods Enzymol.*, 1994, **234**, 294.
18. S. Liang, X. Bian, J. Ma, M. Arogunjo, A. A. Deorukhkar, S. Krishnan and H. Xie, *Biomed. Chromatogr.*, 2013, **27**, 58.
19. S. Strohschein, C. Rentel, T. Lacker, E. Bayer and K. Albert, *Anal. Chem.*, 1999, **71**, 1780.
20. C. Lauridsen, S. W. Leonard, D. A. Griffin, D. C. Liebler, T. D. McClure and M. G. Traber, *Anal. Biochem.*, 2001, **289**, 89.
21. R. Zhang, W. Shen and X. Wei, *et al.*, *Anal. Methods*, 2016, **8**, 7341.
22. M.-J. Lee, W. Feng, L. Yang, Y.-K. Chen, E. Chi, A. Liu and C. S. Yang, *J. Food Drug Anal.*, 2018, **26**, 318.
23. S. P. Yap, T. Julianto, J. W. Wong and K. H. Yuen, *J. Chromatogr. Biomed. Sci. Appl.*, 1999, **735**, 279.
24. H. L. Che, D. M. Tan, P. Meganathan, Y. L. Gan, G. Abdul Razak and J. Y. Fu, *Int. J. Anal. Chem.*, 2015, **2015**, 357609.
25. P. Meganathan, R. S. Jabir and H. G. Fuang, *et al.*, *Sci. Rep.*, 2015, **5**, 13550.
26. J. Fuchs, S. Weber, M. Podda, N. Groth, T. Herrling, L. Packer and R. Kaufmann, *Free Radical Biol. Med.*, 2003, **34**, 330.
27. E. Bartosińska, J. Jacyna, A. Borsuk-De Moor, M. Kaliszan, W. J. Kruszewski, Z. Jankowski and D. Siluk, *Talanta*, 2018, **176**, 108.
28. I. D. Desai, *Methods Enzymol.*, 1984, **105**, 138.
29. K. Nesaretnam, P. A. Gomez, K. R. Selvaduray and G. A. Razak, *Asia Pac. J. Clin. Nutr.*, 2007, **16**, 498.
30. A. Kamal-Eldin, S. Görgen, J. Pettersson and A.-M. Lampi, *J. Chromatogr. A*, 2000, **881**, 217.

31. A. Kato, A. Gapor, K. Tanabe, M. Yamaoka and H. Mamuro, *J. Jap. Oil Chem. Soc.*, 1981, **30**, 590.
32. T. S. Shin and J. S. Godber, *J. Am. Oil Chem. Soc.*, 1993, **70**, 1289.
33. B. Butinar, M. Bučar-Miklavčič, C. Mariani and P. Raspor, *Food Chem.*, 2011, **128**, 505.
34. J. N. Thompson and G. Hatina, *J. Liq. Chromatogr.*, 1979, **2**, 327.
35. W. Tsuzuki, R. Yunoki and H. Yoshimura, *Lipids*, 2007, **42**, 163.
36. S. P. Yap, K. H. Yuen and A. B. Lim, *J. Pharm. Pharmacol.*, 2003, **55**, 53.
37. I. Ikeda, Y. Imasato, E. Sasaki and M. Sugano, *Int. J. Vitam. Nutr. Res.*, 1996, **66**, 217.
38. B. Abuasal, P. Sylvester and A. Kaddoumi, *Drug Metab. Dispos.*, 2010, **38**, 939.
39. M. Okabe, M. Oji, I. Ikeda, H. Tachibana and K. Yamada, *Biosci. Biotechnol. Biochem.*, 2002, **66**, 1768.
40. L. Deng, Y. Peng, Y. Wu, M. Yang, Y. Ding, Q. Chen and Q. Fu, *Lipids Health Dis.*, 2014, **13**, 66.
41. N. Akaho, J. Takata and T. Fukushima, *et al.*, *Drug Metab. Dispos.*, 2007, **35**, 1502.
42. C. M. Compadre, A. Singh and S. Thakkar, *et al.*, *Drug Dev. Res.*, 2014, **75**, 10.
43. A. H. G. Rasool, K. H. Yuen, K. Yusoff, A. R. Wong and A. R. A. Rahman, *J. Nutr. Sci. Vitaminol.*, 2006, **52**, 473.
44. S. Alqahtani, A. Alayoubi, S. Nazzal, P. W. Sylvester and A. Kaddoumi, *AAPS J.*, 2013, **15**, 684.
45. N. Miyoshi, Y. Wakao, S. Tomono, M. Tatemichi, T. Yano and H. Ohshima, *J. Nutr. Biochem.*, 2011, **22**, 1121.
46. G. A. Ledet, S. Biswas and V. P. Kumar, *et al.*, *Int. J. Mol. Sci.*, 2017, **18**, 28.
47. J. Pham, A. Nayel, C. Hoang and T. Elbayoumi, *Drug Delivery*, 2016, **23**, 1514.
48. P. S. Goh, M. H. Ng, Y. M. Choo, N. B. Amru and C. H. Chuah, *Molecules*, 2015, **20**, 19936.
49. J. Y. Fu, W. Zhang, D. R. Blatchford, L. Tetley, G. McConnell and C. Dufès, *J. Controlled Release*, 2011, **154**, 20.
50. H. Ali, A. B. Shirode, P. W. Sylvester and S. Nazzal, *Colloids Surf., A*, 2010, **353**, 43.
51. D. M. Tan, J. Y. Fu, F. S. Wong, H. M. Er, Y. S. Chen and K. Nesaretnam, *Nanomedicine*, 2017, **12**, 2487.
52. S. P. Yap, K. H. Yuen and J. W. Wong, *J. Pharm. Pharmacol.*, 2001, **53**, 67.
53. S. Fairus, R. M. Nor, H. M. Cheng and K. Sundram, *Nutr. J.*, 2012, **11**, 5.
54. S.-F. Chin, N. Aini and A. Hamidm, *et al.*, *Nutrition*, 2008, **24**, 1.
55. E. C. Heng, S. A. Karsani, M. Abdul Rahman, N. A. Abdul Hamid, Z. Hamid and W. Z. Wan Ngah, *Eur. J. Nutr.*, 2013, **52**, 1811.
56. J.-Y. Fu, H.-L. Che, D. M.-Y. Tan and K.-T. Teng, *Nutr. Metab.*, 2014, **11**, 5.

57. K. Nesaretnam, P. A. Gomez, K. R. Selvaduray and G. A. Razak, *Asia Pac. J. Clin. Nutr.*, 2007, **16**, 498.
58. V. Patel, C. Rink and G. M. Gordillo, *et al.*, *J. Nutr.*, 2012, **142**, 513.
59. M. Tasaki, T. Umemura and T. Inoue, *et al.*, *Toxicology*, 2008, **250**, 143.
60. U. S. Food and Drug Administration, GRAS Notices GRN No. 307, Palm oil-derived tocols with tocotrienols and α-tocopherol as the principal components, Washington, DC, 2010.
61. P. Meganathan and J. Y. Fu, *Int. J. Mol. Sci.*, 2016, **17**, 1682.
62. Y. L. Gan, J. Y. Fu and O. M. Lai, *et al.*, *Sci. Rep.*, 2017, **7**, 11542.
63. G. M. Springett, K. Husain and A. Neuger, *et al.*, *EBioMedicine*, 2015, **2**, 1987.
64. S. L. Oo, P. Chang and K. E. Chan, *Nutr. Res.*, 1992, **12**, S217.
65. A. Sheppard, J. A. T. Pennington and J. L. Weihrauch, in *Vitamin E in Health and Disease*, ed. L. Packer, Marcel Dekker Inc, New York, 1993.
66. H. Nakamura, F. Furukawa, A. Nishikawa, M. Miyauchi, H. Y. Son, T. Imazawa and M. Hirose, *Food Chem. Toxicol.*, 2001, **39**, 799.
67. S. G. Yu, A. M. Thomas, A. Gapor, B. Tan, N. Qureshi and A. A. Qureshi, *Lipids*, 2006, **41**, 453.
68. D. T. Tan, H. T. Khor, W. H. Low, A. Ali and A. Gapor, *Am. J. Clin. Nutr.*, 1991, **53**, 1027S.
69. S. P. Yap, K. H. Yuen and J. W. Wong, *J. Pharm. Pharmacol.*, 2001, **53**, 67.
70. E. C. Heng, S. A. Karsani, M. Abdul Rahman, N. A. Abdul Hamid, Z. Hamid and W. Z. Wan Ngah, *Eur. J. Nutr.*, 2013, **52**, 1811.
71. M. L. Wahlqvist, Z. Krivokuca-Bogetic, C. S. Lo, B. Hage, R. Smith and W. Luldto, *Nutr. Res.*, 1992, **12**, S181.
72. A. A. Qureshi, N. Qureshi and J. J. Wright, *et al.*, *Am. J. Clin. Nutr.*, 1991, **53**, 1021S.
73. A. A. Qureshi, S. A. Sami, W. A. Salser and F. A. Khan, *Atherosclerosis*, 2002, **161**, 199.
74. K. H. Yuen, J. W. Wong, A. B. Lim, B. H. Ng and W. P. Choy, *Funct. Foods Health Dis.*, 2012, **3**, 106.
75. R. P. Mensink, A. C. van Houwelingen, D. Kromhout and G. Hornstra, *Am. J. Clin. Nutr.*, 1999, **69**, 213.
76. E. Magosso, M. A. Ansari and Y. Gopalan, *et al.*, *Nutr. J.*, 2013, **12**, 166.
77. A. C. Tomeo, M. Geller, T. R. Watkins, A. Gapor and M. L. Bierenbaum, *Lipids*, 1995, **30**, 1179.
78. K. Nesaretnam, K. R. Selvaduray, G. Abdul Razak, S. D. Veerasenan and P. A. Gomez, *Breast Cancer Res.*, 2010, **12**, R81.
79. V. Patel, C. Rink and G. M. Gordillo, *et al.*, *J. Nutr.*, 2012, **142**, 513.
80. Y. Gopalan, I. L. Shuaib and E. Magosso, *et al.*, *Stroke*, 2014, **45**, 1422.
81. Y. Gan, O. Lai and B. Chew, *et al.*, *J. Oil Palm Res.*, 2016, **28**, 34.
82. E. Serbinova, S. Khwaja and J. Catudioc, *et al.*, *Nutr. Res.*, 1992, **12**, S203.
83. Y. J. Suzuki, M. Tsuchiya, S. R. Wassall, Y. M. Choo, G. Govil, V. E. Kagan and L. Packer, *Biochemistry*, 1993, **32**, 10692.

84. S. U. Weber, J. J. Thiele, N. Han, C. Luu, G. Valacchi, S. Weber and L. Packer, *Free Radical Biol. Med.*, 2003, **34**, 170.
85. Y. Yoshida, E. Niki and N. Noguchi, *Chem. Phys. Lipids*, 2003, **123**, 63.
86. Y. Kanaya, T. Doi and H. Sasaki, *et al.*, *Diabetes Res. Clin. Pract.*, 2004, **66**, S157.
87. A. H. Rasool, K. H. Yuen, K. Yusoff, A. R. Wong and A. R. A. Rahman, *J. Nutr. Sci. Vitaminol.*, 2006, **52**, 473.
88. N. Ma and N. Nna, *Clin. Exp. Hypertens.*, 1999, **21**, 1297.
89. C. L. Powell, J. A. Swenberg and I. Rusyn, *Cancer Lett.*, 2005, **229**, 1.
90. G. J. Fisher, S. Kang and J. Varani, *et al.*, *Arch. Dermatol.*, 2002, **138**, 1462.
91. W. N. Yap, *J. Cosmet. Dermatol.*, 2017, **17**, 555.
92. V. K. Singh, V. Srinivasan and R. Toles, *et al.*, *Proceedings of Human Factors and Medicine Research Task Group NATO RTG 099 Radiation Bioeffects and Countermeasures*, 2005, AFRRI CD 05-02, pp. 16.1.
93. V. K. Singh, R. L. Shafran, W. E. Jackson, T. M. Seed and K. S. Kumar, *Exp. Mol. Pathol.*, 2006, **81**, 55.
94. M. Berbée, Q. Fu, M. Boerma, J. Wang, K. S. Kumar and M. Hauer-Jensen, *Radiat. Res.*, 2009, **171**, 596.
95. S. Kulkarni, S. P. Ghosh and M. Satyamitra, *et al.*, *Radiat. Res.*, 2010, **173**, 738.
96. S.-F. Chin, N. A. A. Hamid and A. A. Latiff, *et al.*, *Nutrition*, 2008, **24**, 1.
97. C. K. Sen, S. Khanna, S. Roy and L. Packer, *J. Biol. Chem.*, 2000, **275**, 13049.
98. S. Khanna, S. Roy, H. Ryu, P. Bahadduri, P. W. Swaan, R. R. Ratan and C. K. Sen, *J. Biol. Chem.*, 2003, **278**, 43508.
99. S. Khanna, L. Parinandi Narasimham, R. Kotha Sainath, S. Roy, C. Rink, D. Bibus and C. K. Sen, *J. Neurochem.*, 2010, **112**, 1249.
100. S. Khanna, S. Roy and A. Slivka, *et al.*, *Stroke*, 2005, **36**, 2258.
101. C. Rink, G. Christoforidis and S. Khanna, *et al.*, *J. Cereb. Blood Flow Metab.*, 2011, **31**, 2218.
102. The Vitamin E in Neuroprotection Study Investigators, *JAMA Neurol.*, 2018, **75**, 444.
103. P. Hunter, *EMBO Rep.*, 2012, **13**, 968.
104. M. Kaileh and R. Sen, *J. Am. Coll. Nutr.*, 2010, **29**, 334S.
105. K. Nesaretnam and P. Meganathan, *Ann. N. Y. Acad. Sci.*, 2011, **1229**, 18.
106. S. J. Wu, P. L. Liu and L. T. Ng, *Mol. Nutr. Food Res.*, 2008, **52**, 921.
107. A. Kuhad and K. Chopra, *Neuropharmacology*, 2009, **57**, 456.
108. A. S. Nazrun, M. Norazlina, M. Norliza and S. I. Nirwana, *Adv. Pharmacol. Sci.*, 2012, **2012**, 7.
109. L. Zhao, I. Kang and X. Fang, *et al.*, *Int. J. Obes.*, 2014, **39**, 438.
110. W.-Y. Wong, L. C. Ward, C. W. Fong, W. N. Yap and L. Brown, *Eur. J. Nutr.*, 2017, **56**, 133.
111. K. S. Ahn, G. Sethi, K. Krishnan and B. B. Aggarwal, *J. Biol. Chem.*, 2007, **282**, 809.

112. A. Radhakrishnan, D. Tudawe, S. Chakravarthi, G. S. Chiew and N. Haleagrahara, *Exp. Ther. Med.*, 2014, **7**, 1408.
113. AOCS, Official Methods and Recommended Practices of the American Oil Chemists' Society 5th edition A.O.C.S. official method ce 8-89 (reapproved 1997) determination of tocopherols and tocotrienols in vegetable oils and fats by HPLC, American Oil Chemists' Society, Champaign, IL, 1997.
114. F. Mangialasche, A. Solomon and I. Kåreholt, *et al.*, *Exp. Gerontol.*, 2013, **48**, 1428.

CHAPTER 3

The Behaviour of Vitamin E in Membranes

J. ATKINSON*[a], D. MARQUARDT[b] AND T. HARROUN[c]

[a]Brock University, Department of Chemistry, 1812 Sir Isaac Brock Way, St. Catharines, Ontario, L2S3A1, Canada; [b]University of Windsor, Department of Chemistry and Biochemistry, 401 Sunset Ave., Windsor, Ontario, N9B3P4, Canada; [c]Brock University, Department of Physics, 1812 Sir Isaac Brock Way, St. Catharines, Ontario, L2S3A1, Canada
*E-mail: jatkinson@brocku.ca

3.1 Introduction

The study of tocopherols and tocotrienols in the vitamin E family has the ultimate goal of describing their effect on organism health, and so the literature is replete with nutritional studies attempting to link diets and vitamin doses to indicators of animal health. However, a more complete understanding of the action of α-tocopherol (α-Toc), the most important form of vitamin E in mammals, also requires a molecular level description. Since the majority of the non-adipose pool of vitamin E is found in tissue cellular membranes,[1] it is critical to understand the behavior of α-tocopherol in phospholipid bilayer membranes.

There are several quality reviews on the biophysical investigation of α-Toc in membranes.[2–8] In this review we will focus on recent (since ~2008) biophysical studies on the membrane behavior of α-Toc as well as biochemical investigations where membrane structure was a consideration.

Food Chemistry, Function and Analysis No.11
Vitamin E: Chemistry and Nutritional Benefits
Edited by Etsuo Niki

Published by the Royal Society of Chemistry, www.rsc.org

3.2 Membrane Localization, Stabilization, and Fluidity

Two main questions are asked when enquiring how α-Toc resides in membranes: (1) the orientation and depth of the phenol (chromanol) group, and (2) the preferred location of α-Toc in particular lipid mixtures or phases. A previous review[6] has summarized many of the early results for the position of the chromanol from fluorescence and quenching probes, concluding that the chromanol hydroxyl group (a phenol) likely resides approximately at the same depth as the glycerol ester groups in phosphatidylcholine bilayers such as DOPC[9] (1,2-dioleoyl-*sn*-glycero-3-phosphocholine) and DSPC (1,2-distearoyl-*sn*-glycero-3-phosphocholine), but is closer to the choline group in mixed chain lipids like POPC (1-palmitoyl-2-oleoyl-*sn*-glycero-3-phosphocholine). Results from neutron diffraction studies unambiguously show that the phenol of α-Toc is located high in the bilayer near the lipid–water interface.[10]

3.2.1 Transverse Location of Vitamin E

Recent studies using probe molecules have predominately used deuterated α-Toc and neutron diffraction to determine the transverse location of tocopherol in model membranes. These studies examined the location of α-Toc in multi-bilayer stacks deposited on a solid substrate when the bilayers were hydrated using a saturated salt solution which generated a relative humidity of ~94%. In a series of publications, the location of α-[5-$^2H_3$]tocopherol (α-Toc-5_{d3}) was determined in DPPC (1,2-dipalmitoyl-*sn*-glycero-3-phosphocholine), POPC, DOPC, PAPC (1-palmitoyl-2-arachidonoyl-*sn*-glycero-3-phosphocholine), and DAPC[10] (1,2-diarachidonoyl-*sn*-glycero-3-phosphocholine), DMPC[11] (1,2-dimyristoyl-*sn*-glycero-3-phosphocholine), POPS (1-palmitoyl-2-oleoyl-*sn*-glycero-3-phospho-L-serine), POPE (1-palmitoyl-2-oleoyl-*sn*-glycero-3-phosphoethanolamine), egg sphingomyelin (SM),[12] SOPC (1-stearoyl-2-oleoyl-*sn*-glycero-3-phosphocholine), and SDPC (1-stearoyl-2-docosahexaenoyl-*sn*-glycero-3-phosphocholine).[13] The SOPC and SDPC experimental results were further augmented with molecular dynamic (MD) simulations. With the exception of DMPC (discussed later), the αToc-5_{d3} label is relatively invariant in the phospholipid environment and located at or near the lipid–water interface, regardless of headgroup or fatty acid chain unsaturation. On average, the αToc-5_{d3} is located approximately 2–5 Å above the glycerol moiety of the lipid bilayer towards the aqueous phase. This is slightly higher than the location observed in the MD simulation. Comparison of diffraction and MD data on the same lipid system (SOPC and SDPC) showed the C5 methyl group of α-Toc to be ~2 Å above the lipid carbonyl groups for the experimental, whereas simulations gave a location ~4 Å below the carbonyl groups.[13]

Earlier MD simulations performed on various PC and PE lipids yielded the hydroxyl oxygen of α-Toc near (within ~1–4 Å) the *sn*-1 of the carbonyl carbon of the phospholipid.[14,15] This location was demonstrated for simulations performed at both 280 K and 310 K. In the same study, it was observed that α-Toc is driven up to 8 Å deeper into the hydrocarbon core at an elevated temperature of 350 K. These high temperature results should be viewed with caution; contrary to the MD results, experimentally determined bilayer thicknesses and carbonyl depths decrease with increasing temperature.[16]

Using phospholipids labeled with a fluorescent quenching doxyl moiety, Ausili *et al.* were able to reproduce, qualitatively, the probe-free results of Marquardt *et al.* Specifically, they report that the location of α-Toc is independent of the degree of unsaturation of the fatty acid chain.[17] Discrepancies observed between the studies may stem from using a dynamic quenching probe. For example, the doxyl transverse-diffusion in the bilayer is 10-times faster for 14-doxyl than 5-doxyl, and the effective label location difference between 5-doxyl and 14-doxyl is ~5 Å.[18] Other complicating influences could be from the absolute concentrations of quencher present, which were up to 20 mol%.

It is remarkable that very different approaches to determining the location of α-Toc in phospholipid bilayers seem to arrive at a similar conclusion. The literature argues over angstrom differences in the phenolic headgroup position; however, a membrane environment is so dynamic that these differences ultimately do not change many of the biologically relevant conclusions drawn in these works. The main conclusion is that α-Toc is located near the lipid–water interface, a location that is consistent with its antioxidant mechanism. At the core of this antioxidant mechanism is the synergism between α-Toc intercepting water-borne radicals and the mobility of PUFA chains,[19] which can "snorkel" to the surface of the bilayer, allowing α-Toc to terminate lipid peroxyl radicals centered on the PUFA chain.[10] The original suggestion of a bobbing or floating lipid peroxyl radical was made by Barclay and Ingold[20,21] and later molecular dynamics simulations appear to support this model,[22] but recent, more comprehensive, work has calculated the partition coefficients of the peroxyl radical and hydroperoxide species and clearly showed that the radical species (ROO$^•$) is about two orders of magnitude more hydrophobic than the hydroperoxide (ROOH) and, on average, spends more of its time deeper in the bilayer.[23] The "floating peroxyl radical" is thus a mis-appreciation of (as stated earlier) the true dynamic nature of a phospholipid bilayer, where even a lipid peroxyl radical can be expected to range widely in location, from the bilayer normal to close to the headgroups. Once formed, the more polar LOOH molecule is prone to translocation across the bilayer and even intermembrane transfer.[24]

3.2.2 DMPC – The Exception

Despite the biological irrelevance of DMPC, the location of α-Toc in DMPC bilayers has long served as a model membrane system and has been plagued with inconsistent observations.

Curiously, α-Toc is observed to be high in the bilayer when the intrinsic fluorescence of tocopherol is measured in the presence of the quenching 5-doxyl stearic acid.[25] However, when examined either directly or indirectly in the absence of molecular probe molecules, α-Toc is observed deep in the hydrophobic core of the bilayer.[11,25,26] For example, neutron diffraction studies, using deuterated α-Toc, have reported α-Toc at the center of a DMPC bilayer. Quasi-elastic neutron scattering (QENS) studies have further postulated that α-Toc resides deep in the hydrophobic core of a DMPC bilayer.[27] Sharma *et al.*[27] observed a two-fold effect of α-Toc on DMPC dynamics: α-Toc had a fluidizing effect at low temperature and a stiffening effect at high temperature. These observed dynamics, coupled with the lack of an observed impact on lipid lateral diffusion, is best explained by α-Toc being positioned deep in the membrane. Further, it was observed that α-Toc could not be oxidized by waterborne O_2^- or H_2O_2 unless lipid peroxides were also included,[26] strongly suggesting that tocopherol was away from the lipid–water interface. Finally, FT-IR investigations of the C=O and P=O stretches of DMPC bilayers containing α-Toc showed no change even at α-Toc concentrations as high as 20 mol%, indicating that there were no hydrogen bonds between the OH-group of α-Toc with the lipid headgroups.[25] However, systems containing 25 mol% α-Toc or more do yield changes to the C=O and P=O stretches.[28]

These discrepancies very likely arise from conclusions drawn on data from different experimental systems. Much of the work that reports α-Toc being distant from the lipid headgroups was performed in systems containing 20 mol% α-Toc or less[11,25,27] and the work reporting α-Toc near the headgroups used concentrations of α-Toc >20 mol%, as well as elevated concentrations of doxyl-containing probes (4–20 mol%).[28]

3.2.3 Vitamin E Diffusion

3.2.3.1 Lateral

As with the location of α-Toc, its rate of lateral diffusion is also contested in the literature. Over the past three decades, a number of studies have strived to determine the lateral diffusion rate of α-Toc. There is very little agreement among published diffusion coefficients, which range from 1.8 (±0.1) × 10^{-7} cm^2 s^{-1} (POPC)[29] to 4.8 × 10^{-6} cm^2 s^{-1}[30] in fluid membranes primarily composed of POPC (egg PC). Within this range falls the diffusion coefficient of α-Toc in fluid DPPC (at 52 °C), which has been reported to be 2.5 × 10^{-7} cm^2 s^{-1}[31] (Figure 3.1). In all these cases, the diffusion was determined by lateral diffusion-controlled fluorescence quenching, a strategy that relies on the introduction of a quenching molecule to the system.

Sonnen *et al.* report that there are no notable differences between the diffusion of α-, β-, γ- and δ-tocopherol, which could be considered surprising as one would expect a dependence on the chromanol ring substitution due to steric differences. Similarly, no difference in diffusion coefficients were observed between tocopherols and tocotrienols.[32]

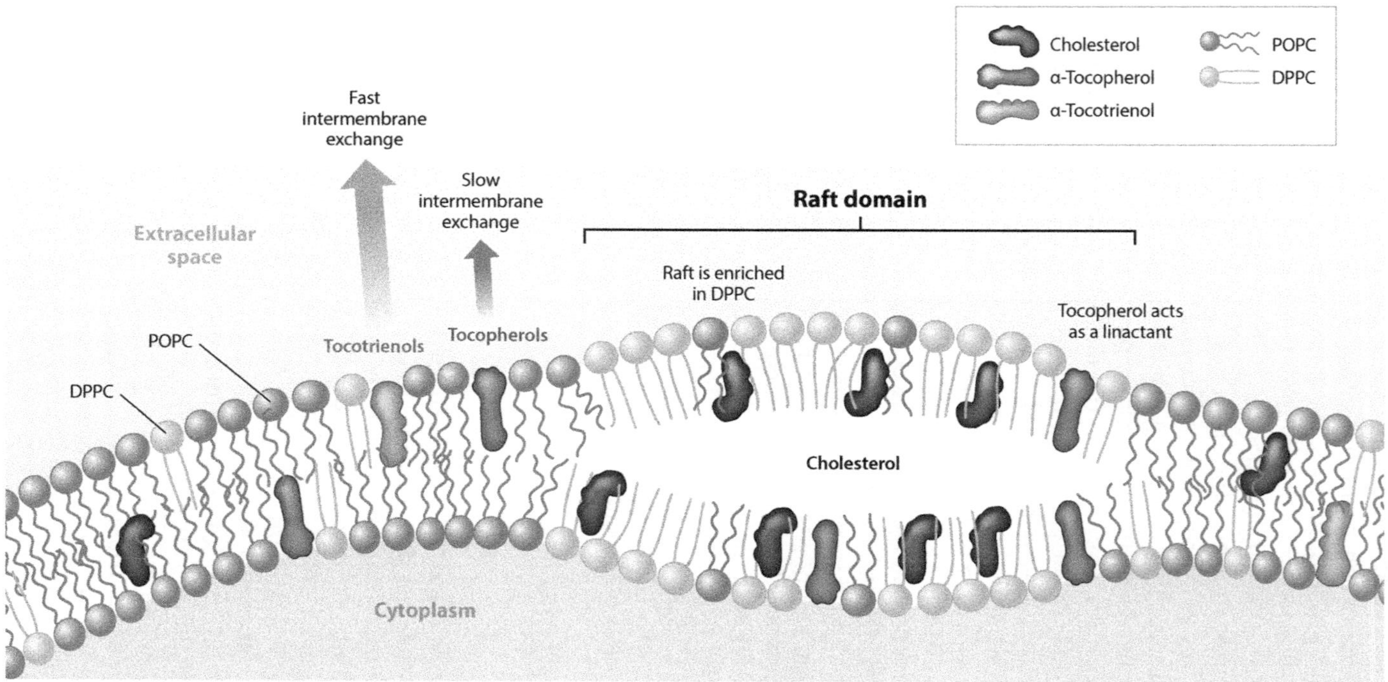

Figure 3.1 Cartoon representation of the orientation and behavior of tocopherols and tocotrienols in phospholipid membranes. The insertion of cholesterol into the raft domain separates phospholipid spacing and thickens the membrane. The normal occurrence of saturated sphingomyelin lipids has been omitted for clarity. The thinner mixed-phospholipid bilayer is shown with partial mono-unsaturated composition (DOPC) but would also contain other polyunsaturated lipids. Scales and bilayer position are only representative. (Illustration by Fiona R. Martin.)

Interestingly, it has been reported that α-Toc does not significantly alter the lateral diffusion of the phospholipid itself.[32] Such an observation is consistent with the dynamics work of Sharma *et al.*[27] Further, it has been stated that α-Toc does not diffuse exceedingly quickly compared to phospholipids.[29]

3.2.3.2 Transverse Diffusion

Experimentally, the transverse diffusion (flip-flop, translocation) of α-Toc has been vastly overlooked. There are a handful of publications inferring the translocation rate of α-Toc to be quite slow, on the order of hours.[33–35] Leng *et al.* report unpublished data where α-Toc in vesicles showed complete loss of α-Toc fluorescence on both sides of the bilayer within minutes of exposure to the oxidant potassium ferricyanide.[13,36] These results are in sharp contrast to the long translocation times reported decades ago, and the reasons for this remain unknown.

MD simulations have given a picture of α-Toc translocation. Qin and Yu determined that α-Toc translocation events only occur at elevated temperatures (350 K) in DPPC, DPPE, POPC, and POPE, and they remark that the translocation of α-Toc is much slower than cholesterol, likely due to a more polar head compared to cholesterol.[14] A subsequent MD study reports multiple translocation events in fluid DMPC, DMPE, DSPC, and DSPE over 200 ns simulations.[15]

Contrary to some previous results, Leng *et al.*[13] report that the translocation half time ($t_{1/2}$) of α-Toc is 0.11 μs for SDPC and 1.8 μs for SOPC at 310 K, after observing multiple translocation events in 180 ns simulations (Figure 3.1). Further, the result suggests a translocation rate dependent on the degree of lipid unsaturation, since the $t_{1/2}$ for SOPC > SDPC.[13]

Despite discrepancies amongst studies to determine the absolute translocation rates of α-Toc, both experimentally and by simulation, both approaches yield translocation rates faster than phospholipids, which have been reported to have $t_{1/2}$ on the order of days to weeks.[37]

3.2.4 Behavior of Non-α-tocopherols in Bilayers

Because of its biological importance, much more attention has been paid to α-Toc in biophysical studies, but the growing interest in non-α-tocopherols (essentially only γ-tocopherol and δ-tocopherol since β-tocopherol is a much rarer form of the vitamin) has prompted work with these forms. The laboratory of Quinn has contributed fundamental insights into the effect of α-Toc in lipid bilayers,[2,5,7,38,39] including the ability of α-Toc to abolish the pre-transition temperature, lower and broaden the main transition temperature, T_m, in saturated,[40,41] and unsaturated (DOPC) phospholipids,[42] and the promotion of the "ripple phase".[8,41,42] Until recently, however, little attention has been paid to non-α-tocopherols. Kamal *et al.*[43] reported the results of X-ray diffraction studies on hydrated aligned multi-bilayers of DPPC containing α-, γ-, and δ-Toc. By varying the temperature and amounts of tocopherols in the bilayers, they constructed phase diagrams for these mixtures. All of the tocopherols tested promoted the ripple phase (P_{β}) above ~10 mol% tocopherol concentration in a manner similar to cholesterol,[44–47] although the effect of γ-tocopherol diminished after about 30 mol%. It has been shown by fluorescence quenching experiments that α-Toc, which has a diffusion coefficient well within the range of other lipids in fluid membranes (10^{-8}–10^{-7} cm^2 s^{-1}),[29,32] retains significant lateral diffusion through the ripple phase of DPPC, even though this phase exists below the main phase transition temperature, T_m, of the gel phase for this saturated phospholipid.[31] It is not known whether tocopherols (or tocotrienols) promote the formation of the ripple phase *in vivo*. Very little biophysical work has been done with tocotrienols.

3.3 Tocopherol and Lipid "Rafts"

Before any discussion of the role of tocopherols in "lipid rafts", the difficulties associated with this term must first be admitted.[48,49] The concept of membrane lipid sorting in epithelial cells,[50–52] and later the isolation of detergent resistant membranes (DRMs),[53–55] brought a new dimension to the analysis of this putative functional lipid structure. An original description of a lipid raft was a membrane domain rich in cholesterol and sphingolipids that phase separates from other lipid domains and is resistant to detergent solubilization. The introduction of α-Toc into this discussion arose because of the biological activities of vitamin E, which appear not to rely on antioxidant action. However, a non-antioxidant mechanism of action has been elusive, and the disruption of signaling mechanisms mediated by raft-resident proteins was an attractive possibility.

By measuring the membrane dipole potential (MDP) of egg-PC model vesicles and cells containing the membrane-residing fluorescent probe, 1-(3-sulfonatopropyl)-4-[β[2-(di-*n*-octylamino)-6-naphthyl]vinyl] pyridinium betaine (di-8-ANEPPS), Davis *et al.*[56] studied how tocopherol and tocopherol succinate modified ligand binding to P-glycoprotein. The MDP exists between the hydrophobic interior of the membrane and the waters of hydration associated with the lipid headgroups[57,58] and this potential is thought to control the affinity of P-glycoprotein for its ligands. They noticed that when PC vesicles contained 10 mol% cholesterol, the MDP increased but substitution with 10 mol% α-Toc or 7-ketocholesterol decreased the dipole potential. They hypothesized that cholesterol-rich domains (rafts), which increase the MDP, are disrupted by the presence of α-Toc (and α-tocopherol succinate), resulting in a lessened packing density and a decreased MDP. This is, however, an indirect means for showing tocopherol association with raft domains.

In 2009, the Lagrost laboratory reported[59,60] that α-tocopherol, but not γ-tocopherol, prevented apoptosis in A7R5 smooth muscle cells exposed to the oxysterol 7-ketocholesterol. 7-Ketocholesterol-dependent apoptosis could be inhibited by pre- or co-exposure of cells to α-Toc, but not after prior exposure to oxysterol. By isolating DRM fractions, they showed that addition of 100 μM α-Toc to cells one hour before exposure to 20 μg mL^{-1} 7-ketocholesterol significantly impaired the incorporation of 7-ketocholesterol into DRMs. When exposed to 100 μM α-Toc, approximately 25% of the total cell tocopherol was found in the DRM fraction, but this was easily displaced by a second exposure to 1% Triton, suggesting that tocopherol was rather loosely associated with the DRM raft.

α-Toc has also been shown to impair 7-ketocholesterol[61]- and 7-β-hydroxycholesterol[62]-induced apoptosis in cultured 158N murine oligodendrocytes. The dose of α-Toc is rather high (400 μM) but was shown to preferentially occupy DRMs when co-exposed to cells with 7-ketocholesterol (20 μg mL^{-1}). Related work by the same group[63] showed that both α- and γ-tocopherol partially protected murine microglial BV-2 cells from 7-ketocholesterol-induced mitochondrial/peroxisomal dysfunction, although no direct measurements of DRMs were made.

The observations that α-Toc protected against 7-ketocholesterol-initiated apoptosis, and that α-Toc was easily removed from DRMs by a dilute detergent, might be explained by the recent work of Muddana *et al.*[64] who used course-grained molecular dynamics simulations to model the behavior of α-Toc in mixed lipid systems comprised of DPPC and dilinoleoyl phosphatidylcholine (called DUPC in the paper; the "U" standing for "unsaturated"). They concluded that α-Toc behaves as a linactant, residing predominantly at the boundaries between phases of saturated (DPPC) and unsaturated (DUPC) lipids. This was also reflected in their experimental results using giant unilamellar vesicles (GUVs). GUVs were made from DOPC–DPPC–cholesterol (1 : 1 : 0.7) containing the fluorescent probe DiI-C_{12} (1,1′-didodecyl-3,3,3′,3′-tetramethylindocarbocyanine perchlorate), which preferentially partitions to liquid phases when gel–liquid phases coexist, or to the l_d regions when two liquid phases coexist. Inspection by fluorescence microscopy of GUVs made from this lipid mixture showed clearly distinguishable lipid phases. Addition of 10–20 mol% of α-Toc produced uniformly labeled GUVs showing no phase separation. In summary, their results suggest that the preference for α-Toc to reside at domain boundaries increases the miscibility of the existing phases resulting in their mutual dissolution (Figure 3.1). This could explain how α-Toc inhibits 7-ketocholesterol-dependent apoptosis, possibly by promoting the dissolution of 7-ketocholesterol-containing rafts and inactivation of the raft-dependent phosphatidylinositol 3-kinase/Akt signaling pathway,[65] thus keeping the pro-apoptotic protein Bad phosphorylated and active (although see later for a discussion of the reactivity of tocopherol quinones with Akt). Interestingly, γ-tocopherol did not protect against the effects of 7-ketocholesterol-induced apoptosis when apoptotic endpoints were assessed by Western blots for phosphorylated Bad,[59] but did when the measure was peroxisome activation.[63] The reason(s) for this are unknown, but the lack of biophysical studies on the effects of tocols other than α-Toc on lipid phase behavior makes speculation difficult. It also remains unclear whether tocopherols, despite being found in or around DRMs, are not still acting as antioxidants, as there is some evidence that oxysterol treatment of cells can induce oxidative stress.[66–68]

α-Toc protects yeast cells against the membrane perturbing effects of the anti-cancer lysophospholipid edelfosine.[69] This antitumor lipid induces sterol internalization and displaces the proton pump, Pma1p, from lipid microdomains. It was postulated that the protective effect was due to the matching conical shape of α-Toc with the inverted conical-shaped lysolipid, which stabilized the membrane and inhibited Pma1p internalization. The antioxidant action of α-Toc was stated to be necessary as methylation of the 6-OH group of α-Toc to generate the methyl ether could not protect against the action of edelfosine. However, loss of the free phenolic group on α-Toc would very likely disturb the orientation of the chromanol away from the headgroups of the phospholipids, thus altering its behavior as a conical-shaped lipid.

α-Tocotrienol has also been observed in DRMs (termed microdomains by the authors).[70] Using NIH-3T3 mouse fibroblasts, α-Toc and α-tocotrienol were loaded equally into cells, and DRMs prepared and analyzed for the marker protein flotillin. Treatment of cells with MβCD showed a loss of ~55% of isolated DRM cholesterol. Similar treatment of DRMs recovered from cells incubated with α-Toc showed that only 9% of the tocopherol was removed by MβCD. From the data presented, it is not clear how much of the total cell tocopherol actually resides in the DRM fractions, but the minor loss of α-Toc from DRM on treatment with MβCD is in line with our own observations that MβCD is not nearly as efficient at removing tocopherol as it is cholesterol (Harroun, unpublished observation).

Does α-Toc have a preference for residing in more fluid polyunsaturated phospholipid domains? It would be convenient if it did, as it would be a distinct advantage for an antioxidant to freely move to a lipid phase rich in the very lipids it can protect from peroxidative damage. By measuring the free-energy of mixing (ΔG_{mix}) of binary mixtures of α-Toc with DOPC, POPC, and DPPC, Jurak and Conde[71] showed a more favorable (more negative) value for α-Toc mixing with the twice-unsaturated DOPC than for the fully saturated DPPC, supporting the concept that α-Toc may spontaneously move to more highly unsaturated lipid domains. And yet, recently, Leng and Wassall[72] reported molecular dynamic simulations that showed slightly *lower* favorable binding energy between α-Toc and 1-stearoyl-2-docosahexaenoylphosphatidylcholine (SDPC, 18:0-22:6PC) than with 1-stearoyl-2-oleoylphosphatidylcholine (SOPC, 18:0-18:1PC). This would suggest that the free energy of association between α-Toc and unsaturated phospholipids is insufficient to drive α-Toc to phases rich in PUFAs. In essence, we suggest (as others have discussed[73]) that α-Toc does not first partition to unsaturated lipid phases to act as an antioxidant; rather, radical chain termination during peroxidation of unsaturated lipids is more dependent on the intrinsically disordered acyl chains of unsaturated phospholipids, which bring the peroxyl radicals into close contact with the H-atom-donating phenol of α-Toc.[10]

3.4 The Effect on Membrane-dependent Processes

3.4.1 Tocopherols

It has long been known that α-Toc stabilizes the membranes of red blood cells against hemolysis by agents such as retinol, detergents, hemin, *etc.* (see ref. 6). Hincha *et al.* reported that α-Toc also stabilizes model chloroplast membranes, 90% of which were the plant glycolipids monogalactosyldiacylglycerol (MGDG), digalactosyldiacylglycerol (DGDG), and sulfoquinovosyldiacylglycerol (SQDG). Using vesicle leakage and fusion assays after freezing of the 100 nm vesicles at −20 °C, they found that 1–2% α-Toc in combination with sucrose reduced leakage of vesicle-trapped carboxyfluorescein (CF) by as much as 30%, and reduced vesicle fusion completely.[74] Interestingly, the tocopherol content of chloroplast membranes is about 2%, and when the assays were run with higher amounts of α-Toc, protection from freezing damage was much less efficient. The fluorescent probes MC540 and NBD-PE were also used to assess changes in lipid headgroup behavior. Again, at low concentrations of α-Toc (0.5%), the MC540 dye partitioned less into the lipid headgroup region, which it normally prefers, and the anisotropy of the NBD-PE was at a minimum, reflecting restricted mobility and showing that this concentration of α-Toc had stabilized headgroup packing and reduced the fluidity of the bilayer. In the absence of glucose, however, the stabilizing effect was minimal. The authors concluded that insertion of the prenyl tail of α-Toc between the diacyl lipids increases their separation, expanding the bilayer, which tightens the packing of the headgroups. Even though this work was performed with plant glycolipids, the results are in agreement with the generally accepted view that α-Toc has an ordering effect on the acyl chains of lipids.[38,39] α-Toc is also known to modulate the metabolism of polyunsaturated fatty acids in low-temperature stressed *Arabidopsis thaliana.*[75] The observation of higher levels of α-Toc in cold water fish is also interesting in that it may reflect the protective effect of tocopherol in cold adapted membranes.[76]

The ability of α-Toc to stabilize phospholipid membranes appears also to have a role in muscle repair. Howard *et al.*[77] described how α-Toc can assist the repair of laser-damaged mouse muscle cells *in vitro* and in fresh soleus muscle. Successful membrane repair in C2C12 and HeLa cells could be achieved by exposure to rather high doses of α-Toc (200–2500 μM). Such repair is undoubtedly dependent on α-Toc acting as an antioxidant since 200 μM Trolox also resulted in successful repair, and Trolox partitions to about 20% in the bilayer and the rest in solution.[78,79] Membrane repair depended on the presence of a calcium ion, which is known to be required for successful membrane fusion events.[80,81] The authors also acknowledge that membrane fusion events are spurred by lipids with negative curvature including α-Toc.[82,83]

It has been suggested that depth of the chromanol headgroup penetration into the membrane may be a function of the chromanol methylation status. Fluorescence-quenching studies using different tocols in membranes also containing 5- and 16-doxylstearates (5-DS and 16-DS) showed that α-Toc resides closer to the membrane surface than the less-methylated vitamers. In the same study, lesser methylated tocols were shown to be weaker inhibitors of phospholipase A_2.[84] Interestingly, the strong inhibition of PLA_2 by α-Toc rather than other tocopherols may be related to it being a linactant (see ref. 64) which, by stabilizing the edges of mixed lipid phases, makes substrate access by PLA_2 more difficult. The degree of phospholipid acyl chain unsaturation and/or disorder are factors that likely also affect the depth of the chromanol.

Non-α-tocopherols, especially δ-tocopherol, have been shown to modulate processes at biological membranes. This work focuses primarily on dietary provision of mixed tocopherols with mice and in cell culture and generally does not focus on the mechanism of action.[85–92] Occasionally, however, some work looks at potential mechanisms, such as inhibition of the epidermal growth factor receptor (EGFR) internalization.[93] This study found that at a 50 μM dose, δ-tocopherol was the most active form of vitamin E in inhibiting prostate cancer cell growth. γ-Tocopherol was considerably less active, and α-Toc had no effect. δ-Tocopherol attenuated the EGF-induced activation of phosphatidylinositol 3-kinase (PIK3), which in turn stimulated the phosphorylation of Akt on Tyr-308 resulting in cell cycle arrest and apoptosis. (See the section on tocopheryl quinones for a further discussion on the reason for the selective action of vitamin E forms.)

3.4.2 Tocotrienols

The reported biological activities of tocotrienols[94–104] are so diverse that it is hard to imagine that a single mechanism of action exists to explain all observed effects. One might expect, given the number of membrane-resident signaling proteins that are affected (*e.g.* mitogen-activated receptor protein kinases and many others), that the presence of tocotrienols could play a role in changing the enzyme's local membrane environment. The common observation that δ- and γ-tocotrienols have greater activities compared to α-tocotrienol or α-Toc, suggests that the lack of a fully methylated chromanol is a key structural feature. Regretfully, no biophysical data exists (*i.e.* X-ray or neutron diffraction) on the effects of tocotrienols on the phospholipid bilayer structure and dynamics, making it difficult to speculate on their mechanism of action from this perspective.

It is also known that tocotrienols are more rapidly taken up by cell membranes *in vitro*. In human aortic endothelial cells (HAECs), α-tocotrienol accumulates to levels approximately 25–95-fold greater than α-Toc[105] and in human umbilical vein endothelial cells (HUVECs), α-tocotrienol accumulations was favored 10-fold over α-Toc.[106] Sontag and Parker[107] showed that the rates of tocotrienol accumulation in HepG2 cells, as well as total amounts delivered, were substantially higher than for tocopherols, and Saito *et al.*[108] showed a 70-fold higher initial rate of α-tocotrienol uptake in Jurkat E6-1 cells. Similar observations have been made in Caco2 intestinal epithelial cells.[109] Intermembrane transfer of tocotrienols between liposomes is also faster than for tocopherols[110] as is the more rapid arrival of tocotrienols than tocopherols to mitochondria when delivered to HepG2 cells (Figure 3.1).[107]

3.5 Tocol Quinones and Hydroquinones

A possible explanation for some of the biological activities of non-α-tocols may be that some of the activity is due to their respective quinones produced by peroxidative processes after the free phenols (chromanols) are taken up by the cell. The unique biological activity of γ-tocopherol is thought to involve the free 5-position on the aromatic ring that acts as a nucleophile to trap peroxynitrite.[111,112] However, when β-, γ-, and δ-tocols are oxidized to their quinones, electrophilic sites exist for conjugate addition reactions with nucleophiles such as thiols. Elisia and Kitts[113,114] reported that δ-Toc elicited apoptosis in Caco2 and FHs 74 Int cells to a greater extent than γ-Toc, whereas α-Toc was inactive. Activity was assessed by measurement of Nrf-2 activation. They also observed that cellular uptake of non-α-Toc forms was greater than for α-Toc in both cell lines. They speculated that this may, in part, explain the greater biological activity observed for γ-Toc and δ-Toc when compared with α-Toc. But one wonders whether this activity was actually the result of trace amounts of tocopheryl quinones (TQ) being produced during aerobic cell culture and their reaction with the reactive thiols of the Nrf2 protein partner Keap1 (Figure 3.2).[115] We have noted that the rate of reaction of *N*-acetyl-cysteine with δ-tocopheryl and δ-tocotrienyl quinone is ~8-times faster than with γ-tocopheryl and γ-tocotrienyl quinone at pH 7.5. α-Toc is unreactive in this assay (Krueger and Atkinson, unpublished observations). In their studies on the growth inhibition of colon cancer cells, Dolfi *et al.*[116] also observed that while δ- and γ-tocopherol inhibited growth of cells more so than α-Toc, γ- and δ-tocopheryl quinones were 100 and 20-times more active than their respective chromanols. It is possible that much of the literature describing the dose-responsive biological activities of the non-α-tocopherols is obscuring the actual active quinones that are only produced slowly by adventitious oxidation of the parent chromanols during cell culture or in test animals. A relevant example of the potential of quinone reactivity was the discovery of lactoquinomycin and related pyranonaphthoquinones as inhibitors of the kinase AKT/PKB by their reaction with the catalytic activation loop cysteines.[117] The same kinase is described by the Sylvester group in their work with γ- and δ-tocotrienols[118,119] and

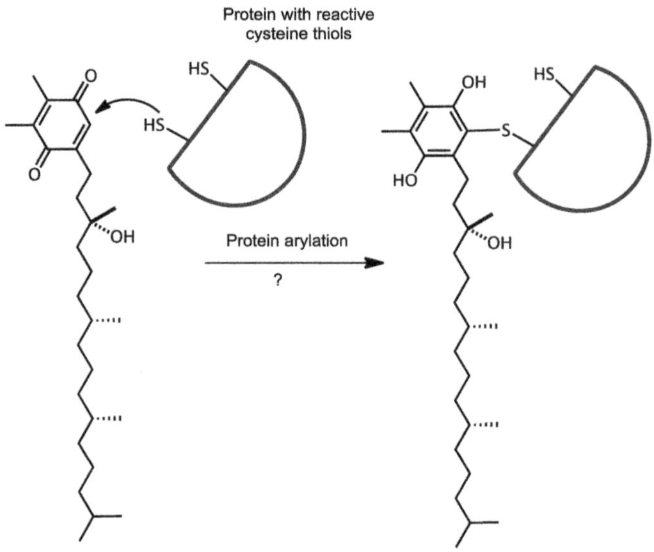

Figure 3.2 Illustration of the potential reactivity of γ-tocopheryl quinone with proteins containing reactive cysteine residues resulting in arylation of the protein.

should probably be re-investigated using pure samples of the appropriate quinones.

Unfortunately, the membrane behaviors of tocopheryl quinones have attracted little attention. The groups of Strzałka and Kruk reported on the spectroscopic properties of prenyllipids such as plastiquinone in monolayers[120,121] and bilayers, and have occasionally included tocopheryl quinone or tocopheryl hydroquinone in their work. They describe that since the quinones have less polar headgroups they can be "squeezed out" of a Langmuir–Blodgett monolayer with less applied collapse pressure than the more polar hydroquinones. Interestingly, both α-TQ and α-tocopheryl hydroquinone had the highest collapse pressure compared to plastiquinones of varying prenyl chain length. This was suggested to be due to the more polar nature of α-TQ (which has the tertiary hydroxy group) and the more polar nature of the hydroquinones in general. By determining fluorescence lifetimes[122] and anisotropy,[123] they also showed that the headgroups of both α-Toc and α-THQ were closer to the membrane surface than plastiquinones and plastiquinols. While the concentration of α-TQ in rat heart mitochondrial membranes[124] and plasma[125] is 100- to 1000-fold less than α-Toc, it has also been shown that α-TQ can be enzymatically reduced to α-THQ[126] (Figure 3.3), an antioxidant that is over 10-times better than α-Toc ($k = 0.268\ M^{-1}\ s^{-1}$ *versus* 0.022) as measured by the fluorescence intensity loss of the antioxidants during Fe^{2+}/H_2O_2-prompted peroxidation in egg lecithin vesicles[127] and five times better in an ascorbate/Fe^{2+} peroxidation assay.[128] *In vitro* experiments with model chloroplast lipids using the azo initiator AMVN show that α-TQ is not an antioxidant.[129]

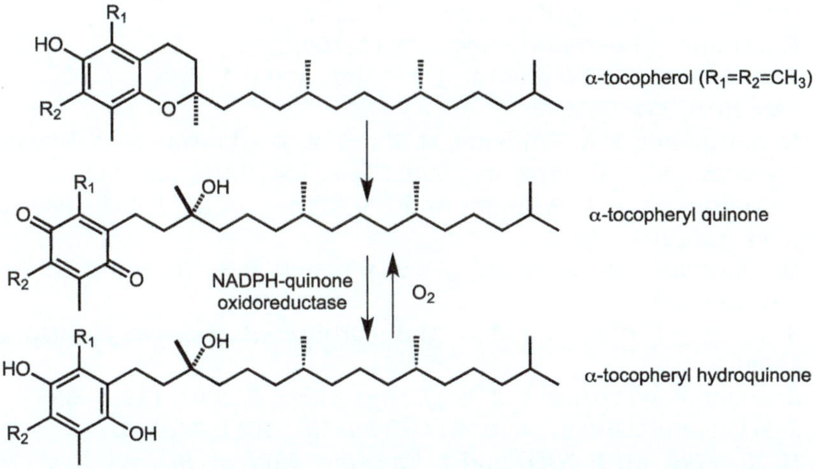

Figure 3.3 Conversion of tocopherols (and tocotrienols) from the chromanols to the quinones and NADPH-dependent enzymatic reduction to hydroquinones. The structure drawn shows α-Toc when both R_1 and R_2 are CH_3; γ-tocopherol when R_1 = H and R_2 = CH_3; and δ-tocopherol when R_1 and R_2 = H.

3.6 Conclusion

The orientation and dynamic behaviour of α-Toc in phospholipid bilayers has been studied for many years. We now appreciate that α-Toc orients in membranes with its phenolic group at the level of the glycerol ester portion of phospholipids just below the hydrated headgroups. However, this is a highly dynamic averaged position and it is now reasonable to conclude that the H-atom-donating phenol of the chromanols is free to access lipid peroxyl radicals at essentially any depth in the bilayer. Yet, the picture remains incomplete, since very little attention and biophysical assessment has been paid to non-α-tocopherols and the tocotrienols. Information on how these forms of the vitamins differ in their membrane behaviour is needed to answer why non-α-tocols, especially γ- and δ-tocopherol and tocotrienols, have such diverse biological activities involving cell signaling, lipid and protein kinase activity and more.

References

1. D. H. Blatt, S. W. Leonard and M. G. Traber, *Nutrition*, 2001, **17**, 799.
2. P. J. Quinn, *Subcell. Biochem.*, 1998, **30**, 319.
3. X. Y. Wang and P. J. Quinn, *Prog. Lipid Res.*, 1999, **38**, 309.
4. P. J. Quinn, *Biochemistry*, 2004, **69**, 58.
5. P. J. Quinn, *Vitam. Horm.*, 2007, **76**, 67.
6. J. Atkinson, R. F. Epand and R. M. Epand, *Free Radical Biol. Med.*, 2008, **44**, 739.

7. P. J. Quinn, *Prog. Lipid Res.*, 2012, **51**, 179.
8. P. J. Quinn, *J. Controlled Release*, 2012, **160**, 158.
9. T. A. Harroun, D. Marquardt, J. Katsaras and J. Atkinson, *J. Phys.: Conf. Ser.*, 2010, **251**, 012039.
10. D. Marquardt, J. A. Williams, N. Kucerka, J. Atkinson, S. R. Wassall, J. Katsaras and T. A. Harroun, *J. Am. Chem. Soc.*, 2013, **135**, 7523.
11. D. Marquardt, J. A. Williams and J. J. Kinnun, *et al.*, *J. Am. Chem. Soc.*, 2014, **136**, 203.
12. D. Marquardt, N. Kucerka, J. Katsaras and T. A. Harroun, *Langmuir*, 2015, **31**, 4464.
13. X. Leng, J. J. Kinnun and D. Marquardt, *et al.*, *Biophys. J.*, 2015, **109**, 1608.
14. S. S. Qin, Z. W. Yu and Y. X. Yu, *J. Phys. Chem. B*, 2009, **113**, 16537.
15. S. S. Qin and Z. W. Yu, *Acta Phys.-Chim. Sin.*, 2011, **27**, 213.
16. N. Kucerka, M. P. Nieh and J. Katsaras, *Biochim. Biophys. Acta*, 2011, **1808**, 2761.
17. A. Ausili, A. M. de Godos, A. Torrecillas, F. J. Aranda, S. Corbalan-Garcia and J. C. Gomez-Fernandez, *Phys. Chem. Chem. Phys.*, 2017, **19**, 6731.
18. A. Kyrychenko and A. S. Ladokhin, *J. Phys. Chem. B*, 2013, **117**, 5875.
19. O. Soubias and K. Gawrisch, *J. Am. Chem. Soc.*, 2007, **129**, 6678.
20. L. R. C. Barclay, *Can. J. Chem.*, 1993, **71**, 1.
21. L. R. C. Barclay and K. U. Ingold, *J. Am. Chem. Soc.*, 1981, **103**, 6478.
22. P. Novikova and P. M. Krasil'nikov, *Biofizika*, 2009, **54**, 675.
23. J. Garrec, A. Monari, X. Assfeld, L. M. Mir and M. Tarek, *J. Phys. Chem. Lett.*, 2014, **5**, 1653.
24. A. W. Girotti, *Free Radical Biol. Med.*, 2008, **44**, 956.
25. K. Fukuzawa, W. Ikebata, A. Shibata, I. Kumadaki, T. Sakanaka and S. Urano, *Chem. Phys. Lipids*, 1992, **63**, 69.
26. K. Fukuzawa, *J. Nutr. Sci. Vitaminol.*, 2008, **54**, 273.
27. V. K. Sharma, E. Mamontov, M. Tyagi and V. S. Urban, *J. Phys. Chem. B*, 2016, **120**, 154.
28. A. Ausili, A. Torrecillas, A. M. de Godos, S. Corbalán-García and J. C. Gómez-Fernández, *Langmuir*, 2018, **34**, 3336.
29. G. Gramlich, J. Zhang and W. M. Nau, *J. Am. Chem. Soc.*, 2004, **126**, 5482.
30. F. J. Aranda, A. Coutinho, M. N. Berberan-Santos, M. J. E. Prieto and J. C. Gómez-Fernández, *Biochim. Biophys. Acta*, 1989, **985**, 26.
31. R. Meyer, A. F. Sonnen and W. M. Nau, *Langmuir*, 2010, **26**, 14723.
32. A. F. P. Sonnen, H. Bakirci, T. Netscher and W. M. Nau, *J. Am. Chem. Soc.*, 2005, **127**, 15575.
33. F. Bellemare and M. Fragata, *J. Membr. Biol.*, 1981, **58**, 67.
34. V. A. Tyurin, V. E. Kagan, E. A. Serbinova, N. V. Gorbunov, A. N. Erin, L. L. Prilipko and T. S. Stoichev, *Bull. Exp. Biol. Med.*, 1986, **102**, 1677.
35. V. E. Kagan, R. A. Bakalova and Z. Z. Zhelev, *et al.*, *Arch. Biochem. Biophys.*, 1990, **280**, 147.
36. X. L. Leng, J. A. Williams and D. Marquardt, *et al.*, *Biophys. J.*, 2014, **106**, 94a.

37. D. Marquardt, F. A. Heberle, T. Miti, B. Eicher, E. London, J. Katsaras and G. Pabst, *Langmuir*, 2017, **33**, 3731.
38. X. Wang and P. J. Quinn, *Prog. Lipid Res.*, 1999, **38**, 309.
39. X. Wang and P. J. Quinn, *Mol. Membr. Biol.*, 2000, **17**, 143.
40. X. Y. Wang, H. Takahashi, I. Hatta and P. J. Quinn, *Biochim. Biophys. Acta*, 1999, **12**, 335.
41. X. Wang, K. Semmler, W. Richter and P. J. Quinn, *Arch. Biochem. Biophys.*, 2000, **377**, 304.
42. X. Wang and P. J. Quinn, *Biochim. Biophys. Acta*, 2002, **1567**, 6.
43. M. A. Kamal and V. A. Raghunathan, *Biochim. Biophys. Acta*, 2012, **1818**, 2486.
44. F. de Meyer and B. Smit, *Proc. Natl. Acad. Sci. U. S. A.*, 2009, **106**, 3654.
45. S. Karmakar and V. A. Raghunathan, *Phys. Rev. Lett.*, 2003, **91**, 098102.
46. K. Mortensen, W. Pfeiffer, E. Sackmann and W. Knoll, *Biochim. Biophys. Acta*, 1988, **945**, 221.
47. P. W. Westerman, M. J. Vaz, L. M. Strenk and J. W. Doane, *Proc. Natl. Acad. Sci. U. S. A.*, 1982, **79**, 2890.
48. P. J. Quinn, *Prog. Lipid Res.*, 2010, **49**, 390.
49. S. Sonnino and A. Prinetti, *Curr. Med. Chem.*, 2013, **20**, 4.
50. K. Simons and G. van Meer, *Biochemistry*, 1988, **27**, 6197.
51. G. van Meer and K. Simons, *J. Cell. Biochem.*, 1988, **36**, 51.
52. K. Simons and J. L. Sampaio, *Cold Spring Harbor Perspect. Biol.*, 2011, **3**, a004697.
53. D. A. Brown and E. London, *Annu. Rev. Cell Dev. Biol.*, 1998, **14**, 111.
54. D. A. Brown and E. London, *J. Biol. Chem.*, 2000, **275**, 17221.
55. E. London and D. A. Brown, *Biochim. Biophys. Acta*, 2000, **1508**, 182.
56. S. Davis, B. M. Davis, J. L. Richens, K. A. Vere, P. G. Petrov, C. P. Winlove and P. O'Shea, *J. Lipid Res.*, 2015, **56**, 1543.
57. H. Brockman, *Chem. Phys. Lipids*, 1994, **73**, 57.
58. L. G. Wang, *Annu. Rev. Biochem.*, 2012, **81**, 615.
59. M. C. Royer, S. Lemaire-Ewing and C. Desrumaux, *et al.*, *J. Biol. Chem.*, 2009, **284**, 15826.
60. S. Lemaire-Ewing, C. Desrumaux, D. Neel and L. Lagrost, *Mol. Nutr. Food Res.*, 2010, **54**, 631.
61. K. Ragot, D. Delmas, A. Athias, T. Nury, M. Baarine and G. Lizard, *Chem. Phys. Lipids*, 2011, **164**, 469.
62. K. Ragot, J. J. Mackrill and A. Zarrouk, *et al.*, *Biochem. Pharmacol.*, 2013, **86**, 67.
63. M. Debbabi, T. Nury and A. Zarrouk, *et al.*, *Int. J. Mol. Sci.*, 2016, **17**, 1973.
64. H. S. Muddana, H. H. Chiang and P. J. Butler, *Biophys. J.*, 2012, **102**, 489.
65. R. Lasserre, X. J. Guo and F. Conchonaud, *et al.*, *Nat. Chem. Biol.*, 2008, **4**, 538.
66. S. Lemaire-Ewing, C. Prunet and T. Montange, *et al.*, *Cell Biol. Toxicol.*, 2005, **21**, 97.
67. G. Lizard, S. Gueldry and O. Sordet, *et al.*, *FASEB J.*, 1998, **12**, 1651.

68. H. Kolsch, M. Ludwig, D. Lutjohann and M. L. Rao, *J. Neural Transm.*, 2001, **108**, 475.
69. T. Bitew, C. E. Sveen, B. Heyne and V. Zaremberg, *J. Biol. Chem.*, 2010, **285**, 25731.
70. T. Nakamura, A. Noma and J. Terao, *SpringerPlus*, 2014, **3**, 550.
71. M. Jurak and J. Minones Conde, *Biochim. Biophys. Acta*, 2013, **1828**, 2410.
72. X. L. Leng, A. Cavazos, B. Ray, M. Ghelfi, J. Atkinson, F. Q. Zhu and S. Wassall, *Biophys. J.*, 2017, **112**, 74a.
73. S. R. Shaikh, S. R. Wassall, D. A. Brown and R. Kosaraju, *Curr. Top. Membr.*, 2015, **75**, 209.
74. D. K. Hincha, *FEBS Lett.*, 2008, **582**, 3687.
75. H. Maeda, T. L. Sage, G. Isaac, R. Welti and D. Dellapenna, *Plant Cell*, 2008, **20**, 452.
76. A. Fujisawa, W. C. Dunlap and Y. Yamamoto, *Comp. Biochem. Physiol., Part B: Biochem. Mol. Biol.*, 2010, **157**, 145.
77. A. C. Howard, A. K. McNeil and P. L. McNeil, *Nat. Commun.*, 2011, **2**, 597.
78. M. Lucio, C. Nunes, D. Gaspar, H. Ferreira, J. L. F. C. Lima and S. Reis, *Food Biophys.*, 2009, **4**, 312.
79. L. Barclay, J. D. Artz and J. J. Mowat, *Biochim. Biophys. Acta*, 1995, **1237**, 77.
80. G. Q. Bi, J. M. Alderton and R. A. Steinhardt, *J. Cell Biol.*, 1995, **131**, 1747.
81. A. Draeger, K. Monastyrskaya and E. B. Babiychuk, *Biochem. Pharmacol.*, 2011, **81**, 703.
82. A. Bradford, J. Atkinson, N. Fuller and R. P. Rand, *J. Lipid Res.*, 2003, **44**, 1940.
83. M. A. Churchward, T. Rogasevskaia, D. M. Brandman, H. Khosravani, P. Nava, J. K. Atkinson and J. R. Coorssen, *Biophys. J.*, 2008, **94**, 3976.
84. A. Grau and A. Ortiz, *Chem. Phys. Lipids*, 1998, **91**, 109.
85. G. Lu, H. Xiao and G. X. Li, *et al.*, *Carcinogenesis*, 2010, **31**, 687.
86. J. Ju, X. Hao and M. J. Lee, *et al.*, *Cancer Prev. Res.*, 2009, **2**, 143.
87. F. Guan, G. Li and A. B. Liu, *et al.*, *Cancer Prev. Res.*, 2012, **5**, 644.
88. G. X. Li, M. J. Lee, A. N. B. Liu, Z. H. Yang, Y. Lin, W. C. J. Shih and C. S. Yang, *Cancer Prev. Res.*, 2011, **4**, 404.
89. X. Zheng, X. X. Cui and T. O. Khor, *et al.*, *Cancers*, 2011, **3**, 3762.
90. J. D. Lambert, G. Lu, M. J. Lee, J. Hu, J. Ju and C. S. Yang, *Mol. Nutr. Food Res.*, 2009, **53**, 1030.
91. S. Das Gupta, J. Y. So and B. Wall, *et al.*, *Mol. Carcinog.*, 2014, **54**, 916.
92. A. Barve, T. O. Khor and S. Nair, *et al.*, *Int. J. Cancer*, 2009, **124**, 1693.
93. H. Wang, J. Hong and C. S. Yang, *Mol. Carcinog.*, 2015, **55**, 1728.
94. M. M. Kanchi, M. K. Shanmugam, G. Rane, G. Sethi and A. P. Kumar, *Drug Discovery Today*, 2017, **22**, 1765.
95. Q. Jiang, *Adv. Nutr.*, 2017, **8**, 850.
96. T. Eitsuka, N. Tatewaki, H. Nishida, K. Nakagawa and T. Miyazawa, *Int. J. Mol. Sci.*, 2016, **17**, 1605.
97. P. W. Sylvester, *Clin. Transl. Med.*, 2014, **3**, 30.

98. M. T. Ling, S. U. Luk, F. Al-Ejeh and K. K. Khanna, *Carcinogenesis*, 2012, **33**, 233.
99. R. S. Wong and A. K. Radhakrishnan, *Nutr. Rev.*, 2012, **70**, 483.
100. R. Kannappan, S. C. Gupta, J. H. Kim and B. B. Aggarwal, *Genes Nutr.*, 2012, **7**, 43.
101. B. B. Aggarwal, C. Sundaram, S. Prasad and R. Kannappan, *Biochem. Pharmacol.*, 2010, **80**, 1613.
102. T. Miyazawa, A. Shibata and P. Sookwong, *et al.*, *J. Nutr. Biochem.*, 2009, **20**, 79.
103. C. Constantinou, A. Papas and A. I. Constantinou, *Int. J. Cancer*, 2008, **123**, 739.
104. P. W. Sylvester, *Vitam. Horm.*, 2007, **76**, 329.
105. Y. Naito, M. Shimozawa and M. Kuroda, *et al.*, *Atherosclerosis*, 2005, **180**, 19.
106. N. Noguchi, R. Hanyu, A. Nonaka, Y. Okimoto and T. Kodama, *Free Radical Biol. Med.*, 2003, **34**, 1614.
107. T. J. Sontag and R. S. Parker, *J. Lipid Res.*, 2007, **48**, 1090.
108. Y. Saito, Y. Yoshida, K. Nishio, M. Hayakawa and E. Niki, *Ann. N. Y. Acad. Sci.*, 2004, **1031**, 368.
109. W. Tsuzuki, R. Yunoki and H. Yoshimura, *Lipids*, 2007, **42**, 163.
110. Y. Yoshida, E. Niki and N. Noguchi, *Chem. Phys. Lipids*, 2003, **123**, 63.
111. S. Christen, A. A. Woodall, M. K. Shigenaga, P. T. Southwell-Keely, M. W. Duncan and B. N. Ames, *Proc. Natl. Acad. Sci. U. S. A.*, 1997, **94**, 3217.
112. Q. Jiang, J. Lykkesfeldt, M. K. Shigenaga, E. T. Shigeno, S. Christen and B. N. Ames, *Free Radical Biol. Med.*, 2002, **33**, 1534.
113. I. Elisia and D. D. Kitts, *Biofactors*, 2013, **39**, 663–671.
114. I. Elisia and D. D. Kitts, *Mol. Cell. Biochem.*, 2015, **404**, 123.
115. A. Uruno and H. Motohashi, *Nitric Oxide*, 2011, **25**, 153.
116. S. C. Dolfi, Z. Yang, M. J. Lee, F. Guan, J. Hong and C. S. Yang, *J. Agric. Food Chem.*, 2013, **61**, 8533.
117. L. Toral-Barza, W. G. Zhang and X. Huang, *et al.*, *Mol. Cancer Ther.*, 2007, **6**, 3028.
118. P. W. Sylvester and R. V. Tiwari, *Autophagy Curr. Trends Cell. Physiol. Pathol.*, 2016, 89.
119. P. W. Sylvester and N. M. Ayoub, *Anticancer Agents Med. Chem.*, 2013, **13**, 1039.
120. J. Kruk, K. Strzalka and R. M. Leblanc, *Biochim. Biophys. Acta*, 1992, **1112**, 19.
121. J. Kruk, K. Strzalka and R. M. Leblanc, *Biochim. Biophys. Acta*, 1993, **1142**, 6.
122. J. Kruk, B. Mysliwa-Kurdziel, M. Jemiola-Rzeminiska and K. Strzalka, *Photochem. Photobiol.*, 2006, **82**, 1309.
123. M. Jemiola-Rzemimska, J. Kruk and K. Strzalka, *Chem. Phys. Lipids*, 2003, **123**, 233.
124. W. Gregor, K. Staniek, H. Nohl and L. Gille, *Biochem. Pharmacol.*, 2006, **71**, 1589.

125. H. Kanazawa, C. Miyata, Y. Nagata, S. Urano and Y. Matsushima, *Chem. Pharm. Bull.*, 2000, **48**, 1462.
126. D. Siegel, E. M. Bolton, J. A. Burr, D. C. Liebler and D. Ross, *Mol. Pharmacol.*, 1997, **52**, 300.
127. J. Kruk, M. Jemiolarzeminska and K. Strzalka, *Chem. Phys. Lipids*, 1997, **87**, 73.
128. A. Bindoli, M. Valente and L. Cavallini, *Biochem. Int.*, 1985, **10**, 753.
129. B. Nowicka, J. Gruszka and J. Kruk, *Biochim. Biophys. Acta*, 2013, **1828**, 233.

CHAPTER 4

Chemical Reactivity and Cellular Uptake of Tocopherols and Tocotrienols

YOSHIRO SAITO*[a] AND YASUKAZU YOSHIDA*[b]

[a]Systems Life Sciences Laboratory, Department of Medical Life Systems, Faculty of Life and Medical Sciences, Doshisha University, Kyotanabe, Kyoto 610-0394, Japan; [b]Health Research Institute, National Institute of Advanced Industrial Science and Technology (AIST), Takamatsu, Kagawa 761-0395, Japan
*E-mail: ysaito@mail.doshisha.ac.jp, yoshida-ya@aist.go.jp

4.1 Introduction

Both tocopherols (T) and tocotrienols (T3) act as radical-scavenging antioxidants, and the reactivities of α, β, γ, and δT and the corresponding T3 isoforms toward free radicals are similar; however, T3 has unique physical characteristics derived from its side chain. This chapter focuses on the chemical and physical properties of vitamin E isoforms, particularly on the differences between T and T3. Furthermore, this section addresses the importance of the cellular uptake of T and T3 regarding their cytoprotective activity in cultured cells.

Food Chemistry, Function and Analysis No.11
Vitamin E: Chemistry and Nutritional Benefits
Edited by Etsuo Niki

Published by the Royal Society of Chemistry, www.rsc.org

4.2 Reactivities Toward Free Radicals

The action and activity of T and T3 as radical-scavenging antioxidants have been studied and compared by many investigators.[1–6] The corresponding α-, β-, γ-, and δ-isoforms of T and T3 exert a similar antioxidant activity in solution, as the side chain located at position 2 of chromanols has a small effect on the reactivity of the phenolic O–H bond toward free radicals. There are several reports of the reactivities of T and T3 toward free radicals.[1–6] It has been observed that the corresponding T and T3 have the same reactivity toward free radicals. The aryloxyl radical-scavenging rate constants of T and T3 were measured in ethanol solution *via* stopped-flow spectrophotometry.[5] The rate constants of α, β, γ, and δT showed good agreement with those of the corresponding α, β, γ, and δT3. The relative reactivities of the antioxidants toward peroxyl radicals, which are chain-carrying species in lipid peroxidation, can be clearly and quantitatively assessed *via* a competition method using reference compounds. In these methods, the manner in which the test compound protects a reference compound from degradation or consumption by exposure to a controlled free radical flux is evaluated. The results of typical examples carried out in organic solution using pyrogallolsulfonephthalein (Pyrogallol Red) as a reference compound are shown in Figure 4.1A.[1] The peroxyl radicals were generated from the azo initiator at a constant rate.[7] The generation of free radicals at a controlled and constant rate is essential for kinetics studies. The results show clearly that the reactivities decrease in the order of $\alpha > \beta = \gamma > \delta$ and that the corresponding T and T3 have the same reactivities (Figure 4.1B). It should be noted that T and T3 gave the same induction period (lag phase) (Figure 4.1C), suggesting that the stoichiometry number—*i.e.*, the number of free radicals scavenged by each T and T3 molecule—is the same. The measurement of the temporal disappearance of T and T3 induced by free radicals also revealed similar reactivities for the corresponding T and T3 isoforms in both homogeneous solution and in liposomal membranes.[8,9] These results show clearly that the corresponding isoforms of T and T3 have the same chemical reactivities toward free radicals.

4.3 Antioxidant Activities

The results of *in vitro* studies of the relative antioxidant activities of T and T3 performed in different laboratories are not always consistent. The antioxidant activity against lipid peroxidation has been evaluated using several markers or parameters, such as conjugated diene, lipid hydroperoxides, thiobarbituric acid-reactive substances (TBARS), oxygen consumption, and induction period. It has been observed that the corresponding T and T3 isoforms reduced the rate of oxidation to the same level for the same duration.[1,2,9] The results shown in Figure 4.1C confirm that αT and αT3 exerted the same antioxidant activities against lipid peroxidation in solution. The antioxidant activities in this system decreased in the order of $\alpha > \beta = \gamma > \delta$,

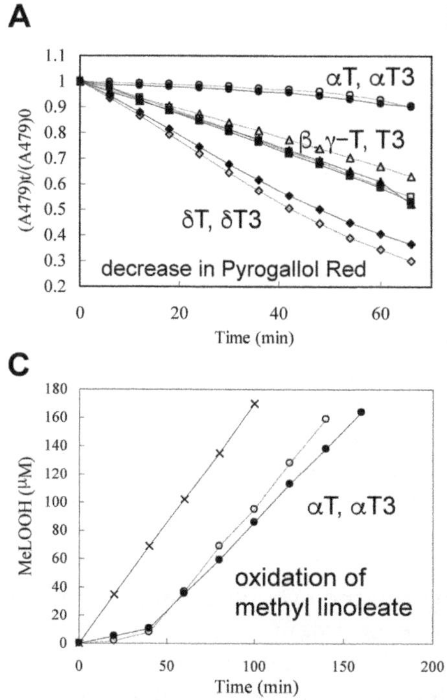

Figure 4.1 Reactivities of T and T3 toward free radicals in solution. (A) Effect of α (circle), β (triangle), γ (square), and δ (diamond) isoforms of T (solid) and T3 (open) (15 μM) on the consumption of Pyrogallol Red (15 μM) induced by AMVN (2.5 mM) in ethanol at 37 °C under air. (A475)t/(A479)0 denotes the ratio of absorbance at 479 nm caused by PR at time *t* to that of the start of the experiment. (B) Consumption of T and T3 induced by AMVN (10 mM) in acetonitrile at 37 °C under air. The symbols are the same as those in (A). (C) Inhibition by αT (solid circles) and αT3 (open circles) (5 μM) of the oxidation of methyl linoleate (24.4 mM) induced by AMVN (0.84 mM) in acetonitrile at 37 °C. Methyl linoleate hydroperoxide (MeLOOH) was measured in the absence (crosses) and presence (circles) of antioxidant. Adapted from ref. 1 with permission from Elsevier, Copyright 2003.

which is in good agreement with the relative reactivities observed toward peroxyl radicals. Similar antioxidant activities were reported for αT and αT3 and for γT and γT3 against the oxidation of cholesterol in aqueous dispersions containing Tween-20 that were initiated by the water-soluble radical initiator 2,2′-azobis(2-amidinopropane) dihydrochloride (AAPH).[10]

Conversely, the antioxidant activities of T and T3 in membranes appear to be more complicated. Servinova *et al.*[11] reported that αT3 exerted antioxidant activity against iron–ascorbate- and iron–NADPH-induced lipid peroxidation in rat liver microsomal membranes 40–60 times higher and protection of cytochrome P-450 against oxidative damage 6.5 times better than did αT. It was also reported that γT3 and αT3 were more effective than αT in inhibiting the *in vitro* oxidative damage of both lipids and proteins induced by iron–ascorbate in rat brain mitochondria.[12] Furthermore, it was reported that T3 incorporated into human low-density lipoprotein (LDL) reacted with peroxyl radicals and suppressed its oxidation as efficiently as the corresponding T isomers.[8]

We compared the antioxidant activities of T and T3 in phosphatidylcholine (PC) liposomal membranes. αT and αT3 suppressed the formation of PC hydroperoxides to the same extent as that observed in the oxidation of β-linoleoyl-γ-palmitoyl PC (PLPC) liposomal membranes (Figure 4.2A).[1] The competition between αT3 and αT for scavenging peroxyl radicals in the membranes was studied in two different membranes: PLPC liposomes and dimyristoyl (14:0) PC liposomes. The lipophilic radical initiator 2,2′-azobis(2,4-di-methylvaleronitrile) (AMVN) was incorporated into the same vesicles to generate free radicals within the membranes. It was found that αT and αT3 were consumed at the same rate in PLPC and 14:0 PC liposomal membranes (Figure 4.2C). These data show that αT and αT3 scavenge peroxyl radicals and suppress lipid peroxidation with the same efficacy in PC liposomal membranes.

The action of the T and T3 isoforms within the PC liposomal membranes was studied by fluorescence quenching experiments using 2,3-diazabicyclo[2,2,2]-oct-2-ene and fulorazophone L.[12] The quenching of these azoalkanes proceeds by hydrogen atom transfer, as does the scavenging of reactive free radicals by T and T3. In agreement with previous reports, the rate decreased in the order of $\alpha > \beta = \gamma > \delta$, and the rate constants for the corresponding T and T3 were the same both in solution and in liposomal membranes.[12] Furthermore, the lateral diffusion coefficients were also compared for the α and δ forms of T and T3, and no marked effects of the nature of the isoprenoid side chain on the diffusion were observed. These results suggest that the difference between T and T3 regarding their antioxidant action in membranes, if present, is small.

4.4 Action of T and T3 as Reductants

Metal ions decompose hydrogen peroxide and hydroperoxides to give oxygen radicals, eventually leading to oxidative stress. In general, transition metal ions at a higher valency state, such as Fe(III) and Cu(II), decompose peroxides more slowly than do those at a lower valency state, such as Fe(II) and Cu(I). Therefore, a reductant can act as a prooxidant. For example, ascorbate (vitamin C) reduces Fe(III) to yield Fe(II), and the combination of iron–ascorbate has been used often to initiate lipid peroxidation. αT also reduces Fe(III) to Fe(II)[13] and Cu(II) to Cu(I).[14] Thus, under certain circumstances, αT may act as a prooxidant.[13–16] Furthermore, it was found that αT3 reduces Cu(II) to Cu(I) as rapidly as does αT.[1] αT and αT3 are oxidized by Cu(II) to yield α-tocopheryl quinone (αTQ) and α-tocotrienyl quinone (αT3Q), respectively. Under these circumstances, αT and αT3 may act as a prooxidant.

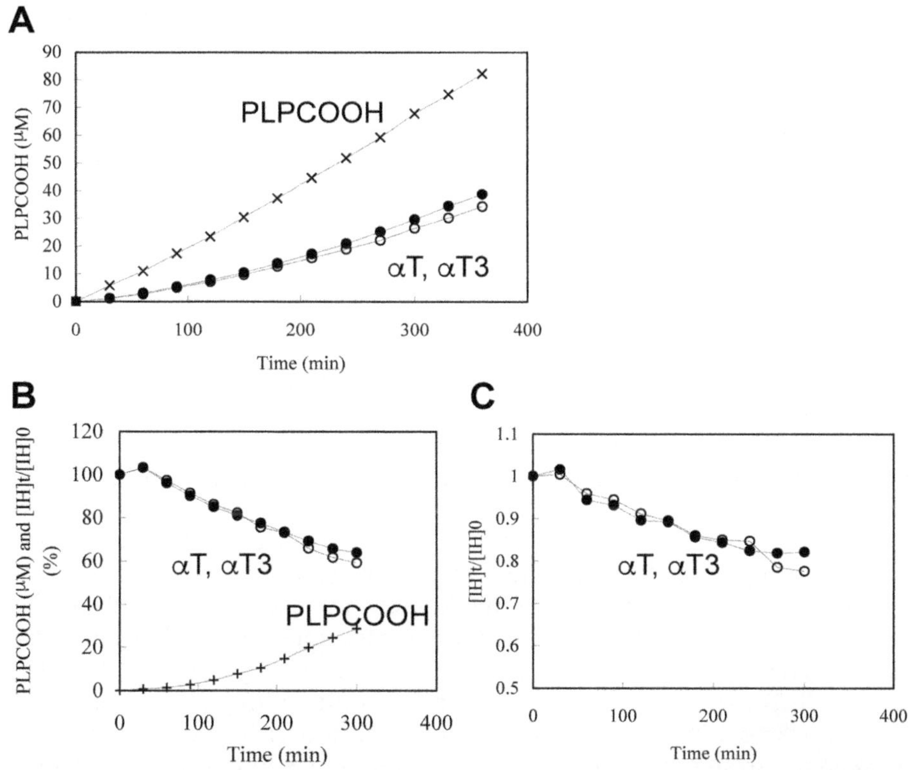

Figure 4.2 Antioxidant effects of αT and αT3 in membranes. (A) Inhibition of the oxidation of PLPC liposomal membranes by αT (solid circles) or αT3 (open circles). PLPC (2.83 mM) vesicles were oxidized with MeO-AMVN (0.20 mM) in the absence (crosses) and presence of antioxidant (5 μM) at 37 °C under air. PLPC hydroperoxides (PLPCOOH) were measured *via* HPLC, as described in ref. 1. (B) and (C) Consumption of antioxidant and accumulation of PLPC hydroperoxides were followed by the oxidation of PLPC (B) and 14 : 0 PC (C) (2.83 mM) liposomal membranes induced by 0.20 mM MeO-AMVN in the presence of both αT (solid circles) and αT3 (open circles) at 37 °C. Adapted from ref. 1 with permission from Elsevier, Copyright 2003.

Interestingly, in contrast to αT and αT3, the β, γ, and δ forms of T and T3 did not reduce Cu(II) at an appreciable rate. They did not yield the corresponding TQ or T3Q nor exert a prooxidant effect in the oxidation of methyl linoleate micelles in aqueous dispersions.[1] In support of these observations, it was found that the oxidation potentials of αT and αT3 were significantly lower than those of γT and γT3. The observation that the T and T3 isoforms have a similar oxidation potential also suggests that the corresponding T and T3 isoforms have similar chemical reactivities toward free radicals. It may be worth noting that metal ions are sequestered by specific proteins *in vivo*, which stabilizes metal ions and makes it difficult for them to be reduced.

Therefore, the prooxidant action observed for antioxidants *in vitro* may not be important *in vivo*.

4.5 Physical Effects of T and T3 on Membranes

The effects of the incorporation of T and T3 on the physical properties of membranes have been studied using chemiluminescence and fluorescence probes.[1,17–20] αT induces a decrease in membrane fluidity.[18] It was reported that both αT and αT3 affected membrane order to a similar extent, but αT3 imposed a more pronounced motional anisotropy on the membrane.[20] The effects of vitamin E homologues on membrane properties were studied using the spin probes 5- and 16-doxylstearic acid (5-NS and 16-NS, respectively).[1] The ESR spectra of the two spin probes, 5-NS and 16-NS, were indistinguishable from each other in homogeneous solution. However, after their incorporation into liposomal membranes, they exhibited different ESR spectra. Both the correlation times for 16-NS and order parameters for 5-NS were increased regardless of the isoform of T and T3 in both multilamellar and unilamellar liposomal membranes. Interestingly, T increased the correlation time of 16-NS more significantly than did the corresponding T3, while T and T3 exerted similar effects for order parameters of 5-NS, suggesting that T increased the rigidity of the membrane more significantly than T3, while both T and T3 exerted similar effects on the fluidity of the liposomal membranes near the water–lipid interface.[1] This is ascribed to the difference between the straight phytyl side chain and bending isoprenyl side chain of T and T3, respectively.

4.6 Incorporation of T3 and T into Membranes

Information on the rate of incorporation of T and T3 into membranes has been reported.[1] T and T3 were added into aqueous suspensions of unilamellar vesicles containing either 5-NS or 16-NS, and the change in ESR parameters was monitored. The ESR parameters (S(t) and τ(t)) increased with time, and they were compared with those of the membranes into which the spin probe and antioxidant were incorporated simultaneously beforehand ($S(\infty)$ and $\tau(\infty)$). When all isoforms are incorporated into the membranes, the ratio should be 1.0, and a faster incorporation of T or T3 into the membrane should yield a larger ratio of the ESR parameters. Cholesterol was also studied, for comparison. Both the correlation times and order parameters were higher when T3, rather than T, were added into the aqueous suspensions after some specific time of incubation, suggesting that T3 was more readily incorporated into the liposomal membranes than was T.

To elucidate inter-membrane mobility, the oxidation of multilamellar PLPC liposomes was induced by the lipid-soluble radical initiator AMVN incorporated into PLPC liposomal membranes in the presence of either αT or αT3, which was inserted into 14:0 PC liposomal membranes separately.

The formation of PLPC hydroperoxides was suppressed by αT and αT3 localized in the different membranes, and the inhibiting effect of αT3 was more significant than that of αT. Another experiment showed that αT3 migrated from 14 : 0 PC membranes to PLPC membranes faster than did αT. The difference in the mobilities of αT and αT3 is small within the membranes but the inter-membrane mobility of αT3 is higher than that of αT.

4.7 Cellular Uptake and Distribution of Tocopherols and Tocotrienols

T and T3 have a similar radical-scavenging activity; however, the biological efficacies of these vitamin E isoforms are different. The prominent cytoprotective effects of αT3 against oxidative-stress-induced cell death have been reported.[21–25] The accelerated cellular uptake of T3 compared with T has been demonstrated in several studies.[9,21–25] The time-dependent cellular uptake of αT and αT3 in human T lymphoma Jurkat cells shows that the initial rate of cellular uptake of αT3 is approximately 70-fold higher than that of αT (Figure 4.3A).[24] A concentration-dependent study revealed an accelerated incorporation of αT3, which was 6.5- and 2.2-fold higher than that of αT after incubation for 24 h and 72 h, respectively (Figure 4.3A). Similar results were obtained in primary neuronal cells; namely, a longer preincubation time increased both the cellular content and the cytoprotective effects of T more significantly than it did those of T3, the effect of preincubation being relatively small for T3.[21] The apparent higher capacity of T3 compared with T is ascribed to the faster uptake of T3 into cells *in vitro* compared with T. Furthermore, the unique properties of αT3, particularly for c-Src kinase and 12-lipoxygenase, have been demonstrated and might contribute further to the strong cytoprotective effect of αT3.[23,25] Vitamin E content in plasma is regulated by the αT transfer protein, which specifically facilitates αT transfer to the ATP-binding cassette transporter A1 in the hepatic plasma membrane, resulting in the acceleration of αT secretion to the plasma.[26] The αT transfer protein possesses a higher affinity for αT than it does for other isoforms, and this property increases the bioavailability of α-T in humans compared with αT3. The cellular content of vitamin E is determined not only by cellular uptake but also by metabolism, such as cytochrome P-450 activity. Cytochrome P-450 family 4 subfamily F member 2 (CYP4F2) activity toward αT is limited relative to the other forms of vitamin E, and the half-life of other isoforms is shorter than that of αT.[27] Collectively, based on physiological content, bioavailability, and metabolism, it is thought that in the isoform of vitamin E, αT plays an important physiological role in the cytoprotective effects against oxidative stress.

The cellular distribution and localization of an antioxidant are important factors that determine its effects. A subcellular fractionation analysis of αT- and αT3-treated Jurkat cells revealed a similar cellular distribution of these antioxidants, which are mainly enriched in the microsomal fraction

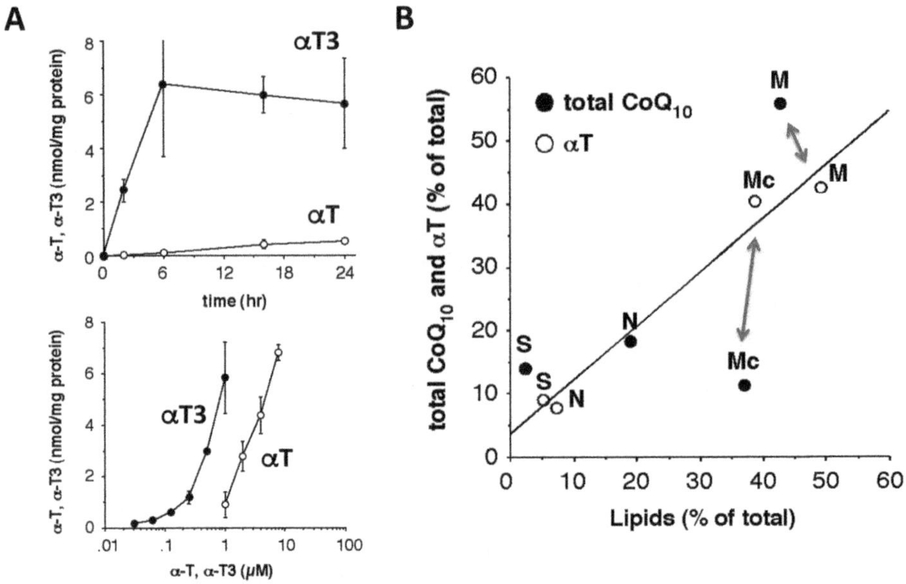

Figure 4.3 Cellular uptake and distribution of αT and αT3. (A) Cellular uptake of αT and αT3 into Jurkat cells. The cells were cultured in serum-free medium with αT and αT3 for the indicated times, and the cellular vitamin E content was measured. Upper panel, 1 μM αT3 (closed circles) or 1 μM αT (open circles) for the indicated times. Lower panel, variable amounts of αT3 (closed circles) or αT (open circles) for 24 h. The means ± SD are shown ($n = 3$). (B) Cellular content of αT, coenzyme Q_{10} (CoQ_{10}) and total lipid in subcellular fractions. PC12 cells cultured with 10 μM αT (open circles) or CoQ_{10} (closed circles) for 24 h were fractioned to the nuclear fraction (N), mitochondrial fraction (M), microsomal fraction (Mc), and cytosolic fraction (S), and the contents of these antioxidants and total lipids were measured. The means are shown ($n = 2$). The correlation of αT with total lipids is shown ($y = 3.75 + 0.85\,x$, $R^2 = 0.974$). Figure A from ref. 24 with permission from John Wiley and Sons, Copyright © 2006, John Wiley and Sons. Figure B adapted from ref. 28 with permission from Elsevier, Copyright 2009.

containing endoplasmic reticulum and plasma membrane.[28] The cellular distribution of these vitamin E isoforms is directly proportional to the lipid distribution in Jurkat cells and rat pheochromocytoma PC12 cells (Figure 4.3B). It is interesting to note that coenzyme Q (CoQ), which is an electron transporter and a lipid-soluble antioxidant, is distributed particularly in the mitochondrial fractions, without any relationship with the lipid distribution.[28] For the antioxidative function of vitamin E, its presence in biological membranes, where lipid radicals are formed, is significant, and the co-distribution of lipids and vitamin E in cells is considered to be biologically reasonable. It is thought that the localization of T in a polyunsaturated fatty-acid-rich non-raft domain, but not in a cholesterol-rich raft domain, is significant for its biological function.[29]

4.8 Cytoprotective Effects of Tocopherols and Tocotrienols

The cytoprotective effects of T and T3 against cell death induced by several stressors have been demonstrated.[21–25] We have reported previously that vitamin E and the iron chelator deferoxamine (DFO) completely blocked selenium deficiency-induced death in Jurkat cells (Figure 4.4).[22] The essential trace element, selenium (Se), forms the active site of antioxidative enzymes, such as glutathione peroxidases (GPxs) and thioredoxin reductases (TRs), as selenocysteine (cysteine analogue in which sulfur is replaced by Se) and is necessary for the biosynthesis of these selenoenzymes.[30] Se deficiency decreases the levels of cellular selenoproteins, including GPx4, and increases reactive oxygen species (ROS) levels. In particular, it induces lipid peroxidation, finally leading to cell death. This cell death shows the character of ferroptosis, which is a non-apoptotic cell death that depends on

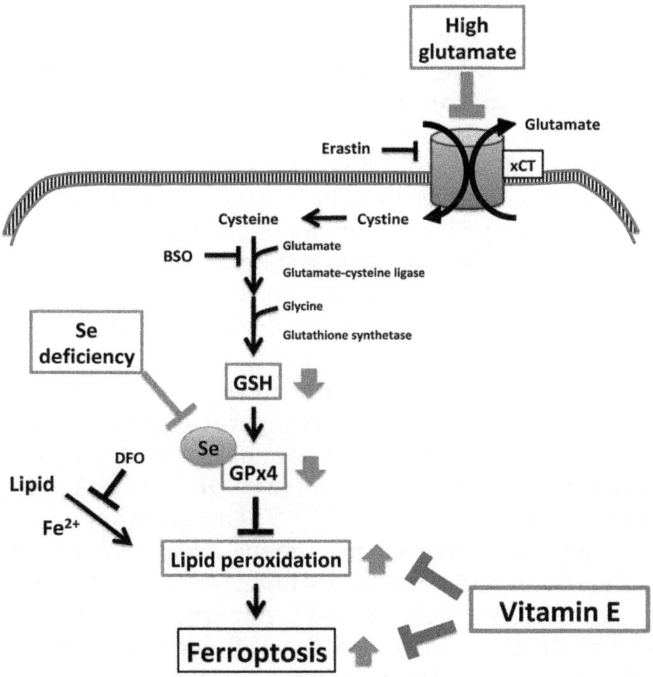

Figure 4.4 Cytoprotective effects of vitamin E against cell death induced by a selenium deficiency and high glutamate. A Se deficiency induces a decrease in the selenoenzyme glutathione peroxidase 4 (GPx4) and induces lipid peroxidation and cell death. A high level of glutamate inhibits the activity of the xc-high-affinity cysteine/glutamate exchange transport system xCT and results in a decrease in cellular cysteine availability, which leads to a decrease in cellular glutathione (GSH) and the execution of cell death. Vitamin E inhibits the lipid peroxidation and cell death induced by Se deficiency and high glutamate levels.

lipid peroxidation and iron (Figure 4.4).[31] GPx4, a major reductant of phospholipid hydroperoxide, has been identified as a critical regulator of ferroptosis, and vitamin E is an effective inhibitor of ferroptosis (Figure 4.4).[32] In the case of Se deficiency-induced Jurkat cell death, the decrease in GPx4 is considered to initiate iron-dependent lipid peroxidation, leading to cell death. It has been known that Se and vitamin E show compensative effects and that a deficiency of both elements causes massive injury. It is thought that under physiological conditions, both αT in lipoproteins and Se in selenoprotein P, which is a major supplier of Se in plasma,[33] help to suppress the induction of lipid peroxidation and oxidative stress. It is interesting to note that both lipoproteins and selenoprotein P are incorporated *via* lipoprotein receptors.[34]

Vitamin E isoforms have been shown to exhibit prominent cytoprotective effects against glutamate-induced neuronal cell death.[21,23,25] Glutamate is an excitatory neurotransmitter in the central nervous system and is known as a contributor to neuronal cell death in neurodegenerative diseases and pathological conditions such as ischemic insults and traumatic brain damage. It is known that high levels of glutamate inhibit the activity of xCT, which is the core subunit of the xc-high-affinity cystine/glutamate exchange transport system.[35] Cystine uptake is one of the important determinants of cellular glutathione (GSH) levels, and inhibition of xCT *via* high glutamate treatment induces GSH depletion and oxidative stress. Erastin has been identified as an inhibitor of xCT, as well as an inducer of ferroptosis.[36] It is thought that ferroptosis is induced by high glutamate levels (Figure 4.4).[36] Vitamin E isoforms inhibit glutamate-induced ROS generation and lipid peroxidation but not GSH depletion; therefore, it is thought that the cytoprotective effects of vitamin E isoforms against glutamate-induced neuronal cell death are mediated by its role as an antioxidant (Figure 4.4).

Furthermore, it has been reported previously that αT exhibits cytoprotective effects against cumene hydroperoxide, monochloramine, the radical initiator AAPH, 4-hydroxy-2-nonenal (4-HNE), 7-ketocholesterol (7-ketoCh), and 24S-hydroxycholesterol (24*S*-OHCh) but not against hydrogen peroxide and arsenic trioxide.[24,37–40] The addition of αT effectively inhibited the cell death induced by cumene hydroperoxide, monochloramine, and AAPH in an antioxidative manner, while the effects of αT against 4-HNE and 7-ketoCh were partial, and it remains unclear whether αT protects against cell death *via* an antioxidative function.[24,37,38] Conversely, the effects of αT against 24*S*-OHCh-induced necroptosis-like cell death are completely independent of its antioxidative function and might involve the physical effects of the increase in rigidity at the membrane interior.[39,40] In the case of 24*S*-OHCh-induced cell death, a relatively high amount of αT (>10 μM) is necessary to inhibit cell death, and αT3 is not effective. Examination of these issues might aid in the understanding of the involvement of antioxidative functions in the cytoprotective effects of vitamin E isoforms.

4.9 Different Biological Action of Tocopherol Quinones

The oxidation of vitamin E generates several oxidized products with different functions. The different biological actions of tocopheryl quinones (TQs), an oxidized product of T, are known (Figure 4.5).[41,42] In general, all quinones are redox-cycling agents that generate ROS, while partially substituted quinones can function as arylating agents that react with cellular nucleophiles, such as thiols, thereby forming covalently linked quinone–thiol Michael adducts.[41] αTQ is a fully substituted quinone—namely, a non-arylating quinone—while γTQ is a partially substituted and electrophilic arylating quinone (Figure 4.5). Treatment of rat pheochromocytoma PC12 cells with αTQ and γTQ resulted in cytoprotective effects against cell death induced by oxidative stress.[42] In αTQ- and γTQ-treated PC12 cells, both forms of tocopheryl hydroquinone (TQH_2) were detected. TQH_2, a reduced form of TQ, can act as a radical-scavenging antioxidant, while TQ itself is not considered to be an effective antioxidant.[42] In the case of γTQ, but not αTQ, a significant increase in GSH was observed, suggesting the involvement of arylating properties in the γTQ-induced cytoprotective effect.[41–44] γTQ did not induce any considerable changes in the activity of glutamate-cysteine ligase (GCL, late-limiting

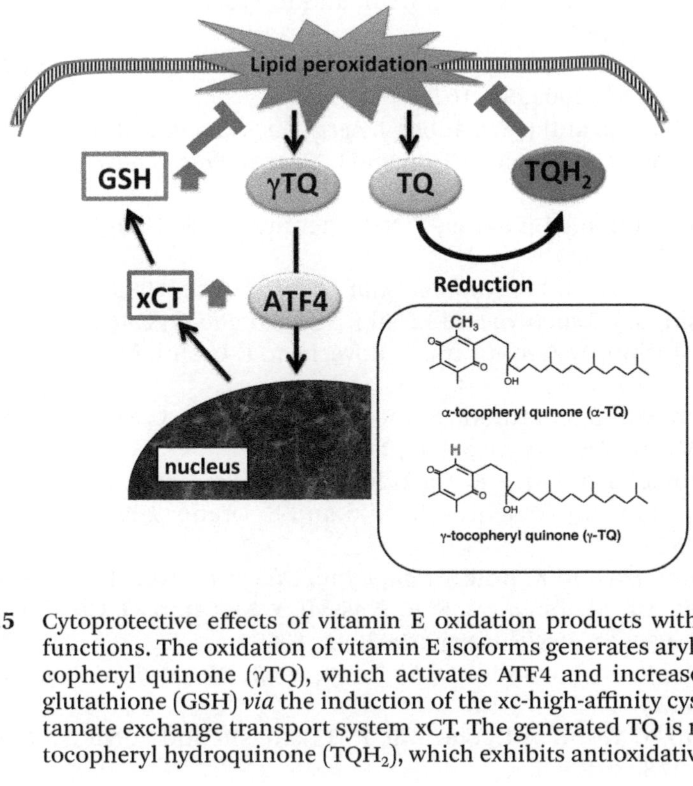

Figure 4.5 Cytoprotective effects of vitamin E oxidation products with different functions. The oxidation of vitamin E isoforms generates arylating γ-tocopheryl quinone (γTQ), which activates ATF4 and increases cellular glutathione (GSH) *via* the induction of the xc-high-affinity cysteine/glutamate exchange transport system xCT. The generated TQ is reduced to tocopheryl hydroquinone (TQH_2), which exhibits antioxidative activity.

enzyme in GSH synthesis), whereas it induced xCT, to increase cysteine availability, *via* the transcriptional activation of the activating transcription factor 4 (ATF4). These observations indicate that γTQ acts as a signal messenger to induce the upregulation of intracellular GSH synthesis (Figure 4.5).[42] It is thought that the generation of γTQ, but not αTQ, under oxidative stress serves as a feedback signal to trigger cellular responses that increase the antioxidative system.

References

1. Y. Yoshida, E. Niki and N. Noguchi, *Chem. Phys. Lipids*, 2003, **123**, 63.
2. Y. Yoshida, Y. Saito, L. S. Jones and Y. Shigeri, *J. Biosci. Bioeng.*, 2007, **104**, 439.
3. L. Müller, K. Theile and V. Böhm, *Mol. Nutr. Food Res.*, 2010, **54**, 731.
4. K. Fukuzawa, A. Ouchi, S. Nagaoka, A. Shibata and K. Mukai, *J. Nutr. Sci. Vitaminol.*, 2014, **60**, 443.
5. A. Ouchi, S. Nagaoka, T. Suzuki, K. Izumisawa, T. Koike and K. Mukai, *J. Agric. Food Chem.*, 2014, **62**, 8101.
6. K. Mukai, A. Ouchi, T. Abe, K. Murata, K. Nakagawa and T. Miyazawa, *J. Agric. Food Chem.*, 2014, **62**, 11901.
7. E. Niki, *Methods Enzymol.*, 1990, **186**, 100.
8. C. Suarna, R. L. Hood, R. T. Dean and R. Stocker, *Biochim. Biophys. Acta*, 1993, **1166**, 163.
9. N. Noguchi, R. Hanyu, A. Nonaka, Y. Okimoto and T. Kodama, *Free Radical Biol. Med.*, 2003, **34**, 1614.
10. Z. Xu, N. Hua and J. S. Gadber, *J. Agric. Food Chem.*, 2001, **49**, 2077.
11. E. Servinova, V. Kagan, D. Han and L. Packer, *Free Radical Biol. Med.*, 1991, **10**, 263.
12. A. F. P. Sonnen, H. Bakirci, T. Netscher and W. N. Nau, *J. Am. Chem. Soc.*, 2005, **127**, 15575.
13. K. Yamamoto and E. Niki, *Biochim. Biophys. Acta*, 1988, **958**, 19.
14. Y. Yoshida, J. Tsuchiya and E. Niki, *Biochim. Biophys. Acta*, 1994, **1200**, 85.
15. M. Maiorino, A. Zamburlini, A. Roveri and F. Ursini, *FEBS Lett.*, 1993, **330**, 174.
16. N. Yamashita, M. Murata, S. Inoue, M. J. Burkitt, L. Milne and S. Kawanishi, *Chem. Res. Toxicol.*, 1998, **11**, 85.
17. M. Yamaoka and M. J. H. Carrillo, *Chem. Phys. Lipids*, 1990, **55**, 295.
18. M. Takahashi, J. Tsuchiya, E. Niki and S. Urano, *J. Nutr. Sci. Vitaminol.*, 1988, **34**, 25.
19. M. Yamaoka and K. Komiyama, *J. Jpn. Oil Chem. Soc.*, 1989, **38**, 478.
20. Y. J. Suzuki, M. Tsuchiya, S. R. Wassall, Y. M. Choo, G. Govil, V. E. Kagan and L. Packer, *Biochemistry*, 1993, **32**, 10692.
21. Y. Saito, K. Nishio and Y. O. Akazawa, *et al.*, *Free Radical Biol. Med.*, 2010, **49**, 1542.
22. Y. Saito, Y. Yoshida, T. Akazawa, K. Takahashi and E. Niki, *J. Biol. Chem.*, 2003, **278**, 39428.

23. C. K. Sen, S. Khanna, S. Roy and L. Packer, *J. Biol. Chem.*, 2000, **275**, 13049.
24. Y. Saito, Y. Yoshida, K. Nishio, M. Hayakawa and E. Niki, *Ann. N. Y. Acad. Sci.*, 2004, **1031**, 368.
25. S. Khanna, S. Roy, H. Ryu, P. Bahadduri, P. W. Swaan, R. R. Ratan and C. K. Sen, *J. Biol. Chem.*, 2003, **278**, 43508.
26. N. Kono, U. Ohto, T. Hiramatsu, M. Urabe, Y. Uchida, Y. Satow and H. Arai, *Science*, 2013, **340**, 1106.
27. M. G. Traber, *J. Lipid Res.*, 2013, **54**, 2295.
28. Y. Saito, A. Fukuhara and K. Nishio, *et al.*, *J. Nutr. Biochem.*, 2009, **20**, 350.
29. J. Atkinson, T. Harroun, S. R. Wassall, W. Stillwell and J. Katsaras, *Mol. Nutr. Food Res.*, 2010, **54**, 641.
30. V. M. Labunskyy, D. L. Hatfield and V. N. Gladyshev, *Physiol. Rev.*, 2014, **94**, 739.
31. B. R. Stockwell, J. P. Friedmann Angeli and H. Bayir, *et al.*, *Cell*, 2017, **171**, 273.
32. W. S. Yang, R. Sri Ramaratnam and M. E. Welsch, *et al.*, *Cell*, 2014, **156**, 317.
33. R. F. Burk and K. E. Hill, *Biochim. Biophys. Acta*, 2009, **1790**, 1441.
34. Y. Mita, K. Nakayama and S. Inari, *et al.*, *Nat. Commun.*, 2017, **8**, 1658.
35. H. Sato, M. Tamba, T. Ishii and S. Bannai, *J. Biol. Chem.*, 1999, **274**, 11455.
36. S. J. Dixon, K. M. Lemberg and M. R. Lamprecht, *et al.*, *Cell*, 2012, **149**, 1060.
37. R. Piga, Y. Saito, Z. Chen, Y. Yoshida and E. Niki, *Arch. Biochem. Biophys.*, 2005, **436**, 101.
38. R. Piga, Y. Saito, Y. Yoshida and E. Niki, *Neurotoxicology*, 2007, **28**, 67.
39. T. Nakazawa, Y. Miyanoki, Y. Urano, M. Uehara, Y. Saito and N. Noguchi, *J. Steroid Biochem. Mol. Biol.*, 2017, **169**, 69.
40. K. Yamanaka, Y. Saito, T. Yamamori, Y. Urano and N. Noguchi, *J. Biol. Chem.*, 2011, **286**, 24666.
41. D. G. Cornwell, S. Kim, P. A. Mazzer, K. H. Jones and P. G. Hatcher, *Lipids*, 2003, **38**, 973.
42. Y. Ogawa, Y. Saito, K. Nishio, Y. Yoshida, H. Ashida and E. Niki, *Free Radical Res.*, 2008, **42**, 674.
43. X. Wang, B. Thomas and R. Sachdeva, *et al.*, *Proc. Natl. Acad. Sci. U. S. A.*, 2006, **103**, 3604.
44. A. L. Levonen, D. A. Dickinson, D. R. Moellering, R. T. Mulcahy, H. J. Forman and V. M. Darley-Usmar, *Arterioscler., Thromb., Vasc. Biol.*, 2001, **21**, 1846.

CHAPTER 5

α-Tocopherol Transfer Protein

NOZOMU KONO AND HIROYUKI ARAI*

Graduate School of Pharmaceutical Sciences, the University of Tokyo, 7-3-1 Hongo, Bunkyo-ku, Tokyo 113-0033, Japan
*E-mail: harai@mol.f.u-tokyo.ac.jp

5.1 Introduction

Vitamin E is a major fat-soluble antioxidant that is composed of a chromanol ring and an aliphatic side chain.[1,2] Naturally-occurring forms of vitamin E can be divided into two subgroups: tocotrienols and tocopherols. Tocotrienols have an unsaturated isoprenoid side chain, whereas tocopherols have a saturated phytyl tail. The phytyl tail has three chiral centers that naturally occur in the *RRR* configuration. Tocopherols and tocotrienols each have four isomers (α, β, γ and δ), each designated according to the position and number of the methyl groups on the chromanol ring.[3] α- and γ-Tocopherols are the most abundant vitamin E forms in natural sources,[4] and in particular, α-tocopherol is the main form of vitamin E found in plasma and tissues in mammals.[5] α-Tocopherol transfer protein (α-TTP), an α-tocopherol-selective vitamin E-binding protein expressed in hepatocytes, plays a critical role in maintaining the α-tocopherol level in the body. Here we will review the molecular mechanism and physiological roles of α-TTP-mediated α-tocopherol transport.

Food Chemistry, Function and Analysis No.11
Vitamin E: Chemistry and Nutritional Benefits
Edited by Etsuo Niki

Published by the Royal Society of Chemistry, www.rsc.org

5.2 Vitamin E Transport in the Body

Dietary vitamin E, which is composed of mainly α-tocopherol and γ-tocopherol, is absorbed from the intestine and released into the circulation with chylomicrons[5,6] (Figure 5.1). Chylomicrons are rapidly catabolized in the circulation by lipoprotein lipase, and the resultant chylomicron remnants containing vitamin E are endocytosed by the liver *via* a receptor-mediated process.[7] In the liver, α-tocopherol is selectively resecreted from the liver cells and transferred to lipoproteins such as VLDL.[5,8,9] α-Tocopherol then travels in the circulation with lipoproteins and is delivered to extrahepatic tissues. Thus, discrimination between the different vitamin E forms occurs in the liver.

5.3 Substrate Specificity of α-TTP

α-TTP is a 30 500 Da cytosolic protein that binds α-tocopherol and enhances its transfer between membranes *in vitro*.[10] α-TTP was first described in 1975 by Catignani *et al.*,[11] and was purified and cloned from rat liver by Sato *et al.*[12,13] Rat α-TTP cDNA encodes a 278-amino acid protein that belongs to the Sec14 protein family of evolutionarily conserved lipid binding proteins. α-TTP possesses a lipid-binding Sec14 domain in the C-terminal side, and has no other known functional domains (Figure 5.2).[14] α-TTP is expressed primarily

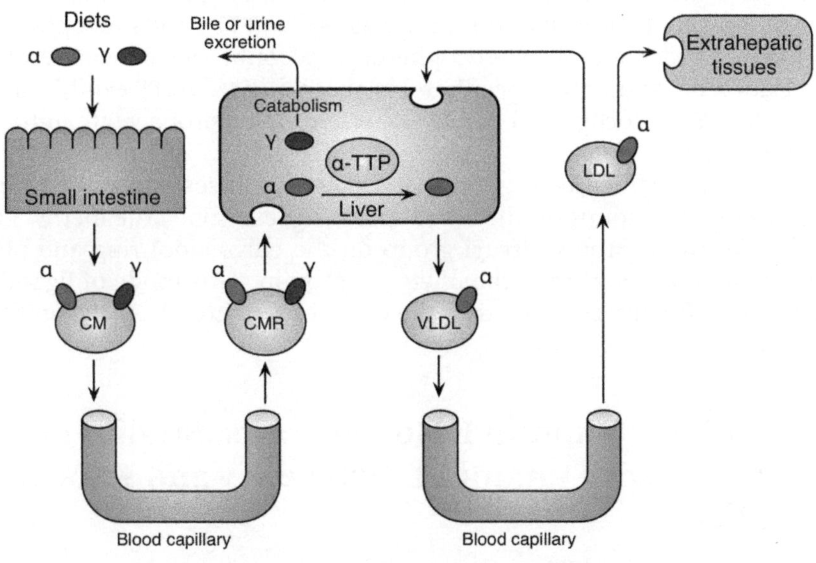

Figure 5.1 Vitamin E transport in the body. α, α-tocopherol; γ, γ-tocopherol; α-TTP, α-tocopherol transfer protein; CM, chylomicron; CMR, chylomicron remnant; VLDL, very low density lipoprotein; LDL, low density lipoprotein. Adapted from ref. 60 with permission from John Wiley and Sons, Copyright © 2014 The Authors.

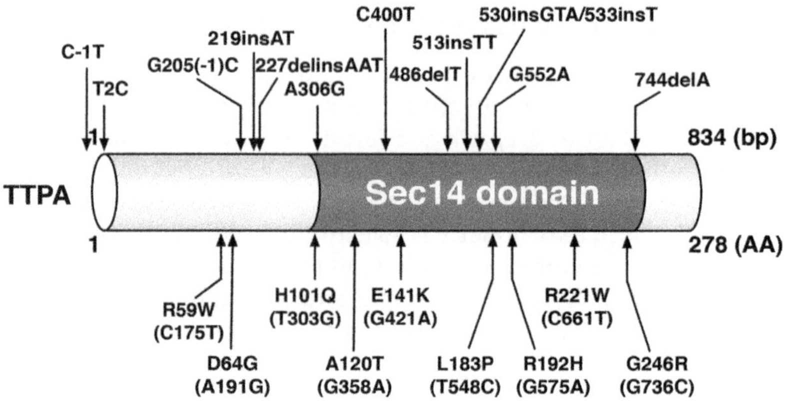

Figure 5.2 α-TTP mutations found in AVED patients. α-TTP consists of only a lipid-binding domain called the Sec14 domain (some domain prediction tools give a hit for a CRAL_TRIO_N domain on the N-terminus of α-TTP). Missense mutations are shown below the panel, and insertions, deletions, and splicing mutations are indicated above the panel. bp, base pairs; AA, amino acids. Reproduced from ref. 60 with permission from John Wiley and Sons, Copyright © 2014 The Authors.

in the liver, where it accounts for about 0.1% (w/w) of the liver cytosolic proteins. The ligand specificity of α-TTP for various tocopherol analogs has been examined using *in vitro* transfer of radiolabeled α-tocopherol between membranes[15] and by Lipidex-based binding assay.[16] α-TTP binds tocopherols in the order of α-tocopherol >> β-tocopherol > γ-tocopherol > δ-tocopherol. The dissociation constant for α-tocopherol is about 25 nM. α-TTP exhibits a weak affinity for *SRR*-α-tocopherol, α-tocotrienol, α-tocopherol acetate, α-tocopherol quinone, and trolox.

The ligand specificities suggest that α-TTP recognizes α-tocopherol by (1) the three methyl groups on the chromanol ring, especially the methyl group at position five, (2) the hydroxyl group on the chromanol ring and (3) the structure and orientation of the phytyl side chain. This mode of ligand recognition was further confirmed by the crystal structure of α-TTP complexed with α-tocopherol.[17,18]

5.4 α-TTP in Vitamin E Homeostasis: Studies of Ataxia with Vitamin E Deficiency and Knockout Mice

The pivotal role of α-TTP in vitamin E homeostasis has been revealed by the discovery that mutations in the α-TTP gene (*TTPA*) are a causative factor of familial vitamin E deficiency, termed ataxia with vitamin E deficiency (AVED).[19,20] AVED is a rare autosomal recessive neurodegenerative disease similar to Friedreich's ataxia that is accompanied by low or undetectable plasma vitamin E levels.[21] In AVED patients, the absorption of α-tocopherol from the intestine and its delivery to the liver are normal, but the secretion of α-tocopherol from the liver is impaired,[22] which results in a very low concentration of vitamin E in the circulation. More than 20 AVED-associated mutations have been discovered (Figure 5.2).[19–21,23–27] Many of the mutations are splice site or truncation mutations, resulting in a defective protein. Several missense mutations have also been found in AVED patients, affecting both alleles, either as a homozygous defect or in combination with one of the other mutations. The causal relationship of α-TTP with vitamin E status and neurological symptoms has been demonstrated by α-TTP-deficient mice.[28,29] Under a normal diet, plasma α-tocopherol levels are undetectable in the homozygous (α-TTP$^{-/-}$) mice, and about half of wild type in the heterozygous mice,[29,30] indicating that α-TTP expression levels directly impact whole-body vitamin E status. Aged α-TTP-deficient mice exhibit AVED-like neurological symptoms, which are almost completely prevented by α-tocopherol supplementation.[31,32]

5.5 Expression of α-TTP in Extrahepatic Tissues

Although α-TTP was first described as a hepatic protein,[13] α-TTP is also expressed in other tissues, such as rat brain, spleen, lung and kidney,[33] pregnant mouse uterus,[34] and the retina.[27]

In the rat brain, α-TTP mRNAs are detected predominantly in the Purkinje layer of the cerebellar cortex, as revealed by *in situ* hybridization histochemistry. The signal is detected in small cells around the Purkinje cells, called Bergmann glial cells, and not in the Purkinje cells themselves. The cell bodies of the Bergmann glia are located between the giant Purkinje cells of the cerebellum. Bergmann glial cells possess a few long processes that end on basement membranes along the blood vessels. Loss of α-TTP in AVED patients causes severe damage to the Purkinje cells.[35] By analogy of the role of α-TTP in hepatocytes, it is speculated that the Bergmann glial cells take up vitamin E from the blood capillaries and secrete it with the help of α-TTP to supply the Purkinje cells and possibly other cerebellar cells (Figure 5.3A). Bergmann glial cells strongly express apoprotein E (ApoE), which may participate in the intercellular transport of vitamin E in the cerebellum.

α-TTP is also highly expressed in Müller cells of the murine retina.[36] The photoreceptor outer segment membrane contains unusually high levels of polyunsaturated fatty acids,[37] and thus is easily damaged by oxidation. AVED patients sometimes show retinitis pigmentosa at a relatively old age. α-TTP expressed in Müller cells may facilitate the transport of vitamin E from the blood capillaries to photoreceptor neurons (Figure 5.3B).

Vitamin E was first described in 1922 as a factor required for normal fetal development in rats and the fetal resorption assay has long been a standard

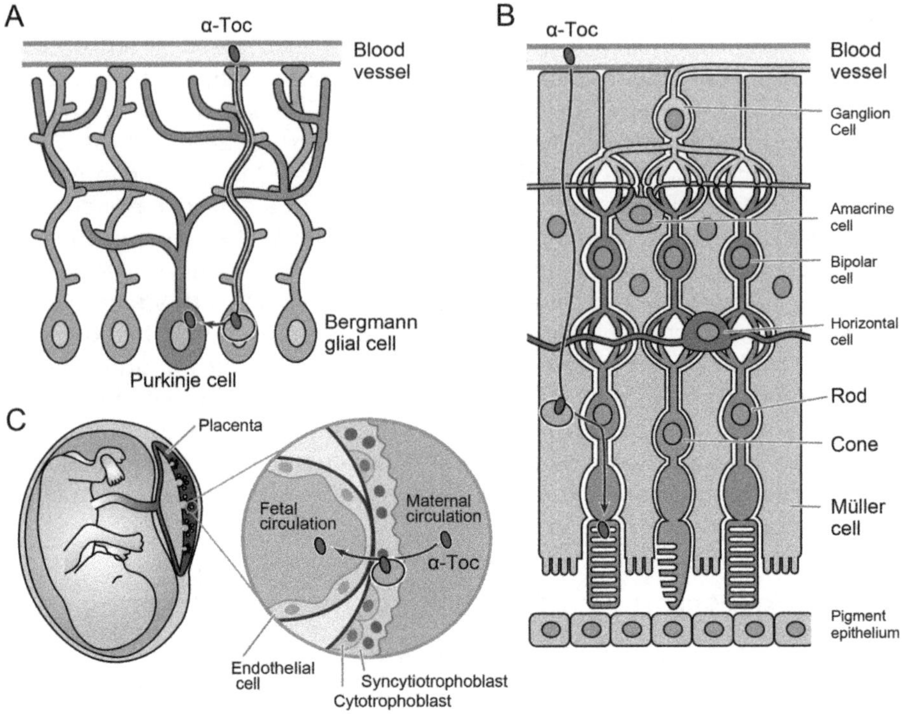

Figure 5.3 α-TTP functions in the cerebellum (A), retina (B), and placenta (C).

test of vitamin E deficiency.[38] In this context, it is interesting to note that the pregnant mouse uterus expresses α-TTP, peaking on day 4.5 post coitum, coinciding with the day of implantation in the mouse.[29,34] At that time, α-TTP is localized in the epithelial cells of the uterus as well as in the inner decidual reaction zone, which later develops into a part of the murine placenta. In humans, intense staining of α-TTP is also seen in term placenta, especially in syncytiotrophoblasts and cytotrophoblasts.[39,40] Pregnant α-TTP-deficient females display severely impaired placentas with a marked reduction of labyrinthine trophoblasts and the embryos died at mid-gestation.[29] Dietary supplementation of excess α-tocopherol or a synthetic antioxidant (BO-653) prevented placental failure and allowed full-term pregnancies in α-TTP-deficient females,[29,41] suggesting that oxidative stress in the labyrinth region of the placenta is protected by vitamin E during development. It has been shown that vitamin E, with a preference for *RRR*-α-tocopherol, crosses the placenta from the mother to the embryo.[42] α-TTP expressed in the uterus may explain the stereospecific transport of tocopherols to the placenta, and up-regulation of α-TTP expression may increase the transport of α-tocopherol to the placenta to prevent deleterious oxidative stress during embryogenesis (Figure 5.3C).

5.6 Intracellular Vitamin E Transport by α-TTP

Based on previous studies of AVED patients, it was suggested that α-TTP catalyzes the intracellular transport of α-tocopherol taken up by the liver cells into nascent VLDL for secretion.[43] To reproduce these processes in a cell culture system, Arita *et al.* devised an assay system using the hepatoma cell line and radiolabeled α-tocopheryl acetate.[44] In this assay, synthetic α-[^{14}C] tocopheryl acetate added to a medium is taken up by hepatoma cells and then hydrolyzed intracellularly by endogenous esterase(s) to produce α-[^{14}C] tocopherol. When α-TTP is expressed in the cells, this tocopherol is efficiently released into the medium. Thus, α-TTP activity in the cells can be measured by counting the radioactivity of α-[^{14}C]tocopherol appearing in the medium. With this system, Arita *et al.* proved that α-TTP facilitated the secretion of α-tocopherol from hepatocytes into the extracellular medium. Unexpectedly, the α-tocopherol secretion catalyzed by α-TTP was not coupled to VLDL secretion because it was not affected by brefeldin A, which inhibits VLDL secretion. On the other hand, the efflux of α-tocopherol from cells to HDL apolipoproteins is mediated by ATP-binding cassette transporter A1 (ABCA1).[45–47] In rat primary hepatocytes and rat hepatoma McARH7777 cells, α-TTP is distributed homogeneously throughout the cytosol, but is localized in endolysosomal compartments upon treatment with chloroquine, an agent that increases the pH in acidic compartments.[36,48] α-TTP was also localized in endolysosomal compartments in human hepatoma HepG2 cells,[46,49] which implies that α-TTP translocates from the cytosol to endolysosomal compartments to acquire α-tocopherol. Collectively, these studies indicate that the most likely mechanism by which α-TTP transports hepatic α-tocopherol is as follows: α-tocopherol associated with circulating plasma lipoproteins (*i.e.* chylomicron remnants) is endocytosed and transported to endolysosomal compartments, where the lipoproteins are hydrolyzed, thereby releasing α-tocopherol. Cytosolic α-TTP picks up α-tocopherol and transports it to the plasma membrane, where it is secreted into the circulation by ABCA1, which is present in the hepatocyte plasma membranes. Consistently, NBD (7-nitrobenz-2-oxa-1,3-diazol-4-yl)-labeled tocopherol (NBD-Toc) was observed to arrive in the lysosome after NBD-Toc in the medium was taken up by cells. From there, NBD-Toc translocates to the cell surface in an α-TTP-dependent manner.[46]

5.7 Role of Phosphoinositides in the Vectorial Transport of α-Tocopherol by α-TTP

The vectorial transport of α-tocopherol to the plasma membrane in the hepatocytes is important because otherwise α-tocopherol would be rapidly degraded by cytochrome P450 in the endoplasmic reticulum,[50] which accounts for more than half of the total membrane of the hepatocytes. Recently, the interaction of α-TTP with phosphoinositide was found to be critical for the intracellular transport of α-tocopherol.[51] The clue for this finding came from α-TTP missense mutations found in AVED patients: among nine disease-associated missense mutations, the arginine missense mutations (R59W, R192H, and R221W) are located in one region on the α-TTP protein surface, which is distinct from the α-tocopherol binding site. R59W α-TTP can bind and transfer α-tocopherol *in vitro*, but cannot stimulate α-tocopherol secretion in cells,[49,51,52] indicating that the R59W mutation impairs an α-TTP function other than α-tocopherol binding. Krugmann *et al.* identified α-TTP as a protein that has an affinity for phosphoinositide-derived matrices.[53] Interestingly, the R59W mutation almost completely abolishes the ability of α-TTP to bind phosphoinositides. The crystal structure of α-TTP complexed with phosphoinositide revealed that the disease-related arginine residues interacted with phosphate groups of the phosphoinositide (Figure 5.4).[51] Phosphoinositides are minor lipid components of biological membranes but play an important role in the organelle targeting of cytosolic proteins. Each type of phosphoinositide has a defined subcellular localization and its polar head group is recognized by a specific protein domain.[54] α-TTP interacts preferentially with two phosphatidylinositol bisphosphates (PIP_2s), phosphatidylinositol 4,5-bisphosphate [$PI(4,5)P_2$] and phosphatidylinositol 3,4-bisphosphate [$PI(3,4)P_2$], both of which are concentrated

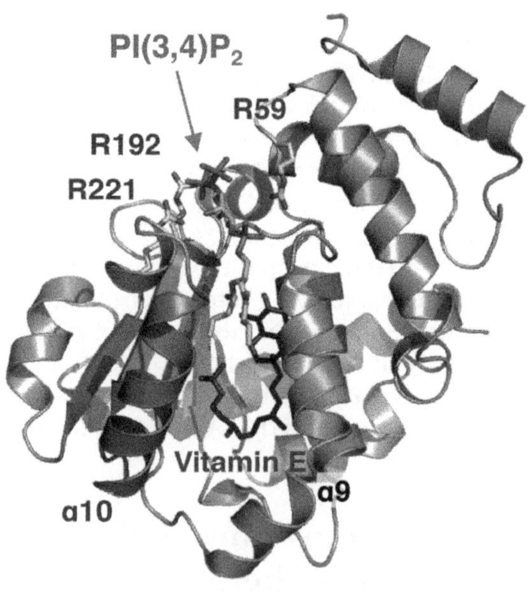

Figure 5.4 Crystal structure of α-TTP–vitamin E–$PI(3,4)P_2$ complex. A ribbon representation of the structure of α-TTP in complex with vitamin E and $PI(3,4)P_2$ (Protein Data Bank (PDB) identifier 3W67; ref. 51). PI(3,4)P2 and vitamin E are shown as stick models. The side chains of R59, R192, and R221 are depicted as stick models. The "lid" helix (α10) is labelled. Adapted from ref. 60 with permission from John Wiley and Sons, Copyright © 2014 The Authors.

in the plasma membrane.[51] In fact, masking or degrading PI(4,5)P_2 inhibits α-TTP-mediated α-tocopherol secretion from cultured hepatocytes.[51,55] These results suggest that α-TTP transfers α-tocopherol to the plasma membrane by targeting these PIP_2.

The crystal structure of α-TTP suggests that the acyl chains of PIP_2 lie near the entrance of the α-tocopherol-binding pocket and contact hydrophobic residues of the "lid" helix (Figure 5.4), whose conformation greatly differs between closed and open states of α-TTP.[17] The position of the lid of the α-TTP-PIP_2 complex is intermediate between the closed and open states, suggesting that the binding of acyl chains of PIP_2 causes the lid of the α-tocopherol binding pocket to open. Experiments with liposomes revealed that α-TTP acts as an α-tocopherol/PIP_2 exchanger.[51] These observations suggest that the ternary complex is a snapshot of a transition state in a lipid exchange reaction in which a PIP_2 in the membrane is swapped for α-tocopherol. Consistently, *in vitro* studies have shown that α-TTP transfers α-tocopherol between membranes through a direct protein–membrane interaction[56–58] and the transfer is promoted by phosphoinositides in the acceptor membrane.[51] Based on these findings, a potential mechanism of intracellular α-tocopherol transport by α-TTP is as follows: α-TTP containing α-tocopherol first interacts with the head group of a PIP_2 in the target membrane through its positively charged surface where disease-related arginine residues are clustered. The PIP_2 is then transferred to α-TTP, which leads to an opening of the lid of the α-tocopherol binding pocket and stimulates the transfer of α-tocopherol to the target membrane. As a result, α-TTP exchanges α-tocopherol for the PIP_2 (Figure 5.5).

5.8 Future Prospects

A substantial body of evidence from clinical and experimental studies has established that α-TTP plays a critical role in maintaining vitamin E levels in the body. Moreover, recent studies have begun to uncover the molecular mechanism underlying the intracellular transport of α-tocopherol by α-TTP. As stated earlier, it is likely that α-TTP targets the plasma membrane by binding to phosphoinositides to release α-tocopherol, and this binding might be impaired by arginine missense mutations associated with AVED. On the other hand, it remains unclear how α-TTP targets endolysosome compartments to pick up endocytosed α-tocopherol. Further studies are needed to identify the target lipids or proteins of α-TTP on endolysosome compartments. In this regard, of interest is the finding that bis(monoacylglycerol) phosphate (BMP), a phospholipid unique to late endosomes, promotes the membrane binding of α-TTP.[59] Moreover, how α-TTP function is regulated by intracellular signaling remains unclear. In the first study to purify α-TTP,[12] two isoforms of α-TTP, which have the same molecular weight but different isoelectric points, were found to exist in the liver. Studies with monoclonal antibodies revealed that both isoforms are produced from the same α-TTP mRNA transcript and one isoform is produced in several days after translation in the hepatocytes,

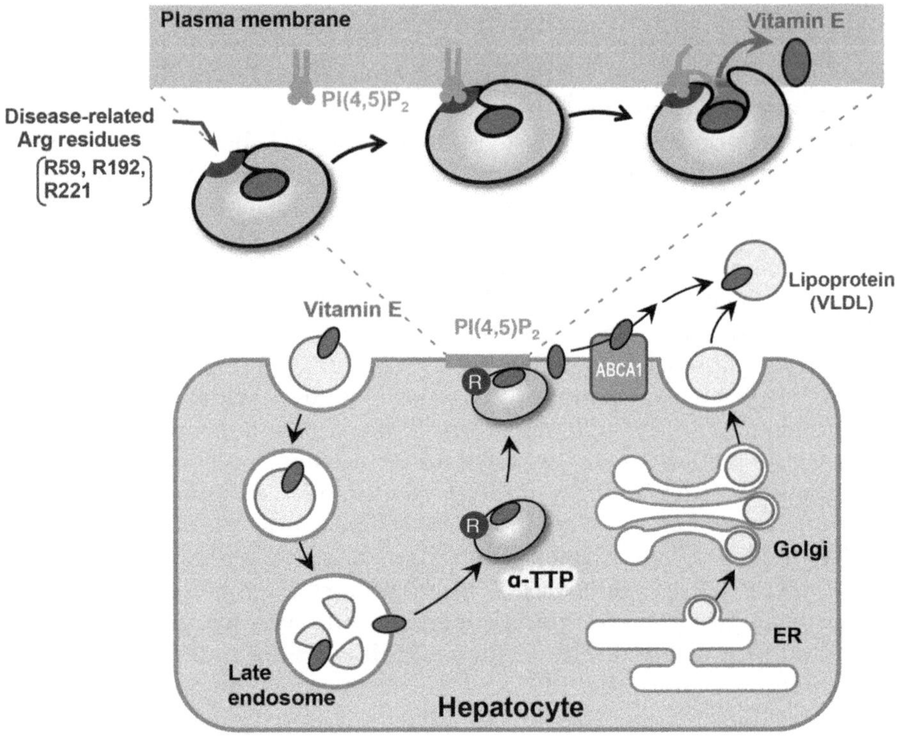

Figure 5.5 A model of intracellular vitamin E transport in hepatocytes. Adapted from ref. 60 with permission from John Wiley and Sons, Copyright © 2014 The Authors.

suggesting the presence of post-translational protein modification(s) of α-TTP (our unpublished observations). Further cell biological and biochemical studies will be required for a full understanding of α-TTP-mediated intracellular α-tocopherol transport and its regulation. Such studies should also provide mechanistic insights into other AVED-associated missense mutations of α-TTP.

References

1. D. J. Mustacich, R. S. Bruno and M. G. Traber, *Vitam. Horm.*, 2007, **76**, 1.
2. E. Niki and M. G. Traber, *Ann. Nutr. Metab.*, 2012, **61**, 207.
3. IUPAC-IUB Joint Commission on Biochemical Nomenclature, *Eur. J. Biochem.*, 1982, **123**, 473.
4. J. G. Bieri and R. P. Evarts, *Am. J. Clin. Nutr.*, 1974, **27**, 980.
5. M. G. Traber and H. J. Kayden, *Am. J. Clin. Nutr.*, 1989, **49**, 517.
6. M. Meydani, J. S. Cohn, J. B. Macauley, J. R. McNamara, J. B. Blumberg and E. J. Schaefer, *J. Nutr.*, 1989, **119**, 1252.
7. A. D. Cooper, *J. Lipid Res.*, 1997, **38**, 2173.

8. M. G. Traber, G. W. Burton, K. U. Ingold and H. J. Kayden, *J. Lipid Res.*, 1990, **31**, 675.
9. M. G. Traber, G. W. Burton and L. Hughes, *et al.*, *J. Lipid Res.*, 1992, **33**, 1171.
10. D. Manor and S. Morley, *Vitam. Horm.*, 2007, **76**, 45.
11. G. L. Catignani, *Biochem. Biophys. Res. Commun.*, 1975, **67**, 66.
12. Y. Sato, K. Hagiwara, H. Arai and K. Inoue, *FEBS Lett.*, 1991, **288**, 41.
13. Y. Sato, H. Arai, A. Miyata, S. Tokita, K. Yamamoto, T. Tanabe and K. Inoue, *J. Biol. Chem.*, 1993, **268**, 17705.
14. K. Saito, L. Tautz and T. Mustelin, *Biochim. Biophys. Acta*, 2007, **1771**, 719.
15. A. Hosomi, M. Arita, Y. Sato, C. Kiyose, T. Ueda, O. Igarashi, H. Arai and K. Inoue, *FEBS Lett.*, 1997, **409**, 105.
16. C. Panagabko, S. Morley and M. Hernandez, *et al.*, *Biochemistry*, 2003, **42**, 6467.
17. R. Meier, T. Tomizaki, C. Schulze-Briese, U. Baumann and A. Stocker, *J. Mol. Biol.*, 2003, **331**, 725.
18. K. C. Min, R. A. Kovall and W. A. Hendrickson, *Proc. Natl. Acad. Sci. U. S. A.*, 2003, **100**, 14713.
19. K. Ouahchi, M. Arita and H. Kayden, *et al.*, *Nat. Genet.*, 1995, **9**, 141.
20. T. Gotoda, M. Arita and H. Arai, *et al.*, *N. Engl. J. Med.*, 1995, **333**, 1313.
21. I. Di Donato, S. Bianchi and A. Federico, *Neurol. Sci.*, 2010, **31**, 511.
22. M. G. Traber, R. J. Sokol, G. W. Burton, K. U. Ingold, A. M. Papas, J. E. Huffaker and H. J. Kayden, *J. Clin. Invest.*, 1990, **85**, 397.
23. L. Cavalier, K. Ouahchi, H. J. Kayden, S. Di Donato, L. Reutenauer, J. L. Mandel and M. Koenig, *Am. J. Hum. Genet.*, 1998, **62**, 301.
24. M. Hoshino, N. Masuda, Y. Ito, M. Murata, J. Goto, M. Sakurai and I. Kanazawa, *Ann. Neurol.*, 1999, **45**, 809.
25. T. Shimohata, H. Date and H. Ishiguro, *et al.*, *Ann. Neurol.*, 1998, **43**, 273.
26. F. Usuki and K. Maruyama, *J. Neurol. Neurosurg. Psychiatry*, 2000, **69**, 254.
27. T. Yokota, T. Shiojiri and T. Gotoda, *et al.*, *Ann. Neurol.*, 1997, **41**, 826.
28. Y. Terasawa, Z. Ladha and S. W. Leonard, *et al.*, *Proc. Natl. Acad. Sci. U. S. A.*, 2000, **97**, 13830.
29. K. Jishage, M. Arita and K. Igarashi, *et al.*, *J. Biol. Chem.*, 2001, **276**, 1669.
30. S. W. Leonard, Y. Terasawa, R. V. Farese and M. G. Traber, *Am. J. Clin. Nutr.*, 2002, **75**, 555.
31. T. Yokota, K. Igarashi and T. Uchihara, *et al.*, *Proc. Natl. Acad. Sci. U. S. A.*, 2001, **98**, 15185.
32. L. Ulatowski, R. Parker, G. Warrier, R. Sultana, D. A. Butterfield and D. Manor, *Neuroscience*, 2014, **260**, 120.
33. A. Hosomi, K. Goto and H. Kondo, *et al.*, *Neurosci. Lett.*, 1998, **256**, 159.
34. D. E. Kaempf-Rotzoll, K. Igarashi and J. Aoki, *et al.*, *Biol. Reprod.*, 2002, **67**, 599.
35. A. Larnaout, S. Belal and M. Zouari, *et al.*, *Acta Neuropathol.*, 1997, **93**, 633.
36. M. Shichiri, N. Kono and Y. Shimanaka, *et al.*, *J. Biol. Chem.*, 2012, **287**, 2926.

37. M. Neuringer, G. J. Anderson and W. E. Connor, *Annu. Rev. Nutr.*, 1988, **8**, 517.
38. H. M. Evans and K. S. Bishop, *Science*, 1922, **56**, 650.
39. D. E. Kaempf-Rotzoll, M. Horiguchi, K. Hashiguchi, J. Aoki, H. Tamai, O. Linderkamp and H. Arai, *Placenta*, 2003, **24**, 439.
40. D. E. Rotzoll, R. Scherling, R. Etzl, H. Stepan, L. C. Horn and J. M. Pöschl, *Eur. J. Obstet. Gynecol. Reprod. Biol.*, 2008, **140**, 183.
41. K. Jishage, T. Tachibe and T. Ito, *et al.*, *Biol. Reprod.*, 2005, **73**, 983.
42. R. V. Acuff, R. G. Dunworth, L. W. Webb and J. R. Lane, *Am. J. Clin. Nutr.*, 1998, **67**, 459.
43. H. J. Kayden and M. G. Traber, *J. Lipid Res.*, 1993, **34**, 343.
44. M. Arita, K. Nomura, H. Arai and K. Inoue, *Proc. Natl. Acad. Sci. U. S. A.*, 1997, **94**, 12437.
45. J. F. Oram, A. M. Vaughan and R. Stocker, *J. Biol. Chem.*, 2001, **276**, 39898.
46. J. Qian, S. Morley, K. Wilson, P. Nava, J. Atkinson and D. Manor, *J. Lipid Res.*, 2005, **46**, 2072.
47. M. Shichiri, Y. Takanezawa and D. E. Rotzoll, *et al.*, *J. Nutr. Biochem.*, 2010, **21**, 451.
48. M. Horiguchi, M. Arita, D. E. Kaempf-Rotzoll, M. Tsujimoto, K. Inoue and H. Arai, *Genes Cells*, 2003, **8**, 789.
49. J. Qian, J. Atkinson and D. Manor, *Biochemistry*, 2006, **45**, 8236.
50. T. J. Sontag and R. S. Parker, *J. Biol. Chem.*, 2002, **277**, 25290.
51. N. Kono, U. Ohto, T. Hiramatsu, M. Urabe, Y. Uchida, Y. Satow and H. Arai, *Science*, 2013, **340**, 1106.
52. S. Morley, C. Panagabko, D. Shineman, B. Mani, A. Stocker, J. Atkinson and D. Manor, *Biochemistry*, 2004, **43**, 4143.
53. S. Krugmann, K. E. Anderson and S. H. Ridley, *et al.*, *Mol. Cell*, 2002, **9**, 95.
54. M. A. Lemmon, *Nat. Rev. Mol. Cell Biol.*, 2008, **9**, 99.
55. S. Chung, M. Ghelfi, J. Atkinson, R. Parker, J. Qian, C. Carlin and D. Manor, *J. Biol. Chem.*, 2016, **291**, 17028.
56. W. X. Zhang, G. Frahm, S. Morley, D. Manor and J. Atkinson, *Lipids*, 2009, **44**, 631.
57. S. Morley, M. Cecchini, W. Zhang, A. Virgulti, N. Noy, J. Atkinson and D. Manor, *J. Biol. Chem.*, 2008, **283**, 17797.
58. W. X. Zhang, V. Thakur and A. Lomize, *et al.*, *J. Mol. Biol.*, 2011, **405**, 972.
59. M. Baptist, C. Panagabko, J. D. Nickels, J. Katsaras and J. Atkinson, *Lipids*, 2015, **50**, 323.
60. N. Kono and H. Arai, *Traffic*, 2015, **16**, 19.

CHAPTER 6

Tocopheryl Phosphate

ANGELO AZZI*

Vascular Biology Laboratory, JM USDA-HNRCA at Tufts University, 711 Washington St., Boston, MA 02111, USA
*E-mail: angelo.azzi@tufts.edu

6.1 Introduction

α-Tocopherol, as its stereoisomer *RRR*, is the only tocopherol that has been shown to be effective in preventing vitamin E deficiency and, until other tocopherols are tested for this capacity, it is the only one that can be termed vitamin E.[1] Soon after its discovery,[2] α-tocopherol was found to have *in vitro* antioxidant properties,[3] a function that was soon extended to *in vivo* situations. More recently, novel properties of α-tocopherol have been discovered, consistent with its cellular and body functions as a modulator of cell signaling and gene expression.[4] Possibly, both radical scavenger and signaling properties can coexist *in vivo*, although more studies are required to provide conclusive evidence of that. In the last 15 years, a substantial amount of research has been carried out to verify if α-tocopherol metabolic products could have pathophysiological roles. In this direction, two research lines have been followed: the first is the short sidechain metabolites of α-tocopherol[5] and the second is its phosphorylated form, α-tocopheryl phosphate (TP) (Figure 6.1).[5] In this chapter, the latter research line, and the chemical and biological properties of TP, will be described in more detail.

Food Chemistry, Function and Analysis No.11
Vitamin E: Chemistry and Nutritional Benefits
Edited by Etsuo Niki

Published by the Royal Society of Chemistry, www.rsc.org

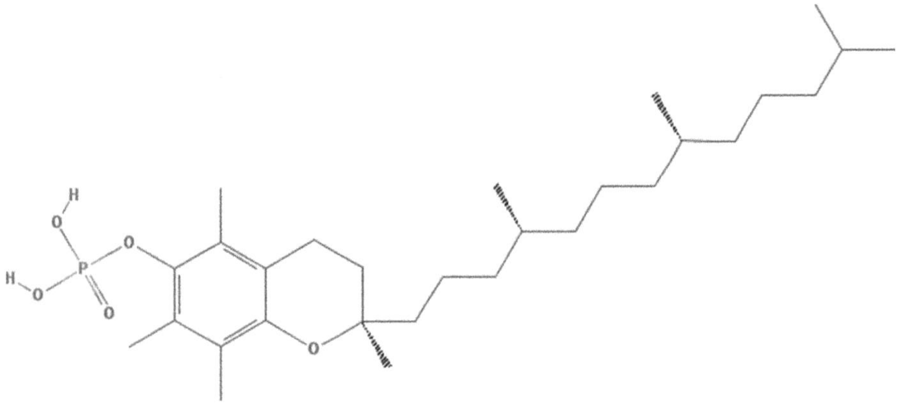

Figure 6.1 Structure of tocopheryl phosphate (TP).

6.2 Synthesis, Extraction and Analysis of TP

In the following paragraphs, some indications will be given of the chemical synthesis of TP, its extraction from organic samples and its quantitative analysis. They have been used successfully in several studies.

6.2.1 Chemical Synthesis of TP

Several protocols have been published describing the chemical synthesis of TP. α-Tocopherol was reacted with phosphoric anhydride at 70–85 °C and in this reaction, besides TP, tocopheryl pyrophosphates were produced. Following hydrolysis at 100 °C with water, the resulting solution contained TP, di-tocopheryl phosphate (T2P) and phosphoric acid.[6] Alkalization with sodium hydroxide produced the sodium salts of TP and sodium phosphate. By adding ethanol, the insoluble sodium phosphate was separated by centrifugation. A further alkalization precipitated the disodium TP.[6] Using a similar protocol, instead of phosphoric anhydride, phosphorous pentoxide was used.[7] In some biological studies, pure TP was used and in others TP (55%) with T2P (30.6%) was used.

6.2.2 Extraction of TP

Biological samples containing TP were homogenized in dichloromethane and centrifuged to separate denatured and precipitated protein. The dichloromethane phase was dried under nitrogen gas, re-dissolved in KOH and incubated at 80 °C. Hexane was then added at room temperature to extract un-esterified α-tocopherol and discarded. The basic water solution containing TP was acidified to pH 1 by HCl to neutralize the phosphate ester charge and then the electrically neutral TP was extracted in hexane and dried under nitrogen gas.[7]

6.2.3 Analyses of Samples Containing TP

TP from biological samples in a THF and NH_3 solution analyzed by ESMS were eluted with THF–H_2O. The samples were analyzed in negative ion mode with the cone voltage set at 40 V. Samples were also analyzed by HPLC. Acetic acid and isopropanol were added to the sample and sonicated. The dissolved sample was injected into a HPLC system with a fluorescence detector, with mobile phase A (0.2% phosphoric acid in isopropanol) and B (water). Detection by fluorescence was done using an excitation wavelength at 297 nm and emission at 319 nm.[7]

6.3 TP Hydrolysis

TP is resistant to both acid and alkaline hydrolysis.[8] Hydrolysis in cellular systems is slow, as shown by the study of Nishio *et al.*[8] indicating that in THP-1 cells *in vitro*, circa 1.8 μM TP was hydrolyzed in 24 h indicating a rate of approximately 1 nM min^{-1}. Nakayama *et al.* reached a similar conclusion in cultured mouse skin,[9] indicating 4% hydrolysis of the added TP, *i.e.* 0.1 nm h mg^{-1} tissue protein. Finally, alkaline phosphatase could hydrolyze TP at an α-tocopherol formation rate of 0.718 μg IU^{-1}.[10] In the study of Negis *et al.* hydrolysis of TP was obtained by calf intestinal phosphatase that gave a value of 0.08 nmol min IU^{-1}.[11] With optimal substrates, 1 IU phosphatase should catalyze the transformation of 1 μmole of substrate per minute. The value obtained, more than 10 000-fold below an optimal phosphatase substrate, indicates how difficult it is for phosphatases to hydrolyze TP. TP uptake and α-tocopherol hydrolysis in THP-1 cells was studied by adding 10 μM TP to cells. After 24 hours, 10 pm per μg protein was taken up and, in the same time, 5% free α-tocopherol appeared.[12] From all the studies mentioned above, it is clear that TP hydrolysis takes place in cells and tissues. Such a reaction is logically necessary to maintain a controlled concentration of TP both in terms of cellular synthesis and hydrolysis.

6.4 Biochemical Studies of TP

TP has been employed as an enzyme inhibitor[13,14] and in animal models of muscular dystrophy:[15–17] these data will not be discussed here, given the lack of follow-up and clear controls, separating the effects of TP from those possibly created by TP chelation of Ca^{2+}.

6.5 TP as Pro-vitamin E

Recently, TP was discovered to be present in human tissues. In this first study, TP was found in rat liver in the order of 11–13 μg g^{-1} liver as compared to α-tocopherol (6–8 μg g^{-1} liver); it was thus described[6] as a pro-vitamin or a storage form of vitamin E. Subsequently, by more accurate measures, we have established that the TP content of liver is in the order of 0.1 μg g^{-1} liver. Given that it is such a low amount, the hypothesis that TP could represent a storage form of α-tocopherol could not be validated. On the other hand, the presence of TP at nanomolar concentrations not only in different human and animal tissues but also in plants, suggests that its function is rather that of an active lipid mediator similar, for instance, to inositol-1,4,5-tris-phosphate.[18,19] In fact, TP has been shown to be involved in signal transduction and gene expression and in exerting modulatory actions at the cellular, tissue and body level.

6.6 Biological Synthesis

To investigate the possible conversion of α-tocopherol to TP, radiolabeled [^{14}C] α-tocopherol was incubated with 3T3-L1 fat cells and was also given orally to a rat. Twenty-four hours later, fat cells and rat liver were extracted to search for the presence of [^{14}C] TP produced *in vitro* and *in vivo*, respectively. [^{14}C] TP was found to be present in the samples as detected by LC-MS and LC-MS/MS analyses.

In a different approach, [^{32} P]-ATP was added to mast cells and human primary coronary artery smooth muscle cells in culture. Due to the known lack of ATP transport, cell membranes were permeabilized by using the detergent digitonin.[20] After a 10 min incubation time, cells were harvested. Analyses carried out by thin layer chromatography after acidic extraction of TP in hexane showed that radioactive [^{32} P]-TP was synthesized in both cell types.[11]

6.7 Absorption of TP

In general, TP is present in lower amounts than α-tocopherol in foodstuffs; in some cases, such as cheese, peanuts and chocolate, the TP level is higher than α-tocopherol. TP can be absorbed. Supplementation of TP in rats daily for 32 days resulted in a significant increase in liver TP (62%) and adipose tissue (50%). In the liver, α-tocopherol was significantly increased in TP-supplemented animals, indicating the conversion of TP to α-tocopherol. Uptake of TP supplements was also documented in humans.[6] In addition, radiolabeled [^{14}C] TP given orally to a rat was recovered in the liver.[7]

Cellular uptake and hydrolysis of TP have also been studied by using a deuterated methyl group, at position five of the chroman ring.[8] In mice fed with a diet containing deuterated TP, α-tocopherol was measured in the plasma, liver, brain, heart, and testis, indicating TP uptake into plasma and tissues followed by TP hydrolysis.[8]

In conclusion, it can be safely accepted that TP can be absorbed and part of it can be hydrolyzed to α-tocopherol. Its presence in tissues can be both a consequence of its absorption with foods as well as its cellular synthesis. Such a presence, at very low concentrations, exists despite the activity of cellular phosphatases.

Absorption of TP to reach tissues implies its transfer through membranes. Due to its highly charged phosphate group at neutral pH, such transport cannot occur *via* simple diffusion but requires a carrier. The same can be concluded from experiments in which α-tocopherol was added to isolated cells in culture. In fact, it has been found that TP crosses the plasma membrane since its transport is decreased by inhibitors such as glibenclamide and probenecid. Both compounds inhibit the ATP binding cassette (ABC) and organic anion transporters (OAT). The latter, responsible for solute uptake, may also be responsible for TP transfer into cells. The growth inhibition provided by OAT inhibitors indicates that the effect of TP is exerted from the inside of the cells and not at their surface.[12,21] Because TP can cross membranes, it is possible that its functions are not limited to the cells where it is synthesized; TP may also have paracrine as well as long distance effects.

6.8 Safety of TP

In vivo toxicological tests using rats, mice and rabbits comprised an oral LD50 study, three 28 day oral repeat-dose studies, two dermal toxicity tests, an ocular irritation test, mutagenic potential tests, and chromosomal aberrations tests. TP with the presence of some T2P was used. The conclusion of the tests was that TP is suitable as a dietary supplement and for skin treatments both in animals and humans, without dose-dependent adverse effects.[22] This was confirmed in other studies with animals where no sign of toxicity was observed[23,24] and *in vitro* using bacterial and mammalian cell cultures at concentrations below 100 μM.[22] TP studies by cell release of LDH and MTT[25] assay, in THP-1, Hep-G2 and Caco-2 cells have shown that TP up to 100 μM does not cause significant cytotoxicity.[8]

6.9 Effect of TP on Proliferation

In an early study, Munteanu *et al.*[26] investigated the effect of TP on the proliferation of rat aortic smooth muscle cells and human THP-1 monocytes. The concentrations of TP used (30 and 60 μM), which are apparently higher than the physiological concentrations, were employed to compensate for chelation of the molecule by the presence of 1.8 mM calcium chloride and 0.8 mM magnesium sulfate in the DMEM buffer. Due to the TP chelation of divalent cations, the free remaining TP is reduced to sub-micromolar concentrations. Similar data have been subsequently reported.[12,19,21] In these experiments, the inhibitory effect of TP was stronger than that of α-tocopherol at the same concentrations. In fact, in the experiment of Negis *et al.*,[12] no proliferation inhibition was observed at α-tocopherol concentrations of 25 and 50 μM while the same concentrations of TP inhibited cell proliferation by 60% and 75%, respectively.

6.10 Effect of TP on Gene Expression and Cell Surface Receptor Localization

6.10.1 Effect of TP on CD36

CD36 (cluster of differentiation 36), encoded by the CD36 gene, is an integral membrane protein at the surface of several cell types. CD36 binds ligands including collagen, thrombospondin-oxidized low-density lipoprotein, native lipoproteins, oxidized phospholipids and long-chain fatty acids. α-Tocopherol has been shown to decrease CD36 expression and oxidized LDL uptake in human monocyte-derived macrophages.[27,28] TP, in a concentration dependent way, decreased CD36 protein expression in THP-1 cells. α-Tocopherol, under the same conditions, had no effect, although its cellular concentration would be higher than the concentration of the α-tocopherol produced from the hydrolysis of TP.[8] In THP-1 monocytes, TP decreased transcription of CD36 in a concentration dependent way, while α-tocopherol had no effect. Parallel to the decrease of CD36 transcription and translation, TP cell treatment resulted in a decrease of fluorescent labelled oxidized LDL (3,3′-dioctadecyloxacarbocyanine perchlorate, oxLDL-DiO) binding and uptake.[26]

Additionally, TP induced CD36 internalization in THP-1 monocytes while α-tocopherol was less efficient. CD36 internalization was inhibited by a specific covalent inhibitor, sulfo-*N*-succinimidyl oleate. At the same time, sulfo-*N*-succinimidyl oleate inhibits TP (but not α-tocopherol) binding and internalization. On the other hand, only the uptake of α-tocopherol was inhibited by Dynasore, an inhibitor of clathrin-mediated endocytosis. Glybenclamide (25 μM), which inhibits TP transport and cell proliferation, did not have any effect on the presence of CD36 on the cell surface, suggesting that cell proliferation and CD36 expression are affected by TP in different manners.[21]

TP induction of the vascular endothelial growth factor (VEGF) promoter activity that occurs *via* CD36/PI3Kγ/Akt was inhibited by sulfo-*N*-succinimidyl oleate; this indicates that the TP association to and transport through CD36 may be the earliest event in TP-induced VEGF increased expression.[28]

6.10.2 Effect of TP on THP-1 Monocyte Gene Expression

Of all the genes analyzed by a Gene Chip, the ones that were most affected by TP or α-tocopherol[21] were studied further in terms of their transcription; the data are reported in Table 6.1. TP produced an upregulation of gene expression and transcription. These effects were more pronounced, and in several cases one order of magnitude higher, than those produced by α-tocopherol. The modulation of several of the genes reported in Table 6.1 goes in the direction of a neuroprotection function.

Table 6.1 Fold changes in THP-1 monocyte genes by α-tocopherol and TP by Gene Chip and quantitative RT-PCR. Data from ref. 21.[a]

Name	Gene symbol	Gene Chip αT/control	Gene Chip αTP/control	RT-PCR TP/control
Sarcoglycan-γ (35 kDa dystrophin-assoc. glycoprotein)	SGCG	1.9	15.0	12.0
Tribbles homolog 3 (Drosophila), stress-induced NIPK kinase	SKIP3/NIPK/TRIB3	1.2	11.3	7.8
DNA-damage-inducible transcript 4	RTP801/REDD1	1.1	14.7	22.5
Stanniocalcin 2	STC2	1.7	11.7	27.9
Microtubule-associated protein 1B	MAP1B	1.0	10.5	23.7
UL16-binding protein 1	ULBP1	2.7	7.6	5.1
Prostate or growth differentiation factor 15	PDF/GDF15	0.8	6.9	21.6
Vascular endothelial growth factor A	VEGFA	1.1	6.8	7.6
Nacht, leucine-rich repeat and PYD-containing 12	NLRP12	0.9	4.4	6.5
Growth factor receptor-bound protein 10	GRP10	1.1	3.5	6.0
Insulin-induced gene 1	INSIG1	0.8	3.4	2.9
Activating transcription factor 3	ATF3	1.3	2.8	2.8
Jun dimerization protein P21SNFT	SNFT	1.6	2.5	1.7

[a]αT/control: fold change produced by αT when compared to solvent control. αTP/control: fold change produced by αTP when compared to solvent control.

6.10.3 Effect of TP on NIH3T3-L1 Gene Expression

In a different study with NIH3T3-L1 pre-adipocytes, TP, but not α-tocopherol, activated a set of genes—TRB3 (Tribbles Homolog 3), Sestrin-2 (SESN2), and Insulin-Induced Gene 1 (INSIG1)—potentially preventing fat accumulation in these cells. However, in differentiated adipocytes, TP produced a transcriptional inhibition of the same genes. These processes cannot be immediately rationalized but may be relevant in the regulation of excess fat accumulation and in the prevention of lipotoxicity.[29]

A comparison of the data in Tables 6.1 and 6.2 shows that, in these two cellular models, TP is far more potent than α-tocopherol. The data also show that several of the genes regulated in the THP-1 monocyte are also regulated in NIH3T3-L1 pre-adipocytes, but many genes are also cell specific.

6.10.4 Comparison Between α-Tocopherol and TP

In a study with THP-1 monocytic leukemia cells, no proliferation inhibition was observed with α-tocopherol concentrations of 25 and 50 μM while the same concentrations of TP inhibited cell proliferation by 60% and 75%, respectively.[12] TP, in a concentration dependent way, decreased CD36 protein expression in THP-1 cells. α-Tocopherol under the same conditions had no effect although its cellular concentration would be higher than the concentration of the α-tocopherol produced from TP hydrolysis.[8] In THP-1 monocytes, TP decreased transcription of CD36 in a concentration dependent way while α-tocopherol had no effect.[26]

Of the 202 genes that were regulated in THP-1 monocytes by TP or α-tocopherol, 101 were upregulated only by TP and six downregulated. α-Tocopherol exclusively upregulated 28 genes and downregulated three. Not many genes were upregulated (35) or downregulated (35) by both compounds.[21] These results clearly indicate that the effect of TP cannot be the consequence of its hydrolysis to α-tocopherol. The two compounds appear to have, to a significant extent, independent functions and only to a lesser extent is their action common.[21]

6.11 Mechanistic Interpretation of TP Action

From the information reported above, it appears that the addition of TP to cells or administration to animals may result in some hydrolysis with production of free α-tocopherol. Therefore, some of the effects of TP may be in fact due to α-tocopherol. However, as discussed previously, several effects of TP cannot be as a consequence of the hydrolysis of α-tocopherol. In particular, the inhibition of cell proliferation by TP is much stronger than by α-tocopherol as well as the modulation of a number of genes. Furthermore, some genes are uniquely modulated by TP. Also, *in vivo*, the effects of TP on atherosclerosis have been shown to be much stronger than those of tocopheryl acetate, which is hydrolyzed more efficiently than TP. These differences between the effects of α-tocopherol and TP prompt questions about the mechanism of TP action at the molecular level. The TP enhanced expression of vascular endothelial growth factor (VEGF) has been taken as an *in vitro* system to dissect the pathway connecting the two events. In cultured HEK293 cells, VEGF-promoter activity was increased by TP.[30] This reaction was sensitive to phosphatidylinositol-3-kinase γ (PI3Kγ) inhibitors. VEGF expression was reduced when the human tocopherol-associated protein 1 (hTAP1/SEC14L2) was overexpressed, suggesting that the protein, known to bind TP and α-tocopherol,[31,32] may have subtracted TP from its activating site. Further results indicate that TP activates PI3Kγ/Akt signaling following its binding to and/or transport by CD36.[33] Although much of the detail needs to be clarified further, the data point to an interaction between TP and the PI3K system-mediated signal transduction and gene expression.

It has been proposed that TP can be an activated form of α-tocopherol.[11] Activation by phosphorylation is known for vitamin B6 and vitamin B1, leading to pyridoxal phosphate and thiamine diphosphate, respectively. Therefore, it would not be an unique event in nature if vitamin E were also phosphorylated and with this became activated. Several data indicate that the effects of TP in some cellular and animal models are more intense than those produced by α-tocopherol, in line with the paradigm of TP being an

Table 6.2 Genes up-regulated in NIH3T3-L1 pre-adipocytes more than 3.4-fold by TP.

Genes	T/C	TP/C	TP/T
CHAC1, Ca/H-Transporter regulator	0.9	13.1	14.7
Hypothetical protein RTP801, Redd1	1.1	14.7	13.5
Inhibin b E	0.8	9.4	11.1
Microfibrillar-associated protein 4	0.4	4.4	11.1
Microtubule-associated protein 1B	1	10.5	10.9
TRB3, SKIP3, Stress-induced NIPK kinase	1.2	11.3	9.3
Prostate differentiation factor mRNA	0.8	6.9	9.1
Sarcoglycan-c (35kD dystrophin-associated glycoprotein) (SGCG)	1.9	15	7.9
Neuroendocrine secretory protein 55	1	8	7.8
Stanniocalcin 2	1.7	11.7	7
cDNA FLJ36044 similar to COP1	0.6	4	6.6
Vascular endothelial growth factor	1.1	6.8	6.5
Protein kinase domains (containing protein similar to)	1.3	8.1	6.3
Asparagine synthase	1.1	6.8	6.2
cDNA DKFZp761M02121, SESTRIN 2	0.9	5.6	6.1
VEGF Vascular endothelial growth factor	0.8	5.1	6
Phosphoserine aminotransferase (PSA)	1	5.7	5.9
Cystathionase and a CHORD containing protein 1 (CHP1) pseudogene	1	5.6	5.6
p8 protein homolog (COM1)	0.8	4.2	5.5
Vascular endothelial growth factor1 isoform 12	1	5.6	5.4
Transmembrane 6 superfamily member 1 (TM6SF1)	0.8	4.4	5.3
Insulin-like growth factor binding protein 3	1.6	7.9	5
INSIG1/insulin-induced gene 1	0.8	3.4	4.4

activated form of α-tocopherol. In some other experiments, TP has shown effects while α-tocopherol has not; these data may be compatible with the possibility that in some cells, the kinase responsible for α-tocopherol phosphorylation is absent. Experiments have shown that some genes are uniquely modulated by α-tocopherol and not by TP. Such data may be consistent with effects produced by α-tocopherol related to its antioxidant or non-antioxidant function but not to its conversion to TP. Finally, there are experiments indicating an antagonism between TP and α-tocopherol that do not seem to be rationalized by the possibilities discussed above. It appears to be clear that more experiments are needed to fully understand the functions of these two related molecules, α-tocopherol and its phosphorylated form, TP.

6.12 Some *In Vivo* Applications of TP

6.12.1 Effects of TP on Atherosclerosis and Inflammation

Feeding a high cholesterol diet to rabbits results, within a brief time, in the development of atherosclerosis, an event which was prevented by TP more efficiently than the acetate derivative.[34] TP diminished the expression of CD36 more than the acetate derivative.[34] Inflammatory markers such as interleukin (IL)-1β, IL-6, IL-8, plasminogen activator inhibitor-1, tumor necrosis factor-α and C-reactive protein were also diminished. Tocopherol acetate only decreased plasma levels of CRP, IL-6 and IL-8.[24] Similar experiments were carried out in ApoE-deficient mice treated with a high fat and cholesterol diet. TP treated animals showed a noticeable reduction in aortic fat deposition and plasma diminution of pro-inflammatory markers IL-6, MCP-1, IL-1β, IFN-γ and TNF-α.[35] TP has been shown to exert protective effects against skin damage; in normal human epidermal keratinocytes after UVB irradiation, it inhibited the production of PGE2 and other pro-inflammatory stimuli.[36]

TP provided treated rats with significant cardioprotection with improved ventricular performance, reduced myocardial infarction size and cardiomyocyte apoptosis. At the molecular level, a MAP kinase-induced death signal appeared to be converted into a survival signal by enhancing anti-apoptotic p42/44 ERK kinase and p38 MAPKβ and reducing pro-apoptotic proteins p38 MAPKα and JNK.[23]

TP has been shown to possess therapeutic potential in the inhibition of apoptosis and increases the migratory capacity of EPCs under high-glucose/hypoxic conditions. It promoted angiogenesis by upregulating Bcl-2, Akt, and vascular endothelial growth factor receptor 2 and decreasing nuclear factor κB p65, p38MAPK, Notch-1, JNK, and Tie-2.[37]

6.12.2 Brain Effects of TP

Treatment with TP persistently decreased peroxiredoxin 2 and glutathione reductase expression and activity in the mouse cerebral cortex and hippocampus. Glutathione content, glutathione peroxidase, and thioredoxin reductase activities were not affected by TP.[38] α-Tocopherol and TP have been shown to interact with the cannabinoid system in the rodent hippocampus.[39] On synaptic transmission, α-tocopherol and TP had opposite actions on both glutamatergic and GABAergic transmission: while α-tocopherol had a potentiating effect, TP was inhibitory. Both effects were mediated by the cannabinoid receptors.[39]

6.12.3 Effect of TP on Tumors

TP has been shown to repress invasion of human fibrosarcoma HT-1080 cells through the basement membrane more efficiently than α-tocopherol. TP decreased cell motility and adhesion to the extracellular matrix. Cortactin, present in the membrane fraction, was markedly inhibited.[40] The inhibitory effects of TP on colon cancer cell proliferation have also been described.[41]

6.12.4 Vehicle for Transdermal Drug Delivery

A TP-gel formulation has been used to transdermally deliver oxycodone to Sprague–Dawley rats.[42,43] A new lipid excipient, a phosphorylated tocopherol mixture, TPM, enhanced the solubilization and oral bioavailability of poorly water soluble CoQ10 in a lipid formulation.[44] A nanocarrier system was developed containing TP, T2P and different lipids/phospholipids (phosphatidylcholine, stearyl amine, Phospholipon 90H, Phospholipon 100H), with and without cholesterol. 5-Fluorouracil as a model compound showed that this system had an acceptable encapsulation efficiency and more than two years' stability.[45]

6.13 Conclusions

TP has been found to be physiologically produced in cells and organisms. Despite this, only α-tocopherol kinase and phosphatase activities have been shown to exist but none of the relative enzymes have been isolated. TP has shown properties that are qualitatively equal to but more intensive than those of α-tocopherol in several models, suggesting that it may act as an activated form of α-tocopherol. In other experiments, it has shown unique properties not shared by the parent molecule. Finally, and most intriguing, antagonistic behaviors of TP and α-tocopherol have been observed. In terms of a molecular target, TP appears to modulate the PI3K system, as seen from cellular and molecular data. In terms of cellular systems, many cells have shown that TP modulates several genes but the affected gene expression was not the same in all cells, indicating cell specificity. Altogether, it can be concluded that many experiments need to be done to fully understand the molecular mechanism and the pathophysiological roles of these two parent molecules, TP and α-tocopherol. The isolation and characterization of the enzymes responsible for phosphorylation and dephosphorylation of α-tocopherol need to be carried out. Studies on the effect of TP on brain cells and in particular on the cerebellar Purkinje cells may result in very interesting conclusions. Following the studies on adipocytes and pre-adipocytes, possible correlations between the amount of TP produced in the body and obesity/lipotoxicity may be the basis of exciting new research. Moreover, the possible phosphorylation of other tocopherols and tocotrienols needs to be investigated. Many intriguing results with these molecules may be the consequence of their possible phosphorylation. Finally, all the recent studies on molecular, cellular and animal models of α-tocopherol and TP, instead of clarifying the molecular targets and the relative mechanisms of the two molecules, have led to, as frequently happens in science, important new questions.

References

1. A. Azzi, *Mol. Aspects Med.*, 2018, **61**, 92.
2. H. M. Evans and K. S. Bishop, *Science*, 1922, **56**, 650.
3. H. S. Olcott and H. A. Mattill, *J. Biol. Chem.*, 1931, **93**, 65.
4. A. Azzi, *Free Radical Biol. Med.*, 2007, **43**, 16.
5. F. Galli, A. Azzi and M. Birringer, *et al.*, *Free Radical Biol. Med.*, 2017, **102**, 16.
6. E. Ogru, R. Gianello, R. Libinaki, *et al.*, *Vitamin E Phosphate: An Endogenous Form of Vitamin E*, Medimond Med. Pub., Englewood, NJ, 2003.

7. R. Gianello, R. Libinaki and A. Azzi, *et al.*, *Free Radical Biol. Med.*, 2005, **39**, 970.
8. K. Nishio, N. Ishida and Y. Saito, *et al.*, *Free Radical Biol. Med.*, 2011, **50**, 1794.
9. S. Nakayama, E. M. Katoh, T. Tsuzuki and S. Kobayashi, *J. Invest. Dermatol.*, 2003, **121**, 406.
10. B. M. Rezk, G. R. Haenen, W. J. Van Der Vijgh and A. Bast, *Biochim. Biophys. Acta*, 2004, **1683**, 16.
11. Y. Negis, J. M. Zingg, E. Ogru, R. Gianello, R. Libinaki and A. Azzi, *IUBMB Life*, 2005, **57**, 23.
12. Y. Negis, M. Meydani, J. M. Zingg and A. Azzi, *Biochem. Biophys. Res. Commun.*, 2007, **359**, 348.
13. J. H. Kay and G. A. Balla, *Proc. Soc. Exp. Biol. Med.*, 1950, **73**, 465.
14. Y. Kuribayashi, M. Katori, M. Majima and K. Yoshida, *Int. J. Tissue React.*, 1996, **18**, 73.
15. O. B. Houchin, *J. Biol. Chem.*, 1942, **146**, 313.
16. O. B. Houchin and H. A. Mattill, *J. Biol. Chem.*, 1942, **146**, 309.
17. J. P. Hummel and D. H. Basinski, *J. Biol. Chem.*, 1948, **172**, 417.
18. V. Luzzi, C. E. Sims, J. S. Soughayer and N. L. Allbritton, *J. Biol. Chem.*, 1998, **273**, 28657.
19. J. M. Zingg, M. Meydani and A. Azzi, *Mol. Nutr. Food Res.*, 2010, **54**, 679.
20. Q. Wang, A. Theriault, A. Gapor and K. Adeli, *Biochem. Biophys. Res. Commun.*, 1998, **246**, 640.
21. J. M. Zingg, R. Libinaki, C. Q. Lai, M. Meydani, R. Gianello, E. Ogru and A. Azzi, *Free Radical Biol. Med.*, 2010, **49**, 1989.
22. R. Libinaki, E. Ogru, R. Gianello, L. Bolton and S. Geytenbeek, *Food Chem. Toxicol.*, 2006, **44**, 916.
23. S. Mukherjee, I. Lekli, M. Das, A. Azzi and D. K. Das, *Biochim. Biophys. Acta*, 2008, **1782**, 498.
24. R. Libinaki, S. Tesanovic and A. Heal, *et al.*, *Clin. Exp. Pharmacol. Physiol.*, 2010, **37**, 587.
25. M. Z. Fiume, *Int. J. Toxicol.*, 2002, **21**, 51.
26. A. Munteanu, J. M. Zingg and E. Ogru, *et al.*, *Biochem. Biophys. Res. Commun.*, 2004, **318**, 311.
27. S. Devaraj, I. Hugou and I. Jialal, *J. Lipid Res.*, 2001, **42**, 521.
28. R. Ricciarelli, J. M. Zingg and A. Azzi, *Circulation*, 2000, **102**, 82.
29. M. Lirangi, M. Meydani, J. M. Zingg and A. Azzi, *Biofactors*, 2012, **38**, 450.
30. J. M. Zingg, A. Azzi and M. Meydani, *J. Cell. Biochem.*, 2015, **116**, 398.
31. A. Stocker, S. Zimmer, S. E. Spycher and A. Azzi, *IUBMB Life*, 1999, **48**, 49.
32. S. Zimmer, A. Stocker, M. N. Sarbolouki, S. E. Spycher, J. Sassoon and A. Azzi, *J. Biol. Chem.*, 2000, **275**, 25672.
33. J. M. Zingg, A. Azzi and M. Meydani, *J. Cell. Biochem.*, 2017, **118**, 1855–1867.
34. Y. Negis, N. Aytan and N. Ozer, *et al.*, *Arch. Biochem. Biophys.*, 2006, **450**, 63.
35. R. Libinaki, A. Vinh, S. Tesanovic-Klajic, R. Widdop and T. Gaspari, *Clin. Exp. Pharmacol. Physiol.*, 2017, **44**(Suppl 1), 107.

36. E. Kato, Y. Sasaki and N. Takahashi, *Bioorg. Med. Chem.*, 2011, **19**, 6348.
37. Z. Wu, X. Zheng, L. Meng, X. Fang, Y. He, D. Li, C. Zheng and H. Zhang, *J. Vasc. Surg.*, 2018, **67**, 1263.
38. M. F. Uchoa, L. F. de Souza and D. B. Dos Santos, *et al.*, *Cell. Mol. Neurobiol.*, 2016, **36**, 1015.
39. N. Crouzin, M. C. de Jesus Ferreira and C. Cohen-Solal, *et al.*, *Free Radical Biol. Med.*, 2011, **51**, 1643.
40. Y. Saitoh, A. Yumoto and N. Miwa, *Int. J. Oncol.*, 2009, **35**, 1277.
41. S. C. Dolfi, Z. H. Yang, M. J. Lee, F. Guan, J. Hong and C. S. Yang, *J. Agric. Food Chem.*, 2013, **61**, 8533.
42. M. T. Smith, B. D. Wyse, S. R. Edwards, M. El-Tamimy, G. Gaetano and P. Gavin, *J. Pharm. Sci.*, 2015, **104**, 2388.
43. P. D. Gavin, L. S. Simon, T. Schlagheck, A. J. Smith and S. Shakib, *Pain Manage.*, 2017, **7**, 243.
44. A. C. Pham, P. Gavin, R. Libinaki, G. Ramirez and B. J. Boyd, *J. Controlled Release*, 2017, **268**, 400.
45. M. R. Mozafari, R. Javanmard and M. Raji, *Int. J. Pharm.*, 2017, **528**, 381.

CHAPTER 7

Novel Functions of Vitamin E Nicotinate

Y. J. SUZUKI*[a], L. MARCOCCI[b], K. R. DUNCAN[a], D. I. SUZUKI[a] AND N. V. SHULTS[a]

[a]Department of Pharmacology and Physiology, Georgetown University Medical Center, Washington, DC 20057, USA; [b]Department of Biochemical Sciences "A. Rossi Fanelli", Sapienza University of Rome, Rome, Italy
*E-mail: ys82@georgetown.edu

7.1 Introduction

Vitamin E plays an important role in protecting biological membranes from oxidative stress.[1] The hydroxyl group of the chromanol ring of vitamin E reduces reactive oxygen species, in particular lipid peroxyl radicals. As a vitamin, this lipophilic molecule needs to be introduced to the body through the diet. While it is rare, a vitamin E deficiency can cause spinocerebellar ataxia, myopathies, peripheral neuropathy, ataxia, skeletal myopathy, retinopathy, and immune dysfunction.[2] Various nuts, seeds, and vegetable oils are good sources of vitamin E.[3] The recommended daily allowance of vitamin E is 15 mg per day.[4] However, it is thought that the intake of higher levels of vitamin E (300–1000 mg per day) may prevent diseases and increase longevity.[5] Thus, people often attempt to increase their dietary intake of vitamin E through vitamin supplements.

Food Chemistry, Function and Analysis No.11
Vitamin E: Chemistry and Nutritional Benefits
Edited by Etsuo Niki

Published by the Royal Society of Chemistry, www.rsc.org

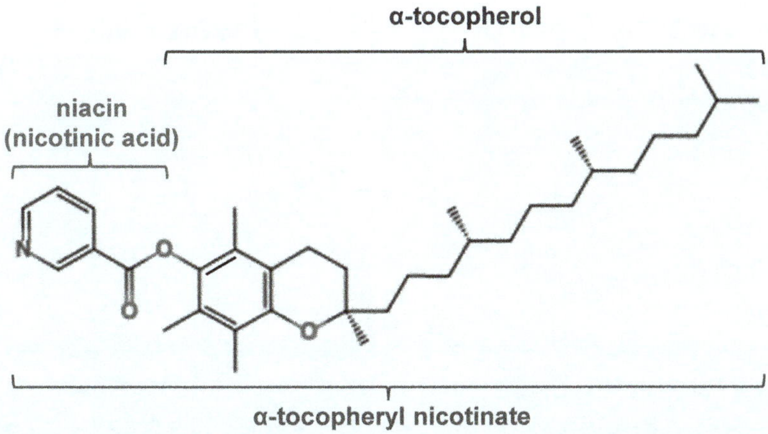

Figure 7.1 Molecular structure of vitamin E nicotinate. α-Tocopheryl nicotinate is composed of α-tocopherol and niacin that are linked through esterification.

Most supplements use the esterified forms of vitamin E because of the better bioavailability.[4] Vitamin E acetate is the most well utilized form of vitamin E ester for dietary supplementation. Other vitamin E esters including succinate, oleate, linoleate, linolenate, palmitate, and nicotinate forms are also available.[5,6] In the vitamin E nicotinate molecule, the hydroxyl group of the chromanol ring of vitamin E is esterified with niacin (Figure 7.1).[5] Niacin (nicotinic acid) together with nicotinamide comprise the vitamin B_3 complex that plays an important role in energy production.[7] We have recently published a review article that summarizes the information available concerning vitamin E nicotinate, particularly in relation to its antioxidant activities.[5] The present chapter focuses on novel functions of vitamin E nicotinate that have recently been discovered in our laboratory.

7.2 Evidence for the Occurrence of Vitamin E Nicotinate in the Biological System

In our previously published study, the role of oxidative stress in right heart failure in response to pulmonary arterial hypertension was investigated.[8] Pulmonary arterial hypertension is a fatal disease due to significant damage to the right heart ventricle. As shown in the transmission electron microscopy images presented in Figure 7.2, the right ventricular muscle from the failing heart due to pulmonary arterial hypertension exhibits marked changes in cellular components (see Figure 7.2 legend for details). We treated rats with ovalbumin and SU5416 to promote severe pulmonary arterial hypertension as described previously.[9] Our goal was to understand the changes in redox status that may occur in the right heart ventricle in response to pulmonary arterial hypertension.[8]

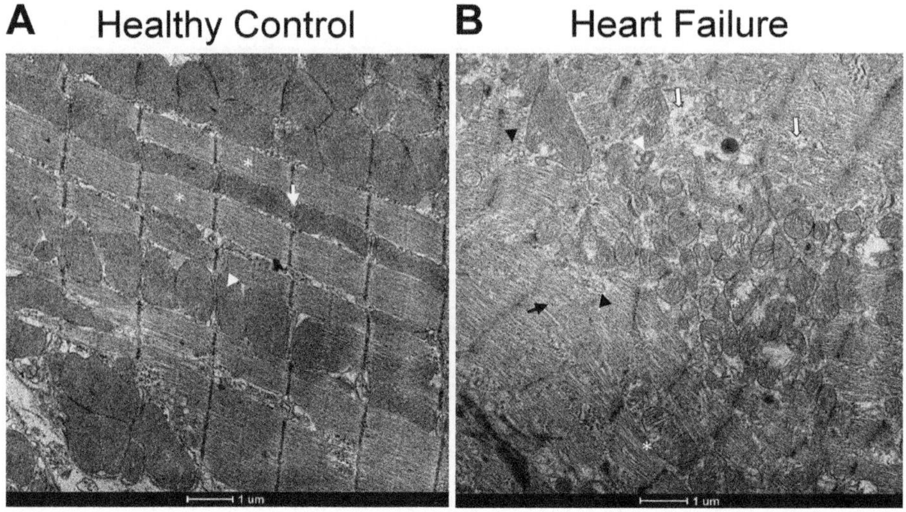

Figure 7.2 Transmission electron microcopy images of the control myocardium and the myocardium undergoing heart failure. (A) The transmission electron microcopy image of the healthy right heart ventricle. The healthy heart muscle has normal cytoarchitectonic structure with continuous, regularly spaced myofibrils (asterisks) alternating with serially arranged and elongated mitochondria (arrow head), along with the dyads and triads, junction of the transverse tubules, and the terminal cisternae of the sarcoplasmic reticulum (arrow). (B) By contrast, the failing right heart ventricle due to pulmonary arterial hypertension exhibits the mismatching of the sarcomere structure with blurred and/or an abject lack of Z-lines, the different orientation of the myofilaments, and the destruction of the contractile filament system such as "mosaic melting" (black arrow) and diffuse and/or focal lysis (white arrows). The reorganization and uneven expansion of the sarcoplasmic reticulum (black arrowheads) and the flattened T-tubules with an electron-dense content in the lumen (white arrowhead) can also be visualized. The shapes and sizes of mitochondria are quite varied with the discontinuous outer membrane, cristae homogenization, and fragmentation, with swollen and electron lucent matrix (asterisks). Magnification ×5500.

Metabolomics analysis identified a number of small molecules that were differentially expressed in the right heart ventricle of rats with pulmonary arterial hypertension compared with controls (Figure 7.3A).[8] We searched for redox-related molecules that are differentially expressed between the control and pulmonary arterial hypertension groups in the database of our metabolomics experiments. We found vitamin E nicotinate to be 30-fold lower in the failing right heart ventricle of rats with pulmonary arterial hypertension compared with healthy controls (Figure 7.3B).

While the decreased levels of antioxidants would be consistent with our overall hypothesis that oxidative stress occurs during right heart failure, the identification of vitamin E nicotinate was quite surprising and exciting for several reasons. First, while vitamin E esters such as acetate and succinate

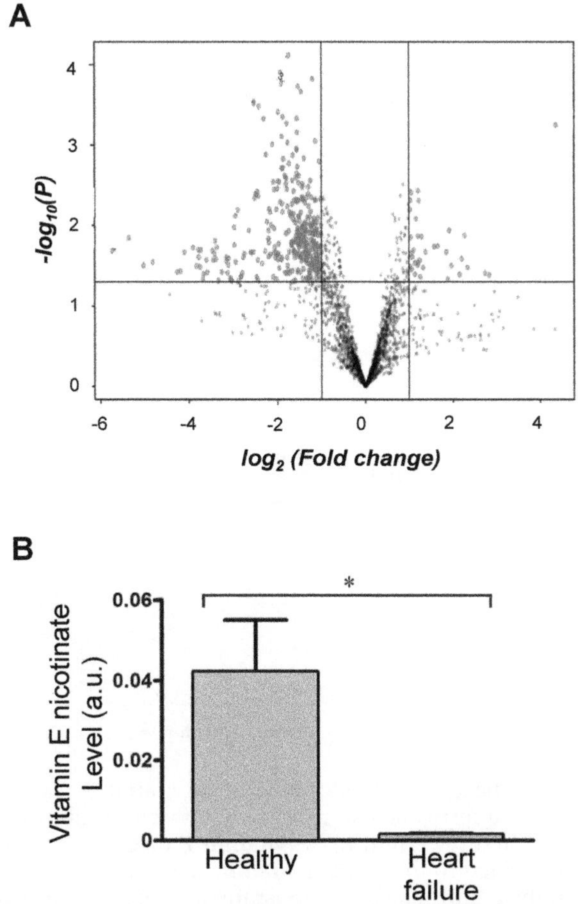

Figure 7.3 Metabolomics studies of the failing heart. Metabolomics analysis was performed in the right heart tissues from control rats and rats with pulmonary arterial hypertension. (A) Volcano plot in the positive ionization mode for heart failure and healthy control. The dot plot represents fold change on the *x*-axis and the statistical significance in *P* value on the *y*-axis. (B) Vitamin E nicotinate levels in control and failing hearts in arbitrary unit (a.u.). The bar graph represents means ± SEM. * denotes that values are significantly different from each other at $P < 0.05$. Adapted from ref. 8 [https://doi.org/10.1371/journal.pone.0176887] under the terms of a CC BY 4.0 license [https://creativecommons.org/licenses/by/4.0/].

forms are well known, we were not aware of vitamin E nicotinate. From the literature, we learned that synthetic vitamin E nicotinate has been used in some cases, including cosmetics. Our recently published review article summarizes the current knowledge about vitamin E nicotinate.[5]

Second, the occurrence of vitamin E nicotinate in the heart tissue was surprising since the animals were fed chow containing vitamin E acetate and

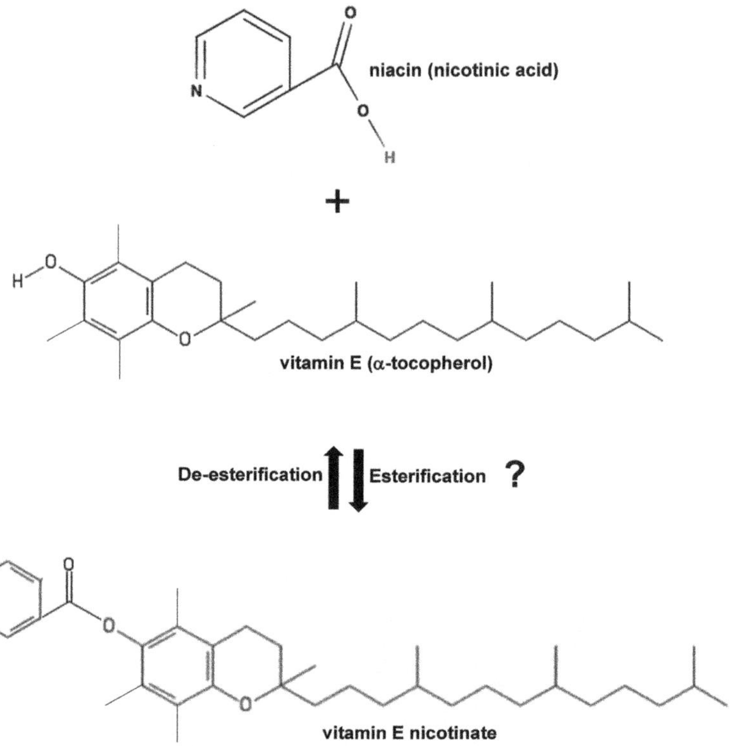

Figure 7.4 De-esterification and esterification of vitamin E nicotinate. Vitamin E nicotinate forms an ester linkage between niacin (nicotinic acid) and α-tocopherol (vitamin E). As vitamin E nicotinate is consumed, it gets de-esterified to form active vitamin E and niacin. We propose that, as vitamin E and niacin get consumed, esterification mechanisms form vitamin E nicotinate in the biological system.

niacin, but not vitamin E nicotinate. If vitamin E nicotinate is consumed, it is likely that it gets de-esterified to form active vitamin E in the reaction depicted in Figure 7.4. However, to our knowledge, there had been no reports on the occurrence of vitamin E nicotinate in the biological system, despite vitamin E nicotinate being simply a combination of vitamin E and niacin, both of which are important functional vitamin molecules in the biological system. Thus, these results revealed that the consumption of vitamin E and niacin results in the formation of vitamin E nicotinate perhaps through some esterification mechanisms (Figure 7.4).

Thirdly, we found that heart failure was associated with decreased vitamin E nicotinate, but not other forms of vitamin E including α-tocopherol.[8] This suggests the exciting possibility that vitamin E nicotinate may function independently from simply serving as a source of active vitamin E. From these studies,[8] we proposed that vitamin E nicotinate is produced in the heart and may play a role in heart failure.[5]

7.3 Evidence for the Occurrence of Vitamin E Nicotinate Signaling

As described earlier, we found that the vitamin E nicotinate expression occurs in the heart and hypothesized that vitamin E nicotinate plays a role in heart failure.[5,8] We also propose that vitamin E nicotinate can function not only through the formation of vitamin E or niacin, but while maintaining the intact vitamin E nicotinate structure. We recently obtained exciting experimental evidence that the intact vitamin E nicotinate structure elicits cellular responses. We treated cultured human vascular smooth muscle cells with either vitamin E nicotinate or vitamin E acetate + niacin and performed metabolomics analysis. Cells were treated for 30 min, washed with phosphate buffered saline, and detached from the plates by trypsin, and the samples were analyzed by mass spectrometry. Analysis of the database produced by the metabolomics experiments revealed that vitamin E nicotinate alters the arachidonoylethanolamine level. Vitamin E nicotinate (TN) was found to be more efficient in increasing the level of arachidonoylethanolamine than vitamin E acetate + niacin separately (T+N) (Figure 7.5). From these results, we propose that the biological system possesses the vitamin E nicotinate receptor that can send signals to produce arachidonoylethanolamine, which is ethanolamine-bound arachidonic acid.

The arachidonoylethanolamine molecule that was increased in response to vitamin E nicotinate could be anandamide (*N*-arachidonoylethanolamine) or virodhamine (*O*-arachidonoylethanolamine) as shown in Figure 7.6. Anandamide is known to function as an agonist for the cannabinoid receptor in the brain.[10] Since our experiments were performed

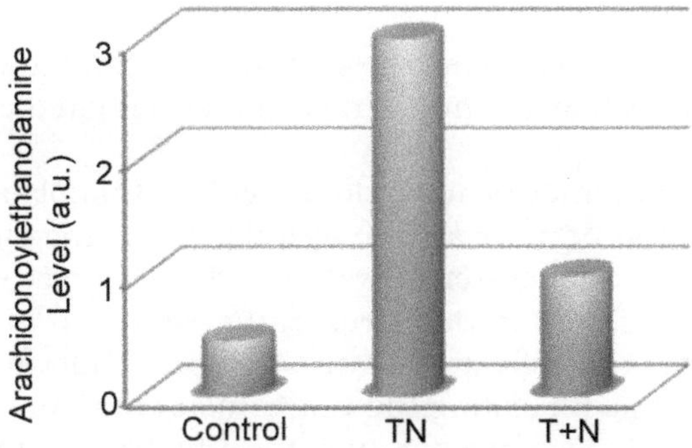

Figure 7.5 Vitamin E nicotinate promotes the formation of arachidonoylethanolamine. Cultured human vascular smooth muscle cells were treated with the vitamin E (α-tocopherol) nicotinate molecule (TN) or the α-tocopherol acetate molecule + the niacin molecule (T+N) for 30 min. Cell samples were subjected to metabolomics analysis using reverse-phase ultra-performance liquid chromatography and quadrupole-time-of-flight mass spectrometry. The bar graph represents means of arachidonoylethanolamine levels in arbitrary units (a.u.).

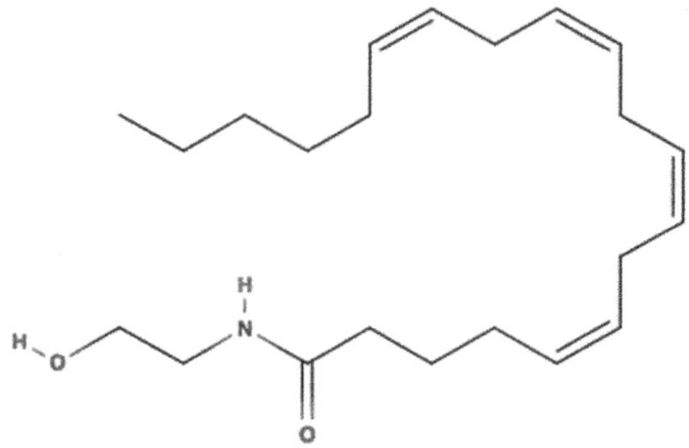

Anandamide
(*N*-arachidonoylethanolamine)

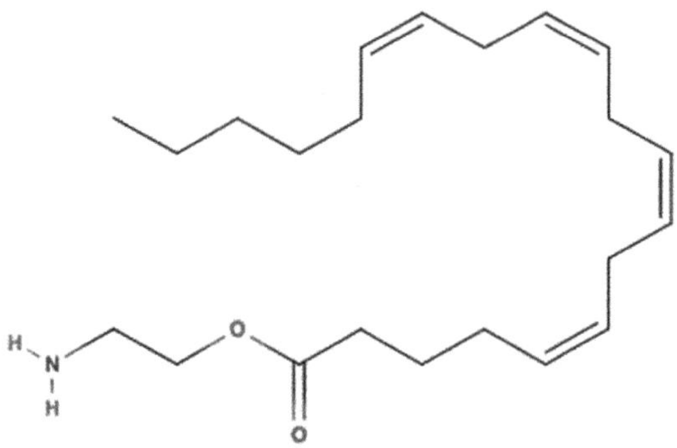

Virodhamine
(*O*-arachidonoylethanolamine)

Figure 7.6 Molecular structures of anandamide and virodhamine.

in vascular smooth muscle cells, these results revealed the possible roles of anandamide as well as vitamin E nicotinate-mediated cell signaling in vascular biology.

Anandamide is the amide of arachidonic acid to ethanolamine, discovered in 1992 in the central nervous system as a ligand for the endocannabinoid CD1 and CD2 receptors.[10] Later, it was found to interact with the TRPV1 receptor and to be present in the peripheral system and in almost all cells where it modulates the cellular pathways for proliferation, inflammation, and vasodilation, thus regulating the functions of the immune system, the cardiovascular system, liver regeneration, the adipose system, and the reproductive system.

Different enzymes have been documented to regulate the synthesis and catabolism of anandamide.[11] Specifically, anandamide is generated from phospholipid *N*-arachidonoyl-phosphatidylethanolamine that is formed through the interaction of membrane arachidonic-acid-containing phospholipids and phosphatidylethanolamine, catalyzed by Ca^{+2}-dependent and Ca^{2+}-independent *N*-acyltransferases.[12] *N*-Arachidonoyl-phosphatidylethanolamine is a substrate for a plethora of lipases that lead to the production of anandamide. Anandamide can be catabolized by oxygenases such as lipoxygenase, cyclooxygenase, and P450 to bioactive molecules like prostaglandins and leukotrienes. Integral membrane fatty acid amide hydrolase and lysosomal *N*-acylethanolamine-hydrolyzing acid amide are also responsible for the catabolism of anandamide that is hydrolyzed to arachidonic acid and ethanolamine in the brain and in the periphery.[13] However, it has been reported that in some tissues (*e.g.*, the liver), fatty acid amide hydrolase can also catalyze the reverse reaction of anandamide synthesis from an excess of arachidonic acid and ethanolamine.[14]

Vitamin E nicotinate may activate these synthetic pathways and/or inhibit the catabolic pathways. As a lipophilic molecule, nicotinated vitamin E might move across the cellular membranes more efficiently than non-esterified vitamin E and modulate the activities of enzymes that regulate the anandamide metabolism. Cyclooxygenase and lipoxygenase that can catabolize anandamide have been shown to be inhibited by vitamin E molecules.[15–17]

Niacin has been shown to bind to the G-protein coupled receptor 109A (GPR109A) and to activate the arachidonic acid cascade to produce prostaglandins, prostacyclin, and thromboxane.[18] Thus, vitamin E nicotinate may bind to the same receptor, leading to a rise in anandamide (Figure 7.7).

Future work should provide mechanistic insights into how the vitamin E nicotinate signaling may be operative.

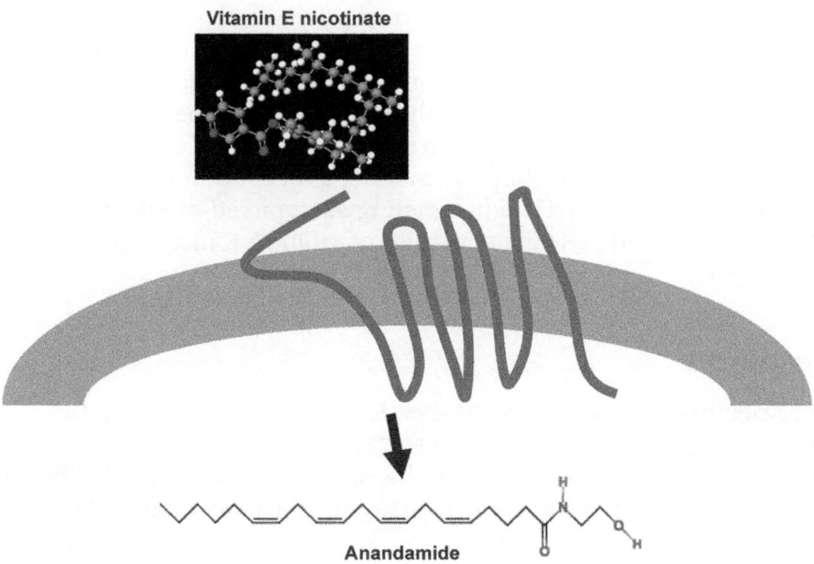

Figure 7.7 Proposed signaling pathway. We propose that vitamin E nicotinate binds to a G-protein coupled receptor and produces intracellular anandamide.

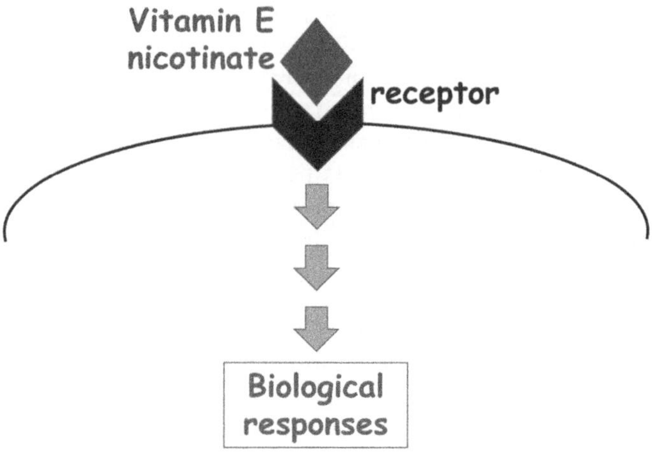

Figure 7.8 Scheme of vitamin E nicotinate signaling. Vitamin E nicotinate binds to a receptor and, in turn, elicits cell signaling.

7.4 Conclusions

Vitamin E nicotinate is not a well-known molecule and is an understudied vitamin E ester. In some situations, especially in cosmetics, synthetic vitamin E nicotinate is used as a source of active vitamin E.[5] While vitamin E nicotinate is comprised of two vitamins that are consumed, whether vitamin E nicotinate ester is formed in the biological system was unknown. We encountered interesting experimental results, suggesting that vitamin E nicotinate occurs in the heart without this ester being supplemented and this molecule may play pathophysiological roles, for example, in the setting of heart failure. Further, we have recently obtained experimental results showing that vitamin E nicotinate activates the formation of intracellular anandamide (Figure 7.7), supporting the exciting concept that the intact vitamin E nicotinate structure elicits cell signaling as depicted in Figure 7.8. These results have provided possible new roles of vitamin E nicotinate in the biological system.

References

1. R. Brigelius-Flohe and M. G. Traber, *FASEB J.*, 1999, **13**, 1145.
2. J. M. Zingg, *Mol. Aspects Med.*, 2007, **28**, 400.
3. J. G. Bieri, L. Corash and V. S. Hubbard, *N. Engl. J. Med.*, 1983, **308**, 1063.
4. Institute of Medicine Panel on Dietary Antioxidants and Related Compounds, *Dietary Reference Intakes for Vitamin C, Vitamin E, Selenium, and Carotenoids*, The National Academies Press, Washington, DC, 2000.
5. K. R. Duncan and Y. J. Suzuki, *Antioxidants*, 2017, **6**, pii: E20.
6. M. Zondlo Fiume, *Int. J. Toxicol.*, 2002, **21**, 51.

7. F. Depeint, W. R. Bruce, N. Shangari, R. Mehta and P. J. O'Brien, *Chem. Biol. Interact.*, 2006, **163**, 113.
8. X. Wang, N. V. Shults and Y. J. Suzuki, *PLoS One*, 2017, **12**, e0176887.
9. S. Mizuno, L. Farkas and A. Al Husseini, *et al.*, *Am. J. Respir. Cell Mol. Biol.*, 2012, **47**, 679.
10. W. A. Devane, L. Hanus and A. Breuer, *et al.*, *Science*, 1992, **258**, 1946.
11. M. Maccarrone, *Front. Mol. Neurosci.*, 2017, **10**, 166.
12. Y. Ogura, W. H. Parsons, S. S. Kamat and B. F. Cravatt, *Nat. Chem. Biol.*, 2016, **12**, 669.
13. I. A. Rahman, K. Tsuboi, T. Uyama and N. Ueda, *Pharmacol. Res.*, 2014, **86**, 1.
14. A. A. Izzo and D. G. Deutsch, *Proc. Natl. Acad. Sci. U. S. A.*, 2011, **108**, 6339.
15. K. A. O'Leary, S. de Pascual-Tereasa, P. W. Needs, Y. P. Bao, N. M. O'Brien and G. Williamson, *Mutat. Res.*, 2004, **551**, 245.
16. E. Lee, M. K. Choi, Y. J. Lee, J. L. Ku, K. H. Kim, J. S. Choi and S.-J. Lim, *Carcinogenesis*, 2006, **27**, 2308.
17. Q. Jiang, *Adv. Nutr.*, 2017, **8**, 850.
18. V. S. Kamanna, S. H. Ganji and M. L. Kashyap, *Int. J. Clin. Pract.*, 2009, **63**, 1369.

CHAPTER 8

Reactive Oxygen Species in Biological Systems

CHRISTINE C. WINTERBOURN*

Department of Pathology and Biomedical Science, University of Otago Christchurch, Christchurch, 2 Riccarton Ave, Christchurch 8011, New Zealand
*E-mail: christine.winterbourn@otago.ac.nz

8.1 Introduction

Free radicals and other reactive oxygen species (ROS) are produced in a wide range of physiological processes. They are capable of inflicting biological damage and are implicated in the pathology of many conditions including cancer, cardiovascular, inflammatory and degenerative diseases. In excess, they can be considered as a deleterious trade-off against the advantages of metabolising oxygen. They are also produced by living organisms to regulate cell signalling pathways such as those involved in growth and differentiation. Reactions of oxidants with antioxidants can both protect against oxidative injury and influence redox-regulated cell signalling pathways. As described throughout this book, vitamin E is best known as an antioxidant. However, antioxidant is a term that is used loosely in the scientific literature and popular press to cover a wide spectrum of activities, only some of which apply to vitamin E. Therefore, to understand the functions of vitamin E, it is essential to appreciate that the terms reactive oxygen species and antioxidant both

Food Chemistry, Function and Analysis No.11
Vitamin E: Chemistry and Nutritional Benefits
Edited by Etsuo Niki

Published by the Royal Society of Chemistry, www.rsc.org

cover a range of chemical species with different properties. Thus all ROS do not undergo identical reactions, and whether a species acts an antioxidant depends on the nature of the oxidant and the environment in which it is produced. This chapter provides an overview of the properties of biologically important oxidants, and their interactions with vitamin E and other antioxidants.

8.2 Vitamin E

As the focus of the book on vitamin E, I will start by highlighting the features of the tocopherol or tocotrienol family members (Figure 8.1) that dictate their antioxidant activity. All contain a chromanol head group linked to a phytyl tail. The chromanol ring is the readily oxidisable redox active centre. This occurs by a one-electron process, in which a hydrogen atom is transferred to the oxidant to form the tocopheroxyl radical. Because only one reducing equivalent is transferred, the oxidants that react with vitamin E are free radicals. Hence, vitamin E analogues are correctly classified as radical scavenging antioxidants. In contrast, they are not good scavengers of two-electron oxidants. Vitamin E is a good scavenger because the

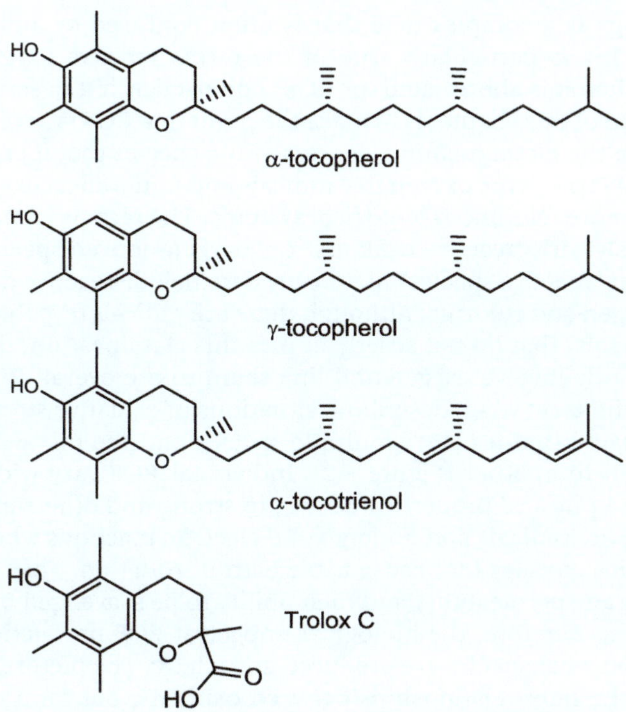

Figure 8.1 Structures of vitamin E family members: α-tocopherol, γ-tocopherol, tocotrienol.

tocopheroxyl radical is relatively stable, it does not react readily with oxygen, and although a weak oxidant it is able to be recycled by reductants like ascorbic acid.[1] In most situations it is able to break chain reactions such as lipid peroxidation. Another important feature is the hydrophobicity of the phytyl chain. This localises vitamin E to lipid environments, including cell membranes and lipoprotein particles. In a complex biological environment, it will scavenge radicals in the lipid phase and be much less effective against those generated in aqueous compartments. However, the polarity of the chromanol group orients it towards the surface of a lipid membrane, and facilitates radical transfer out of the membrane. This transfer to ascorbate underlies the synergism between ascorbate and tocopherols in protecting against lipid peroxidation.

Trolox C is a water-soluble synthetic analogue of vitamin E without the hydrophobic tail. It is convenient for experimentation in aqueous media and has contributed to our knowledge of the redox properties of vitamin E. Trolox C is frequently used experimentally as an "antioxidant". However, because it lacks the tail, it will reside more in the aqueous phase than vitamin E and its radical scavenging profile will not necessarily be the same.

8.3 Reactive Oxygen Species and Antioxidants

Redox biology is a complex field that is often confused by ambiguous terminology. This is particularly true of the term "reactive oxygen species" especially when it is abbreviated to "ROS" and used as if it describes a single entity. An aim of this chapter is to make the point that ROS is a collective term and describe the distinguishing features of the species that it encompasses. The term ROS covers the oxygen free radicals and non-radical oxygen-derived oxidants that are relevant to biological systems. The term is sometimes used synonymously with "reactive oxidants" or "oxygen-derived species", and the broader term "reactive species" may be used to include reactive derivatives of sulfur, nitrogen and chlorine. Although there are radicals (*e.g.* thiyl or carbon centred radicals) that do not strictly fit into this classification, their interactions frequently involve oxygen and link them to the overall ROS network. ROS is a useful term when describing situations of oxidative stress, provided it is appreciated that it covers multiple species and the mix will vary from one situation to another (Figure 8.2). Individual ROS vary widely in their chemical and physical properties. Some are strong and others are weak oxidants. Some are radicals and undergo one-electron reactions whereas others are non-radical species that cause two-electron oxidation. They also vary in their membrane permeability and their ability to be scavenged by individual antioxidants. Therefore, the biological impact of ROS production will vary depending on what species are produced and where. The picture is complex, because of the inter-relationships between oxidants, but for a mechanistic understanding of biological processes, it is important to decipher what specific ROS are involved.

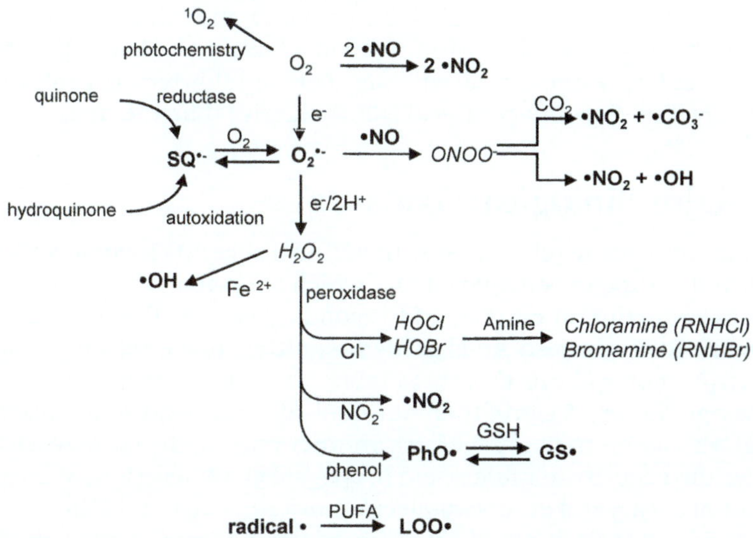

Figure 8.2 Interplay of physiologically relevant reactive oxygen species. The scheme covers most reactive species (or representatives of a class) and selected routes of generation or interconversion. Abbreviations: PUFA, polyunsaturated fatty acid and as in Table 8.1. One-electron oxidants (radicals) are shown in **bold**, two-electron oxidants in *italics*.

Equally important is an understanding of what is meant by "antioxidant". This was initially a chemical term used synonymously with "radical scavenger" to describe species that inhibit radical chain reactions. In current biology it tends to have a much broader meaning. Halliwell and Gutteridge's definition is "any substance that delays, prevents or removes oxidative damage to a target molecule",[2] to which I would make the slight change of replacing "damage" with "modification" to recognise the role of antioxidants in regulating redox processes as well as preventing damage. This broad definition encompasses not only radical scavenging antioxidants but also compounds that react with two-electron oxidants, and these are not necessarily the same. It also covers the network of enzyme systems that break down specific oxidant species, as well as species that inhibit the initiation of a radical reaction (for example metal chelators), and the chaperone and repair mechanisms protect against oxidative injury. Particularly in the context of nutrition and chemoprotection, it also includes compounds that themselves show little reactivity with oxidants, but upregulate cellular expression of antioxidant and detoxification enzymatic pathways. It is clear, therefore, that "antioxidant" as defined covers a diverse array of species with different reactivities and functions. Furthermore, for the antioxidants with redox activity, some scavenge certain ROS but not others, and whether a particular compound acts as an antioxidant or not depends on the context and what species of ROS are involved in the process under study. A prime example is the tocopherols, which are highly efficient radical chain breaking antioxidants but are unreactive with hydrogen peroxide. Another consideration is that radical scavengers act by replacing one radical with another, and this results in antioxidant activity only if the secondary radical is less reactive than the first.

8.4 Major Biological Oxidants

The main biologically relevant ROS (listed in Table 8.1) include superoxide radicals and hydrogen peroxide (H_2O_2), which are primary species generated directly by reduction of oxygen, and secondary species that can be derived from them. ROS can also be classified as either one-electron (radical) or two-electron (non-radical) oxidants (Table 8.1 and Figure 8.2). Radicals are not necessarily more reactive than non-radicals, but because the reaction of a radical with a non-radical generates another radical, the more reactive radicals have the capacity to initiate and propagate chain reactions such as lipid peroxidation. H_2O_2 and other two-electron oxidants do not do this. Likewise, two-electron oxidants generally react poorly or not at all with radical-scavenging antioxidants.

8.4.1 Superoxide

There are many biological sources of superoxide (Figure 8.3). Substantial amounts are formed by mitochondria as a by-product of respiration.[3] It is also generated by various enzymes including members of the NADPH oxidase (NOX) family,[4] which reduce oxygen at the expense of NADPH. Large amounts of superoxide are produced through NOX activity in inflammatory

Table 8.1 Major reactive oxygen species (ROS) in biology.[a]

Radical		Non-radical	
Partially reduced oxygen			
Superoxide	$O_2^{\cdot-}$	Hydrogen peroxide	H_2O_2
Hydroxyl	$^{\cdot}OH$		
Oxygenated species			
Phenoxyl	$PhO^{\cdot}$	Singlet oxygen	$^1O_2$
Alkoxyl	$RO^{\cdot}$	Hydroperoxide	ROOH, PhOOH
Hydroperoxyl	$ROO^{\cdot}$	Hypochlorous acid	HOCl
Semiquinone	$O{=}Ph{-}O^{\cdot-}$ ($SQ^{\cdot-}$)	Hypobromous acid	HOBr
Carbonate	$CO_3^{\cdot-}$	Hypothiocyanous acid	HOSCN
Halogenated species[b]			
		Chloramines	$RNHCl$, $RNCl_2$
		Bromamines	$RNHBr$, $RNBr_2$
Reactive nitrogen species			
Nitric oxide	$NO^{\cdot}$	Peroxynitrite	$ONOO^-$
Nitrogen dioxide	$NO_2^{\cdot}$		

[a]R, alkyl; Ph phenyl.
[b]HOCl, HOBr and HOSCN can also be classified as halogenated (or pseudohalogenated) species.

cells for defence against pathogens[5] whereas lesser amounts are generated in a wide range of other cells in response to stimuli including hormones, cytokines, and growth factors.[4] Ionising or UV radiation, redox cycling xenobiotics and autoxidation of haem proteins such as haemoglobin are all sources of superoxide.[2]

Chemically, superoxide can act as a one-electron oxidant (*e.g.* eqn (8.1)) and reductant (*e.g.* eqn (8.2)).

$$O_2^{\bullet-} + \text{ascorbate} \rightarrow H_2O_2 + \text{ascorbyl radical} \tag{8.1}$$

$$O_2^{\bullet-} + [\text{cytochrome c}]^{3+} \rightarrow O_2 + [\text{cytochrome c}]^{2+} \tag{8.2}$$

Self-oxidation and reduction is the basis of the well-known dismutation reaction (eqn (8.3)). This reaction occurs spontaneously, but it is accelerated by superoxide dismutase (SOD).

$$2O_2^{\bullet-} + 2H^+ \rightarrow H_2O_2 + O_2 \tag{8.3}$$

Superoxide oxidises relatively few biological compounds (Table 8.2). Its protonated form, the perhydroxyl radical (pK_a 5.5) is a stronger and more reactive oxidant that, in contrast to superoxide, is able to initiate lipid peroxidation.[6] Superoxide oxidises α-tocopherol and ascorbate at an appreciable rate.[7,8] It reacts slowly with thiols such as reduced glutathione (GSH) but the thiyl radical can regenerate superoxide so GSH is not an effective superoxide scavenger.[9]

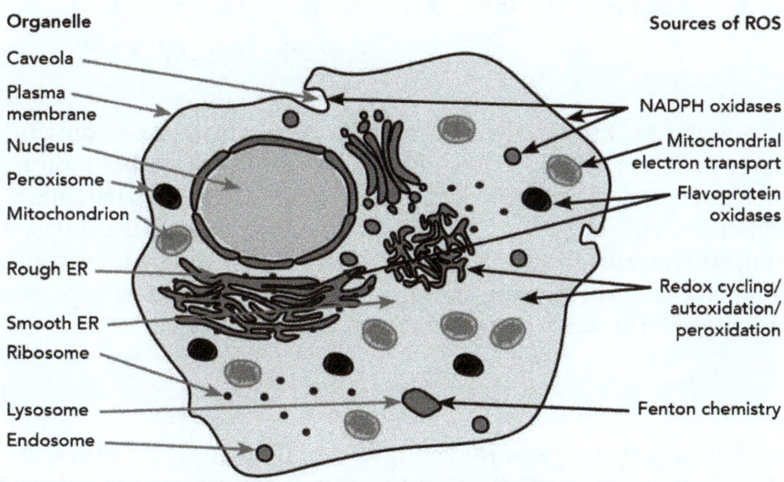

Figure 8.3 Prototype animal cell showing sites of generation of reactive oxygen species.

Table 8.2 One-electron reduction potentials of a selection of biologically relevant free radicals.[a]

Radical couple	Reduction potential (E_0') V
$^•OH, H^+/H_2O$	2.31
$CO_3^{•-}, H^+/HCO_3^-$	1.78
$RO^•/RO^-$	~1.60
$NO_2^•/NO_2^-$	1.04
$O_2^{•-}/2H^+/H_2O_2$	0.94
$PhO^•, H^+/PhOH$	0.90
$CysS^•/CysS^-$	0.92
$ROO^•, H^+/ROOH$	0.77–1.44
$PUFA^•, H^+/PUFA$	0.60
$urate^•, H^+/urate$	0.59
Catechol $SQ^{•-}, H^+$/catechol	0.53
Toc$^•$, H$^+$/Toc	**0.50**
T-O$^•$, H$^+$/T-OH (Trolox C)	**0.48**
Quercetin $SQ^{•-}, H^+$/quercetin	0.33
$Asc^•, H^+/Asc^-$	0.28
$O_2/O_2^{•-}$	−0.33

[a]Species are listed in an approximate order of decreasing reactivity. Standard reduction potentials (E_0') are at pH 7, relative to normal hydrogen electrode, and are from ref. 1,99. Abbreviations or formulae for radicals are from Table 8.1 or as follows: PUFA, polyunsaturated fatty acid bis-allylic-H; Toc, α-tocopherol; Asc, ascorbate. Reduction potentials of different phenoxyl radicals and semiquinones vary depending on substituents. Vitamin E-related species are shown in bold.

A significant biological reaction of superoxide is the oxidation of iron/sulfur proteins such as aconitase. This not only inactivates the enzymes but also releases iron and H_2O_2, which can cause further injury.

$$O_2^{•-} + \text{protein-[4Fe-4S]}^{2+} + 2H^+ \rightarrow \text{protein-[4Fe-4S]}^{3+} + H_2O_2 \rightarrow \text{protein-[3Fe-4S]}^{1+} + Fe^{2+} \quad (8.4)$$

This is considered to be one of the main mechanisms that contribute to superoxide toxicity.[10,11] Superoxide also reacts rapidly with other radicals.[12,13] Its well characterised reaction with nitric oxide ($NO^•$) is particularly fast (rate constant ~10^{10} M^{-1} s^{-1}).[14,15] $NO^•$ can compete with SOD-catalysed dismutation, and interplay with superoxide, and SOD provides a mechanism for regulating $NO^•$ levels. However, the product with $NO^•$ is the strong oxidant, peroxynitrite,[14,16] the reactions of which are discussed later.

$$O_2^{•-} + NO^• \rightarrow ONOO^- \quad (8.5)$$

Other radicals are also favourable targets for superoxide.[13,17] While in some cases the reaction occurs by electron transfer to regenerate the parent compound, with others, such as the tyrosyl and urate radicals, superoxide addition occurs to form a reactive hydroperoxide (as in eqn (8.6) for tyrosine).[18,19]

Superoxide consumes tocopheroxyl radicals during lipid peroxidation,[20] most likely by radical addition.

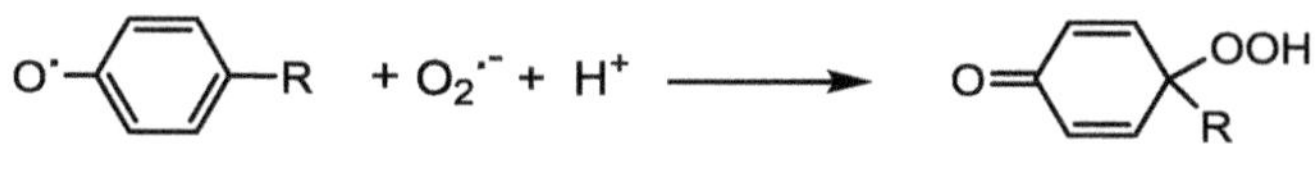

(8.6)

8.4.2 Hydrogen Peroxide

The body encounters H_2O_2 from a variety of endogenous and exogenous sources. Much of it arises from the dismutation of superoxide. It is also generated directly from oxygen by flavoprotein oxidases, many of which reside in peroxisomes, where most of the cellular catalase is also localised and able to limit peroxide release.[21] H_2O_2 is also produced by endoplasmic reticulum-localised oxidase, ERO1, during oxidative folding of exported cell proteins, with one H_2O_2 produced per disulfide formed.[22,23] Although most of the NOX isoforms release superoxide, NOX4 converts it to H_2O_2 before leaving the active site.[24]

Chemically, H_2O_2 is a strong two-electron oxidant, with a standard reduction potential at pH 7.0 (E_0') of 1.32 volts. However, a high activation energy barrier must be overcome to release its oxidising power and it reacts poorly or not at all with most biological molecules.[25] This includes vitamin E and other low molecular weight antioxidants. Although H_2O_2 is a weak one-electron oxidant (E_0' = 0.32 V), it reacts with transition metal centres and these reactions are more favourable than most two-electron oxidations.

Cysteine residues are among the few non-metal targets for H_2O_2. With minor but important exceptions, they react relatively slowly.[26] The reaction occurs with the thiolate anion, so at physiological pH, low pK_a thiols are more reactive. However, low pK_a is not the only determinant of reactivity and a few thiol proteins, most notably the peroxiredoxins, are up to five orders of magnitude more reactive than others.[27,28] The initial product is a sulfenic acid (–SOH), which is generally short-lived and condenses with another thiol to form a disulfide.[29] At higher H_2O_2 concentrations, sulfinic then sulfonic acids can be formed. Many cell regulatory proteins have active site thiols, and the selective oxidation by H_2O_2 is understood to be an important redox signalling mechanism. How this occurs when cells contain much more reactive targets is still the subject of much discussion.[30–34]

H_2O_2 oxidises pyruvate at a comparable rate to cysteine,[35] whereas methionine residues are ~100 times less reactive. A reaction of biological interest is with carbon dioxide to form peroxymonocarbonate.

$$H_2O_2 + CO_2 \rightleftarrows HCO_4^- + H^+ \tag{8.7}$$

Peroxymonocarbonate reacts 200 times faster than H_2O_2 with thiols and methionine.[36,37] The impact of peroxymonocarbonate is limited by the equilibrium in eqn (8.7) lying to the left, and at physiological bicarbonate concentrations, only a few percent of the H_2O_2 is present in this form. However, the reaction can be accelerated by carbonic anhydrase,[37] and some thiol proteins, including protein tyrosine phosphatases, are particularly sensitive to oxidation by peroxymonocarbonate.[38]

H_2O_2 reacts readily with transition metals such as iron and copper to generate more reactive and often biologically damaging species.[11,39] Physiologically, the transition metal centres may be specialised peroxidases, other redox-active metalloproteins, or low molecular weight chelates. The reaction with iron salts is commonly referred to as the Fenton reaction and the product can either be free hydroxyl radicals or a higher oxidation state of the metal:

$$\text{Ligand-Fe}^{2+} + H_2O_2 \rightarrow \text{Ligand-Fe}^{3+} + {}^{\bullet}OH + OH^- \qquad (8.8)$$

An important variation is where the metal is bound to a biological molecule which becomes oxidised by a site-localised mechanism. This can confer selectivity, as observed with copper-catalysed oxidation of DNA and protein oxidation by a bound iron/ascorbate system.[40,41] Although H_2O_2 itself does not initiate lipid peroxidation, it can do so with metal catalysis.[42] Chelation or sequestration of transition metals to limit their redox activity is an important protective mechanism against damage by H_2O_2.[11] Nevertheless, these metals are in a dynamic state in cells with estimates of low molecular weight iron in the micromolar range. To maintain Fenton chemistry at these concentrations, the oxidised metal needs to be recycled. This was once thought to be an important function for superoxide, but reductants such as ascorbate or GSH are much more prominent in this role.

8.4.3 Nitric Oxide and Peroxynitrite

$NO^{\bullet}$ is a relatively unreactive free radical that is produced as a regulatory molecule by the constitutive nitric oxide synthases. It is also produced by inducible nitric oxide synthase for defence against pathogens. $NO^{\bullet}$ reacts mainly with metal centres and superoxide (as noted earlier), as well as with other radicals.[43,44] Although it can react with oxygen to form nitrogen dioxide radicals ($NO_2^{\bullet}$), this reaction will only be efficient when concentrations are high.[45] It may be more relevant in lipophilic environments where both are more soluble.[46] Formation of nitrosothiols (RSNO) on protein thiols is as an important cell regulatory mechanism.[47] This can occur in a reaction between the thiol and NO^+ or between $NO^{\bullet}$ and the thiyl radical, but not as a direct reaction between the thiol and $NO^{\bullet}$.

Peroxynitrite ($ONOO^-$), the product of the reaction of $NO^{\bullet}$ with superoxide, is a strong oxidant that contributes to tissue damage, especially in inflammatory conditions. Although a two-electron oxidant, much of the biological chemistry of peroxynitrite involves radical reactions. As described in detail in recent reviews,[14,16,45] it can break down to hydroxyl radicals and nitrogen dioxide, but physiologically its most favoured reaction is with the carbon dioxide present in biological fluids to form carbonate and nitrogen dioxide radicals:

$$ONOO^- + CO_2 \rightarrow NO_2^{\bullet} + CO_3^{\bullet-} \tag{8.9}$$

Many of its pathophysiological effects are due to these secondary radicals. Peroxynitrite can oxidise thiols directly, but at physiological concentrations of bicarbonate most thiol oxidation is due to the radicals formed in eqn (8.9).[48] An exception is the peroxiredoxins, which are highly reactive with peroxynitrite itself and able to provide antioxidant protection.[49] Nitration of tyrosine residues to give nitrotyrosine, a commonly used biomarker for peroxynitrite, also proceeds *via* a radical mechanism. Either of the products of eqn (8.9) converts tyrosine to its radical which then combines with $NO_2^{\bullet}$ to give the nitrated product.[16] Nitrotyrosine is not exclusive to peroxynitrite, as it can be formed in any system that generates $NO_2^{\bullet}$ radicals, such as nitrite oxidation by a haem peroxidase. Other relevant reactions of peroxynitrite are with metal centres such as in Mn-SOD.[50] It can also induce lipid peroxidation, predominantly *via* its radical products.[51,52] The process can be inhibited by Trolox or vitamin E, which react with the radicals rather than peroxynitrite itself.[51]

8.4.4 Hypohalous Acids

Hypohalous acids are important oxidants in inflammatory situations or infection. They are a double-edged sword as they provide host defence against invading pathogens but can also injure host tissue.[53,54] In mammals, they are generated from their corresponding halides (or in the case of hypothiocyanous acid from the pseudohalide, thiocyanate) by the haem peroxidases, myeloperoxidase, eosinophil peroxidase, lactoperoxidase and peroxidasin, as shown for chloride:

$$H_2O_2 + Cl^- \rightarrow HOCl + OH^- \tag{8.10}$$

The hypohalous acids are two-electron oxidants. Their oxidising ability decreases in the order hypochlorous > hypobromous > hypothiocyanous acid, and therefore their substrate preferences differ.[55–57]

Myeloperoxidase, which is present in neutrophils and to a lesser extent monocytes, is the only mammalian enzyme that produces HOCl. HOCl is highly reactive and strongly microbicidal. Thiols and methionine residues are its most favoured targets, with rate constants greater than 10^7 M^{-1} s^{-1} (Table 8.3).[56,58] HOCl differs from H_2O_2 in oxidising thiols to a sulfenyl chloride rather than a sulfenic acid and forming a greater range of oxidation products. While the sulfenyl chloride can hydrolyse to the sulfenic acid or form a disulfide, it can also cross-link to amines and other nucleophilic centres.[59] The –SCl oxidation product with methionine can hydrolyse to the sulfoxide or react with an amine to form a sulfilimine.[60] HOCl oxidises numerous other substrates and undergoes chlorination reactions. Chloramine formation with amines (eqn (8.11)) is 100-fold slower than thiol oxidation but is important physiologically due to the prevalence of amines.

$$HOCl + RNH_2 \rightarrow RNHCl + H_2O \quad (8.11)$$

Protein and amino acid chloramines are likely to mediate many of HOCl's biological effects.[61,62] Chloramines are less reactive two-electron oxidants than HOCl (Table 8.3) and more selective for sulfur centres. They can also break down to carbonyls, thus contributing to the complex array of reactions that could be responsible for the structural and functional changes seen in proteins exposed to HOCl. Transition-metal-mediated breakdown of chloramines to radicals is likely to explain their modest ability to initiate lipid peroxidation.[52] Tocopherols do react with HOCl[63] but too slowly to compete with its other reactions. Other chlorination reactions, with tyrosine to form chlorotyrosine,[64] unsaturated lipids and NAD^+ to form chlorohydrins,[65,66] and nucleotides to form chlorinated bases,[67] are also slow but give distinctive products that are useful as specific biomarkers.[68,69]

HOBr, the major product of eosinophil peroxidase and peroxidasin, undergoes comparable reactions to HOCl, although there is less of a differential between rates of oxidation and bromination, and brominated products are more prevalent.[56] Tocopherols react faster with HOBr than HOCl but are still not reactive enough to be favourable targets.[70] HOSCN is the preferred product of lactoperoxidase and is produced along with HOCl by myeloperoxidase. It readily oxidises cysteine residues but undergoes few other biological reactions.[57]

8.4.5 Singlet Oxygen

Singlet oxygen is produced in biological systems, primarily in light induced reactions.[71] It is generated in the eye and skin as a result of light exposure and is a source of oxidative damage, and with the use of photosensitisers,

Table 8.3 Relative reactivity of two-electron oxidants based on rate constants for reaction with GSH.

Oxidant	Rate constant[a] ($M^{-1}\ s^{-1}$)
HOCl	3×10^7
HOBr[b]	1×10^7
HOSCN	8×10^4
$ONOO^-$	700
Chloramines	100–700
H_2O_2	0.9

[a]Rate constants are at pH 7 and are from ref. 26, 100–104.
[b]Measured for *N*-acetylcysteine.

during photodynamic cancer therapy. It can also be generated in the absence of light, *e.g.* in the breakdown of lipid hydroperoxides and in the reaction of HOCl with H_2O_2, although compared with other products of HOCl it is likely to be minor.[72] Singlet oxygen is a strong two-electron oxidant that oxidises a range of amino acids, particularly methionine, histidine, tyrosine and tryptophan. It also reacts well with double bonds and is a good initiator of lipid peroxidation. The initial product is commonly an endoperoxide, which breaks down to radicals and other reactive species. Its energy is rapidly quenched in water so does not travel far from its site of generation.

8.4.6 Free Radicals

Free radicals have a fundamental role in lipid peroxidation and the antioxidant activity of vitamin E. As described in more detail elsewhere,[73,74] one-electron oxidation or reduction converts a wide range of endogenous and xenobiotic compounds to their respective radicals. One route is through oxidation by H_2O_2 and a dedicated peroxidase such as myeloperoxidase, or by a similar mechanism involving another redox active metalloprotein such as cytochrome c or haemoglobin. Radical generation from peroxynitrite (eqn (8.9)) has already been discussed. Many compounds form radicals through (predominantly metal catalysed) autoxidation or exposure to UV or ionising radiation, and others such as semiquinones are generated by enzymatic reduction. Importantly, because of the nature of radical reactions, formation of an initiating radical can lead to the generation of others.

Different radicals vary markedly in their reactivity. As illustrated in Table 8.2, the oxidising strength of a radical is given by its one-electron reduction potential. This also reflects its reactivity because the activation energy for radical reactions is generally low. With the proviso that reactant and product concentrations have some influence on the direction of a reaction (as described by the Nernst equation), radicals higher up Table 8.2 oxidise compounds lower down. Hydroxyl radicals are the most oxidising and react with DNA bases, aliphatic and aromatic amino acid residues, sugars and many other biomolecules at near diffusion-controlled rates.[2] Carbonate radicals are capable of oxidising many of these substrates,[75] whereas lower down the table, $NO_2^{\bullet}$ can react with polyunsaturated lipids and the more oxidisable amino acids. Simple phenoxyl radicals such as tyrosyl are more oxidising than polyphenol radicals[76] hence the ability of the latter to act as radical-scavenging antioxidants. Semiquinone radicals generally act as reductants and commonly react with oxygen to produce superoxide.

As noted above for the reactions of $NO^{\bullet}$ with superoxide (eqn (8.5)), an important property of radicals is that they readily react with other radicals. Thus, a radical–radical combination terminates the chain reaction of lipid peroxidation and contributes to effective radical scavenging. For radical scavenging to be a protective function, the product radical needs to be less reactive or damaging than the initial species. This is not necessarily the case. Vitamin E, for example, is a good lipophilic antioxidant when the tocopheroxyl radical can be removed by radical–radical reactions or radical transfer to a co-antioxidant such as ascorbate or ubiquinol.[77,78] However, lipoprotein particles have a low tocopherol concentration that disfavours radical recombination, and if a co-oxidant is not present, tocopheroxyl radicals can abstract a bis-allylic hydrogen and promote lipid peroxidation (tocopherol-mediated peroxidation[79]).

The main physiological aqueous phase radical scavenging antioxidants are ascorbate and GSH. Ascorbate is effective because it reacts with a wide spectrum of radicals and ascorbyl radicals disproportionate in preference to undergoing other reactions. Thiols such as GSH can act as good radical scavenging antioxidants but the mechanism is more complex. Cysteine (and GSH) are reasonably high in Table 8.2, and the thiyl radical is a good enough oxidant to abstract hydrogen from protein residues, initiate lipid peroxidation or form a peroxyl radical with oxygen.[80,81] However, a major physiological decay route for thiyl radicals involves reaction with a thiolate anion to form the disulfide radical anion:

$$GS^{\bullet} + GS^{-} \rightleftarrows GSSG^{\bullet -} \xrightarrow{O_2} GSSG + O_2^{\bullet -} \tag{8.12}$$

This reacts extremely rapidly with oxygen and removes the thiyl radical by driving the equilibrium in the forward direction. Thus, oxygen enables GSH to act as a good radical scavenger.[81] However, the generation of superoxide and H_2O_2 as downstream products has potential consequences. Ascorbate is also able to act synergistically with GSH by scavenging the thiyl radical.[80,82] Synergism is equally relevant for dietary or pharmacological antioxidants. Even though compounds such as polyphenolics and carotenoids are good radical scavengers, dietary intake is unable to increase tissue concentrations sufficiently to impact on total radical scavenging capacity.[2] To have any impact, they would need to accumulate at a site of radical generation, or act synergistically to facilitate radical transfer to more abundant species.

8.5 Compartmentalisation, Diffusion and Identification of Oxidant Targets

Numerous reactions of ROS with isolated biomolecules have been described. However, it does not necessarily follow that all these reactions will be relevant in more complex biological situations where there are multiple targets that must compete for the oxidant. Also, cells and tissues are not homogeneous, but contain compartments with potential diffusion barriers between them. Therefore, the effect of an oxidant may vary depending on whether it acts at the site of generation or is able to diffuse away. The first consideration for diffusion is membrane permeability. This is low for charged species such as superoxide. Thus superoxide produced outside a cell should remain outside and, when generated by mitochondria, it should mostly remain on the side of the membrane from which it is released. Uncharged species, including H_2O_2, can diffuse through cell membranes, whereas hypochlorous (pK_a 7.5) and peroxynitrous (pK_a 6.8) acids penetrate membranes in their protonated form.[83] H_2O_2 diffusion is not completely free, and concentration gradients across membranes exist due to differing consumption rates on either side.[84] Membrane transport of H_2O_2 is facilitated by aquaporins, and this provides a mechanism for regulating intracellular H_2O_2-dependent reactions.[85–87]

An important consideration for the superoxide generated by NADPH oxidases is that the enzyme complexes assemble on membranes, consume NADPH in the cytoplasm and transfer electrons through the membrane to oxygen.[4] Thus, the superoxide produced on the plasma membrane should react extracellularly and that produced on the membrane of organelles such as neutrophil phagosomes or endosomes should remain in the organelle where it is generated. The H_2O_2 formed by dismutation would be less constrained and could re-enter the cytoplasm or diffuse in other directions.

Regardless of membrane barriers, the fate of any ROS depends on how far it diffuses before it is consumed. This is determined by what targets are in its path – the more reactive the targets, the shorter the distance. As nicely demonstrated by Belousov and colleagues,[88,89] for intracellular H_2O_2, it appears that this distance is short. By targeting fluorescently distinguishable variants of the genetically encoded probe, Hyper, to different intracellular sites, they showed that H_2O_2 reacts very close to its site of generation. These observations imply that it is consumed by highly reactive targets in its path.

An assessment of likely targets can be made using a simple kinetic expression based on their concentrations and rate constants for reaction with the oxidant. This kinetic expression has been used to compare thiol proteins as likely cellular targets for H_2O_2.[28,90] This is illustrated in Table 8.4 for a simple system containing estimated cell concentrations of GSH, a typical low pK thiol protein such as the protein tyrosine phosphatase involved in redox signalling, a seleno-glutathione peroxidase and a peroxiredoxin. The highly reactive peroxiredoxin and glutathione peroxidase trap almost all the H_2O_2, with the majority reacting with the more abundant peroxiredoxin. Only a tiny fraction of the H_2O_2 reacts with GSH. The low pK_a thiol protein, with a rate constant more than four orders of magnitude lower that the peroxiredoxin, is unable to compete even with the GSH. How far H_2O_2 can diffuse depends on what targets are present and can be modelled using similar kinetic parameters.[90] Such calculations show that GSH at a typical concentration of several millimoles cannot constrain it within the confines of a cell.[90] For H_2O_2 to act locally, appreciable concentrations of more reactive targets must be present. Only peroxiredoxins and glutathione peroxidases are known to be sufficiently reactive to limit its range to a few microns and other less reactive thiol proteins would not be selectively oxidised at the site of peroxide generation.

Table 8.4 Estimated percentages of hydrogen peroxide reacting with each component in a model homogeneous system containing a typical peroxiredoxin (Prx), a seleno-glutathione peroxidase (GPx), a protein tyrosine phosphatase (PTP) and GSH.[a]

Component	Percentage of H_2O_2 consumed
Prx	85
Gpx	15
GSH	5×10^{-4}
PTP	10^{-5}

[a]Estimated cellular concentrations and rate constants used are GSH 2 mM and 0.89 M^{-1} s^{-1}, Prx 20 μM and 2×10^7 M^{-1} s^{-1}, GPx 1 μM and 6×10^7 M^{-1} s^{-1}, PTP 0.1 μM and 20–160 M^{-1} s^{-1} with details provided in ref. 90. Components are assumed to be in their reduced form.

This type of analysis has implications for redox signalling mechanisms, as oxidation of low pK_a thiol proteins such as protein tyrosine phosphatases has been observed in cells exposed to H_2O_2 or after activation of signalling pathways.[91,92] It implies that other factors must be important in the cellular environment. As discussed elsewhere,[30–32,93] these are not well understood. One possibility is that the thiol or seleno peroxidases become overwhelmed or inactivated at high or sustained H_2O_2 exposure, allowing less favourable reactions to occur. However, while feasible, this seems more applicable to oxidative injury rather than a regulatory process. Alternatively, limits on the recycling mechanisms for the peroxidases, or peroxiredoxin inactivation due to phosphorylation,[32,94] could allow H_2O_2 to increase in concentration and oxidise other targets. However, it seems most likely that an interaction between the H_2O_2 source and target would be needed to localise its oxidation. Examples of localised reactions have been reported.[32,95] An alternative redox signalling mechanism that takes advantage of the high reactivity of the peroxiredoxins is a relay system in which they act as the initial target for H_2O_2 then pass the oxidising equivalents on to another thiol protein.[30,33,96] This was first observed for transcription factor activation in yeast[97] and has now been seen for a number of mammalian systems.[96,98] What seems clear is that multiple mechanisms are likely to operate in the transmission of redox signals but there is still much to be learned in this regard.

Kinetic modelling has also been used to predict diffusion distances and likely targets for reactions of NO, superoxide and peroxynitrite under various conditions[45] and to identify likely targets for HOCl in the neutrophil phagosome.[72] There are provisos as discussed earlier about the impact of non-homogeneity and possible binding interactions, but they provide an important starting point for identifying mechanisms that need to be in place to overcome kinetic constraints. It is also important not to ignore minor reactions if they impart a gain of function to the target. For example, DNA modification may be negligible in terms of the overall reactions of a radical, but by introducing a single mutation, its effects could be far reaching. A minor oxidative modification to a protein could initiate an immune response. In contrast, inhibition of an enzyme by only a few percent is less likely to have functional consequences.

8.6 Conclusions

Free radicals and other ROS are clearly capable of damaging and killing living organisms. This occurs in radiation injury, and numerous cytotoxic chemicals act by generating radicals and releasing excess superoxide and H_2O_2. Animals and plants also produce ROS to combat pathogens. However, many cells generate superoxide or H_2O_2 not as toxins but for the purpose of cell signalling or metabolic regulation. So rather than viewing ROS as strictly damaging and antioxidants as strictly protective, it is now appreciated that there is a continuum of oxidative damage, oxidative stress and redox regulation. Different ROS contribute to these effects in different ways. Some can be directly damaging or cytotoxic. However, lesser damage or exposure to less reactive species can activate stress responses. The ultimate result may still be cell death, for example by apoptosis, but the response may be to increase expression of proteins involved in protective pathways such as antioxidant activity, DNA repair, xenobiotic detoxification and chaperone activity. At the other end of the spectrum, ROS can regulate the activity of oxidant-sensitive enzymes involved in cell metabolism, growth and development. In this context, antioxidant systems are not simply agents for removing ROS and may better be described as reductant systems that regulate oxidant levels in cells and thereby redox metabolism.

While there is still much to discover about ROS in biology, it is simplistic to think just in terms of overall ROS exposure. We need to know what particular species are being generated in any situation, how much, and where the oxidants are located. Extensive information is available on biomolecular modifications caused by individual oxidants, but there are large gaps in our knowledge of mechanisms of oxidation of sensitive targets in cell signalling pathways. Fortunately, recent advances in the development of oxidant-sensitive probes are providing much needed tools for quantifying and localising production of specific ROS in live cells, and these should help answer outstanding questions.

Acknowledgements

Grant support for the author's work has come primarily from the Health Research Council of New Zealand and the Marsden Fund.

References

1. G. R. Buettner, *Arch. Biochem. Biophys.*, 1993, **300**, 535.
2. B. Halliwell and J. M. Gutteridge, *Free Radicals in Biology and Medicine*, Oxford University Press, Oxford, 5th edn, 2015, p. 944.
3. M. P. Murphy, *Biochem. J.*, 2009, **417**, 1.
4. J. D. Lambeth and A. S. Neish, *Annu. Rev. Pathol.*, 2014, **9**, 119.
5. W. M. Nauseef, *Biochim. Biophys. Acta*, 2014, **1840**, 757.
6. J. M. Gebicki and B. H. J. Bielski, *J. Am. Chem. Soc.*, 1981, **103**, 7020.
7. D. E. Cabelli and B. H. J. Bielski, *J. Phys. Chem.*, 1983, **87**, 1809.

8. N. Gotoh and E. Niki, *Biochim. Biophys. Acta*, 1992, **1115**, 201.
9. C. C. Winterbourn and D. Metodiewa, *Arch. Biochem. Biophys.*, 1994, **314**, 284.
10. S. I. Liochev and I. Fridovich, *Free Radical Biol. Med.*, 1994, **16**, 29.
11. J. A. Imlay, *Annu. Rev. Microbiol.*, 2003, **57**, 395.
12. M. Jonsson, T. Lind, T. E. Reitberger, T. E. Eriksen and G. Merenyi, *J. Phys. Chem.*, 1993, **97**, 8229.
13. C. C. Winterbourn and A. J. Kettle, *Biochem. Biophys. Res. Commun.*, 2003, **305**, 729.
14. P. Pacher, J. S. Beckman and L. Liaudet, *Physiol. Rev.*, 2007, **87**, 315.
15. C. Szabo, H. Ischiropoulos and R. Radi, *Nat. Rev. Drug Discovery*, 2007, **6**, 662.
16. G. Ferrer-Sueta and R. Radi, *ACS Chem. Biol.*, 2009, **4**, 161.
17. E. S. Patricio, F. M. Prado, R. P. da Silva, L. A. Carvalho, M. V. Prates and T. Dadamos, *et al.*, *Chem. Res. Toxicol.*, 2015, **28**, 1556.
18. F. Jin, J. Leitich and C. von Sonntag, *J. Chem. Soc., Perkin Trans. II*, 1993, 1583.
19. P. Nagy, A. J. Kettle and C. C. Winterbourn, *J. Biol. Chem.*, 2009, **284**, 14723.
20. L. Bedard, M. J. Young, D. Hall, T. Paul and K. U. Ingold, *J. Am. Chem. Soc.*, 2001, **123**, 12439.
21. N. A. Bonekamp, A. Volkl, H. D. Fahimi and M. Schrader, *Biofactors*, 2009, **35**, 346.
22. T. J. Tavender and N. J. Bulleid, *Antioxid. Redox Signaling*, 2010, **13**, 1177.
23. B. P. Tu and J. S. Weissman, *J. Cell Biol.*, 2004, **164**, 341.
24. T. L. Leto, S. Morand, D. Hurt and T. Ueyama, *Antioxid. Redox Signaling*, 2009, **11**, 2607.
25. C. C. Winterbourn, *Methods Enzymol.*, 2013, **528**, 3.
26. C. C. Winterbourn and D. Metodiewa, *Free Radical Biol. Med.*, 1999, **27**, 322.
27. A. V. Peskin, F. M. Low, L. N. Paton, G. J. Maghzal, M. B. Hampton and C. C. Winterbourn, *J. Biol. Chem.*, 2007, **282**, 11885.
28. G. Ferrer-Sueta, B. Manta, H. Botti, R. Radi, M. Trujillo and A. Denicola, *Chem. Res. Toxicol.*, 2011, **24**, 438.
29. V. Gupta and K. S. Carroll, *Biochim. Biophys. Acta*, 2014, **1840**, 847.
30. C. C. Winterbourn and M. B. Hampton, *Free Radical Biol. Med.*, 2008, **45**, 549.
31. H. S. Marinho, C. Real, L. Cyrne, H. Soares and F. Antunes, *Redox Biol.*, 2014, **2**, 535.
32. S. G. Rhee, H. A. Woo and D. Kang, *Antioxid. Redox Signaling*, 2017, **28**, 537.
33. H. J. Forman, M. Maiorino and F. Ursini, *Biochemistry*, 2010, **49**, 835.
34. S. Stocker, K. Van Laer, A. Mijuskovic and T. P. Dick, *Antioxid. Redox Signaling*, 2018, **28**, 558.
35. J. Vasquez-Vivar, A. Denicola, R. Radi and O. Augusto, *Chem. Res. Toxicol.*, 1997, **10**, 786.

36. D. F. Trindade, G. Cerchiaro and O. Augusto, *Chem. Res. Toxicol.*, 2006, **19**, 1475.
37. E. V. Bakhmutova-Albert, H. Yao, D. E. Denevan and D. E. Richardson, *Inorg. Chem.*, 2010, **49**, 11287.
38. H. Zhou, H. Singh and Z. D. Parsons, *et al.*, *J. Am. Chem. Soc.*, 2011, **133**, 15803.
39. C. C. Winterbourn, *Toxicol. Lett.*, 1995, **82–83**, 969.
40. M. Chevion, *Free Radical Biol. Med.*, 1988, **5**, 27.
41. E. R. Stadtman, *Free Radical Biol. Med.*, 1990, **9**, 315.
42. J. A. Imlay, *Annu. Rev. Biochem.*, 2008, **77**, 755.
43. E. Madej, L. K. Folkes, P. Wardman, G. Czapski and S. Goldstein, *Free Radical Biol. Med.*, 2008, **44**, 2013.
44. M. R. Gunther, B. E. Sturgeon and R. P. Mason, *Toxicology*, 2002, **177**, 1.
45. J. R. Lancaster, Jr, *Chem. Res. Toxicol.*, 2006, **19**, 1160.
46. M. Moller, H. Botti, C. Batthyany, H. Rubbo, R. Radi and A. Denicola, *J. Biol. Chem.*, 2005, **280**, 8850.
47. K. A. Broniowska and N. Hogg, *Antioxid. Redox Signaling*, 2012, **17**, 969.
48. C. Quijano, B. Alvarez, R. M. Gatti, O. Augusto and R. Radi, *Biochem. J.*, 1997, **322**, 167.
49. M. Trujillo, G. Ferrer-Sueta, L. Thomson, L. Flohe and R. Radi, *Subcell. Biochem.*, 2007, **44**, 83.
50. L. A. MacMillan-Crow, J. P. Crow and J. A. Thompson, *Biochemistry*, 1998, **37**, 1613.
51. H. Botti, C. Batthyany, A. Trostchansky, R. Radi, B. A. Freeman and H. Rubbo, *Free Radical Biol. Med.*, 2004, **36**, 152.
52. M. Morita, Y. Naito, T. Yoshikawa and E. Niki, *Redox Biol.*, 2016, **8**, 127.
53. S. J. Klebanoff, *J. Leukocyte Biol.*, 2005, **77**, 598.
54. M. J. Davies, C. L. Hawkins, D. I. Pattison and M. D. Rees, *Antioxid. Redox Signaling*, 2008, **10**, 1199.
55. C. L. Hawkins, D. I. Pattison and M. J. Davies, *Amino Acids*, 2003, **25**, 259.
56. D. I. Pattison and M. J. Davies, *Curr. Med. Chem.*, 2006, **13**, 3271.
57. B. S. Rayner, D. T. Love and C. L. Hawkins, *Free Radical Biol. Med.*, 2014, **71**, 240.
58. C. C. Winterbourn, *Biochim. Biophys. Acta*, 1985, **840**, 204.
59. D. T. Harwood, A. J. Kettle and C. C. Winterbourn, *Biochem. J.*, 2006, **399**, 161.
60. A. V. Peskin, R. Turner, G. J. Maghzal, C. C. Winterbourn and A. J. Kettle, *Biochemistry*, 2009, **48**, 10175.
61. E. L. Thomas, M. B. Grisham and M. M. Jefferson, *Methods Enzymol.*, 1986, **132**, 569.
62. C. C. Winterbourn, A. J. Kettle and M. B. Hampton, *Annu. Rev. Biochem.*, 2016, **85**, 765.
63. D. I. Pattison, C. L. Hawkins and M. J. Davies, *Chem. Res. Toxicol.*, 2003, **16**, 439.
64. N. M. Domigan, T. S. Charlton, M. W. Duncan, C. C. Winterbourn and A. J. Kettle, *J. Biol. Chem.*, 1995, **270**, 16542.

65. C. C. Winterbourn, J. J. van den Berg, E. Roitman and F. A. Kuypers, *Arch. Biochem. Biophys.*, 1992, **296**, 547.
66. W. A. Prutz, R. Kissner, W. H. Koppenol and H. Ruegger, *Arch. Biochem. Biophys.*, 2000, **380**, 181.
67. J. P. Henderson, J. Byun and J. W. Heinecke, *J. Biol. Chem.*, 1999, **274**, 33440.
68. C. C. Winterbourn and A. J. Kettle, *Free Radical Biol. Med.*, 2000, **29**, 403.
69. M. L. Brennan and S. L. Hazen, *Amino Acids*, 2003, **25**, 365.
70. O. Skaff, D. I. Pattison and M. J. Davies, *Chem. Res. Toxicol.*, 2007, **20**, 1980.
71. S. Miyamoto, G. R. Martinez, M. H. Medeiros and P. Di Mascio, *J. Photochem. Photobiol., B*, 2014, **139**, 24.
72. C. C. Winterbourn, M. B. Hampton, J. H. Livesey and A. J. Kettle, *J. Biol. Chem.*, 2006, **281**, 39860.
73. P. J. O'Brien, *Chem. Biol. Interact.*, 2000, **129**, 113.
74. C. C. Winterbourn, in *Encyclopedia of Radicals in Chemistry, Biology & Materials*, ed. C. Chatgilialoglu and A. Studer, Wiley, Chichester, 2012, vol. 3, p. 1259.
75. D. B. Medinas, G. Cerchiaro, D. F. Trindade and O. Augusto, *IUBMB Life*, 2007, **59**, 255.
76. V. E. Kagan, E. A. Serbinova and L. Packer, *Arch. Biochem. Biophys.*, 1990, **280**, 33.
77. R. Yamauchi, *Vitam. Horm.*, 2007, **76**, 309.
78. E. Niki, *Free Radical Biol. Med.*, 2014, **66**, 3.
79. V. W. Bowry, K. U. Ingold and R. Stocker, *Biochem. J.*, 1992, **288**, 341.
80. P. Wardman and C. von Sonntag, *Methods Enzymol.*, 1995, **251**, 31.
81. C. C. Winterbourn, *Free Radical Biol. Med.*, 1993, **14**, 85.
82. B. E. Sturgeon, H. J. Sipe, Jr, D. P. Barr, J. T. Corbett, J. G. Martinez and R. P. Mason, *J. Biol. Chem.*, 1998, **273**, 30116.
83. S. S. Marla, J. Lee and J. T. Groves, *Proc. Natl. Acad. Sci. U. S. A.*, 1997, **94**, 14243.
84. F. Antunes and E. Cadenas, *FEBS Lett.*, 2000, **475**, 121.
85. G. P. Bienert and F. Chaumont, *Biochim. Biophys. Acta*, 2014, **1840**, 1596.
86. M. Bertolotti, S. H. Yim and J. M. Garcia-Manteiga, *et al.*, *Antioxid. Redox Signaling*, 2010, **13**, 1133.
87. E. W. Miller, B. C. Dickinson and C. J. Chang, *Proc. Natl. Acad. Sci. U. S. A.*, 2010, **107**, 15681.
88. N. M. Mishina, P. A. Tyurin-Kuzmin and K. N. Markvicheva, *et al.*, *Antioxid. Redox Signaling*, 2011, **14**, 1.
89. M. E. Matlashov, V. V. Belousov and G. Enikolopov, *Antioxid. Redox Signaling*, 2014, **20**, 1039.
90. C. C. Winterbourn, *Nat. Chem. Biol.*, 2008, **4**, 278.
91. N. K. Tonks, *Cell*, 2005, **121**, 667.
92. S. G. Rhee, Y. S. Bae, S. R. Lee and J. Kwon, *Sci. STKE*, 2000, **2000**, PE1.
93. F. Antunes and P. M. Brito, *Redox Biol.*, 2017, **13**, 1.
94. Z. A. Wood, L. B. Poole and P. A. Karplus, *Science*, 2003, **300**, 650.

95. K. Chen, M. T. Kirber, H. Xiao, Y. Yang and J. F. Keaney, Jr, *J. Cell Biol.*, 2008, **181**, 1129.
96. S. Stocker, M. Maurer, T. Ruppert and T. P. Dick, *Nat. Chem. Biol.*, 2018, **14**, 148.
97. A. Delaunay, D. Pflieger, M. B. Barrault, J. Vinh and M. B. Toledano, *Cell*, 2002, **111**, 471.
98. M. C. Sobotta, W. Liou and S. Stocker, *et al.*, *Nat. Chem. Biol.*, 2015, **11**, 64.
99. S. V. Jovanovic and M. G. Simic, *Ann. N. Y. Acad. Sci.*, 2000, **899**, 326.
100. D. I. Pattison and M. J. Davies, *Chem. Res. Toxicol.*, 2001, **14**, 1453.
101. M. G. Bonini and O. Augusto, *J. Biol. Chem.*, 2001, **276**, 9749.
102. A. V. Peskin and C. C. Winterbourn, *Free Radical Biol. Med.*, 2003, **35**, 1252.
103. P. Nagy, G. N. Jameson and C. C. Winterbourn, *Chem. Res. Toxicol.*, 2009, **22**, 1833.
104. D. I. Pattison and M. J. Davies, *Biochemistry*, 2004, **43**, 4799.

CHAPTER 9

Lipid Peroxidation: Role of Vitamin E

SHANSHAN ZHONG[a,b] AND HUIYONG YIN*[a,b,c]

[a]CAS Key Laboratory of Nutrition, Metabolism and Food Safety, Shanghai Institute of Nutrition and Health, Shanghai Institutes for Biological Sciences (SIBS), Chinese Academy of Sciences (CAS), Shanghai 200031, China; [b]University of Chinese Academy of Sciences, CAS, Beijing 100049, China; [c]School of Life Science and Technology, ShanghaiTech University, Shanghai 200031, China
*E-mail: hyyin@sibs.ac.cn

9.1 Introduction

Lipids are important building blocks of the cellular membrane; they fulfil the function of compartmentalisation and regulate substance and signal transduction between cells and organelles. Fatty acids are essential components of membrane lipids. Fatty acids can be classified based on the carbon chain lengths, degree of unsaturation, and the position of the double bonds. Figure 9.1 illustrates the common nomenclature of fatty acids and chemical structures of polyunsaturated fatty acids (PUFAs). Among the different fatty acids, PUFAs are fatty acids with more than one double bond in their chemical structures.

Food Chemistry, Function and Analysis No.11
Vitamin E: Chemistry and Nutritional Benefits
Edited by Etsuo Niki

Published by the Royal Society of Chemistry, www.rsc.org

Chemical structures	Fatty acids
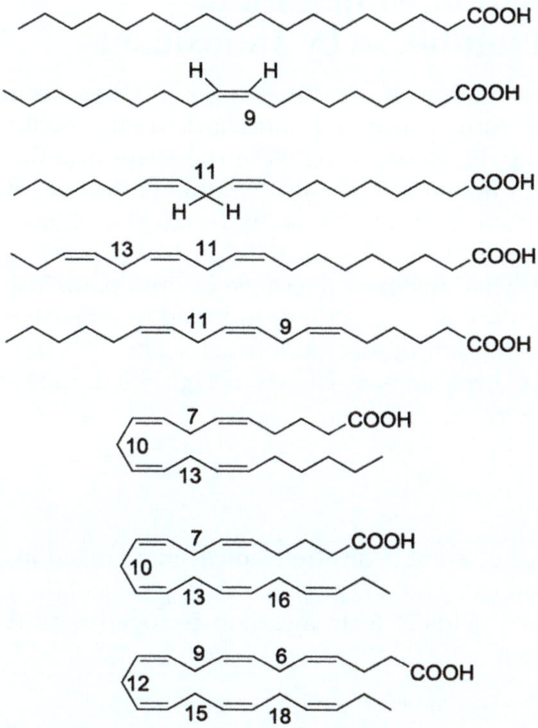	Stearic acid, C18:0
	Oleic acid, C18:1ω9
	Linoleic acid, C18:2ω6
	α-Linolenic acid, C18:3ω3
	γ-Linolenic acid, C18:3ω6
	Eicosatetraenoic acid Arachidonic acid, C20:4ω6
	Eicosapentaenoic acid (EPA), C20:5ω3
	Docosahexaenoic acid (DHA), C22:6ω3

Figure 9.1 Chemical structures and names of common fatty acids.

Oxygen is important for sustaining life on earth, yet it may cause detrimental effects when production of reactive oxygen species (ROS) overwhelms antioxidant systems, a condition which is often referred to as oxidative stress. Excessive generation of ROS initiates free radical lipid peroxidation (LPO), which may lead to damage of the cell membrane, if it is not properly repaired. LPO has been linked to a variety of biological processes such as aging, and diseases including atherosclerosis, cancer, and neurodegenerative diseases including Parkinson's disease (PD) and Alzheimer's disease (AD). LPO is a free radical process, which contains three major elemental reactions: initiation, propagation, and termination.[1–3] LPO can be initiated by a number of processes, such as an enzymatic process, free radicals and redox metals.[4,5] Once it is initiated, the free radical chain propagates until it is eventually terminated by the combining of two radicals, or intercepted by a good hydrogen atom donor such as vitamin E.[6,7]

In this chapter, we will describe the free radical chemical mechanisms of LPO, the role of VitE in LPO, and the biological and disease relevance of LPO. In the end, we will discuss the controversies of antioxidants in LPO and human diseases, highlighting the challenges in this field.

9.2 Chemical Mechanism of Free Radical Lipid Peroxidation: Initiation, Propagation, Termination, and Inhibition by Antioxidants

Historically, free radical chain oxidation of organic compounds has been referred to as autoxidation and research in this area dates back to early 1900.[8] The oxidation of PUFA-containing lipids was initiated in the 1940s and the early work was led by Bolland and Bateman at the British Rubber Producers Research Association.[9] Due to the formation of a peroxyl radical and later hydroperoxides as major products in the oxidation of lipids, this process is often termed lipid peroxidation. The biological relevance of lipid peroxidation has been increasingly recognised since the 1980s followed by extensive studies on the cyclooxygenase oxidation of arachidonic acid. The LPO free radical chain process consists of three steps as shown in Figure 9.2: initiation, propagation, and termination.

9.2.1 Initiation

Generation of a free radical is the key event in any free radical chain reaction, which is referred to as the initiation step (eqn (1)). The initiating radical then abstracts a hydrogen atom in fatty acids to form a carbon-centered radical

Initiation:

$$\mathrm{In{-}In} \longrightarrow \mathrm{In\cdot} + \mathrm{In\cdot} \qquad (1)$$

$$\mathrm{In\cdot} + \mathrm{L{-}H} \xrightarrow{k_{iLH}} \mathrm{In{-}H} + \mathrm{L^{\cdot}} \qquad (2)$$

Propagation:

$$\mathrm{L\cdot} + \mathrm{O_2} \xrightarrow{k_{perox}} \mathrm{L{-}OO\cdot} \qquad (3)$$

$$\mathrm{L{-}OO\cdot} + \mathrm{L{-}H} \xrightarrow{k_p} \mathrm{L{-}OOH} + \mathrm{L^{\cdot}} \qquad (4)$$

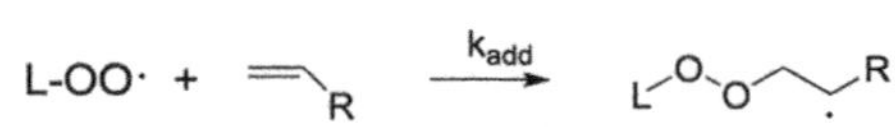
(5)

Termination:

$$2\,\mathrm{L{-}OO\cdot} \xrightarrow{k_t} [\mathrm{L{-}OO{-}OO{-}L}] \qquad (6)$$

$$[\mathrm{L{-}OO{-}OO{-}L}] \longrightarrow \text{Non-radical products (NRP)} + \mathrm{O_2} \qquad (7)$$

Inhibition:

$$\mathrm{L{-}OO\cdot} + \mathrm{ArOH} \xrightarrow{k_{inh}} \mathrm{L{-}OOH} + \mathrm{ArO\cdot} \qquad (8)$$

$$\mathrm{L{-}OO\cdot} + \mathrm{ArO^{\cdot}} \xrightarrow{k_t} \mathrm{NRP} \qquad (9)$$

In, Initiator; L, Lipid; k_{perox}, rate constant for oxygen addition; k_p, rate constant for propagation; k_t, rate constant for termination; k_{add}, rate constant for addition k_{inh}, rate constant for antioxidant inhibition

Figure 9.2 Free radical chain reactions.

($L^{\bullet}$) as shown in eqn (2) in Figure 9.2. For LPO, the initiation can be induced by exogenous sources, such as air pollution, cigarette smoking, ultraviolet exposure, or ionisation irradiation. On the other hand, a variety of intracellular sources can initiate a free radical process that leads to the oxidation of the cellular membrane. One of the major cellular sources for free radical generation is in mitochondria, also known as the power houses for ATP. It is estimated that 1–5% of electrons flowing through the electron transport chain may leak to generate superoxide. In the cytosol, superoxide can be generated from NADPH oxidase, xanthine oxidase, and cytochrome P450, whereas nitric oxide, one of the major reactive nitric species (RNS), can be generated from uncoupled nitric oxide synthase. In addition, free radical chain reactions can be initiated from lipoxygenase enzyme, myeloperoxidase, and prostaglandin H2 synthase (PGHS). For instance, a tyrosyl radical in the active site of PGHS catalyses the cyclisation of arachidonic acid to form PGH2, while the arachidonyl radical escapes from the enzyme active site to initiate the free radical chain reaction of LPO.[10] Similarly, tyrosyl radicals in myoglobin and haemoglobin can also initiate LPO, which has been shown to play an important role in rhabdomyolysis, subarachnoid haemorrhage formation, malaria, and sickle cell diseases.[11] Lastly, a free radical chain reaction can also be initiated from transition metals such as copper and iron, and hydroxyl radicals from gamma irradiation.

9.2.2 Propagation

Once a free radical chain reaction is initiated, the second step is to propagate the chain reactions after the carbon-centered radical $L^{\bullet}$ is quickly added to oxygen to form a peroxyl radical (eqn (3), Figure 9.2). This step occurs at or near the diffusion-controlled rate (k_{perox}) at oxygen pressures above 100 mm Hg. The resulting peroxyl radical, the major species carrying the chain reactions, can abstract a hydrogen atom to form a lipid hydroperoxide (eqn (4)) and a lipid radical; the latter can add a molecular oxygen to form a peroxyl radical. Alternatively, the peroxyl radical may also undergo radical addition to a carbon–carbon double bond; a cyclic peroxide can be formed when the peroxyl radical undergoes a 5-*exo* cyclisation.

Hydrogen atom transfer, transfer of a bis-allylic hydrogen atom in PUFAs to a peroxyl radical, is usually the rate limiting step in the propagation. The rate constant, k_p, for the most common PUFAs (linoleic acids, arachidonic acid, EPA, and DHA) in LPO, is summarised in Table 9.1. In these PUFAs, the C–H bonds at the bis-allylic positions are the weakest and these hydrogen atoms are preferentially abstracted by a peroxyl radical. Notably, the bond dissociation enthalpy (BDE) for these C–H bonds is about 78–80 kcal mol^{-1} and the k_p for methyl linoleate has been measured by independent laboratories to be around 60–62 M^{-1} s^{-1}. The overall oxidation of a particular PUFA is proportional to the number of these bis-allylic C–Hs: linoleate has one such center, arachidonate has three, EPA has four, and DHA has five. Accordingly, the relative propagation rate constants determined for these PUFAs are in a

Table 9.1 The rate constants for hydrogen atom abstraction by a peroxyl radical in PUFAs.

PUFAs	k_p in liposome (n^{-1} s^{-1})[a]	k_p in solution (M^{-1} s^{-1})
Linoleic acid	35 [100]	62 [102]
Arachidonic acid	115 ± 7 [101]	201 ± 12 [101]
Eicosapentaenoic acid (EPA)	145 ± 8 [101]	249 ± 14 [101]
Docosahexaenoic acid (DHA)	172 ± 13 [101]	321 ± 32 [101]

[a]"n", mole fraction of the studied hydrogen donor relative to the sum of acyl chains.

ratio of 1, 3.2, 4.0, and 5.4. In addition to the peroxyl radical, linoleic acid can be oxidised by other ROS with variable k_p. For example, the k_ps for the LA reaction with a $^{\bullet}NO_2$ radical, $CO_3^{\bullet-}$ radical, HOCl, and singlet oxygen are 2×10^5, $>2 \times 10^5$,[12] 18,[13] 1.3×10^5 M^{-1} s^{-1},[14] respectively.[15]

The peroxyl radical can also undergo addition reactions to carbon–carbon double bonds (eqn (5), Figure 9.2). It is well established that peroxyl radicals are much less prone to addition to isolated double bonds compared to conjugated double bonds. In the context of LPO, peroxyl radicals can add to the primary hydroperoxides, which have conjugated double bonds.[16,17] Interestingly, peroxyl radical addition to an adjacent acyl chain with conjugated double bonds was proposed to account for the formation of 4-hydroxy-2-nonenal (4-HNE) from the oxidation of the phospholipids cardiolipins.[18,19] Peroxyl radicals can also undergo *5-exo* cyclisation to form cyclic peroxides when PUFAs have more than two double bonds in the molecule, such as linolenic acids, arachidonic acids,[20–23] EPA and DHA.[24,25]

9.2.3 Termination

The radical chain reactions continue until two peroxyl radicals recombine to form a tetraoxide intermediate (eqn (6), Figure 9.2). This type of tetraoxide intermediate derived from two primary or secondary peroxyl radials is quite unstable and readily decomposes to generate a non-radical product (NRP), including aldehyde, alcohol and molecular oxygen; this reaction is also termed Russell Termination.[26] Notably, the termination step is quite fast and the rate constant for termination (k_t) of two peroxyl radicals can reach 10^7 M^{-1} s^{-1}.[2]

9.2.4 Inhibition by Antioxidants

Alternatively, the free radical chain reactions can be stopped by phenolic antioxidants, such as vitamin E; these compounds are also called free radical chain breakers due to their excellent hydrogen atom donating ability. These antioxidants act by reducing a lipid peroxyl radical to a hydroperoxide and forming a relatively inert phenoxyl radical in the process (eqn (8), Figure 9.2). The k_{inh} for α-TOH ranges from 6×10^3 M^{-1} s^{-1} in phospholipid bilayers to 3×10^6 M^{-1} s^{-1} in organic solvents.[27,28] The α-tocopheroxyl radical can rapidly react with another peroxyl radical ($k_t = 2.5 \times 10^6$ to 3×10^8 M^{-1} s^{-1}) to yield NRP. Therefore, each molecule of α-TOH destroys two radicals and terminates two potential radical chains.[29]

9.3 Free Radical Oxidation of PUFAs: Roles of Vitamin E

As one of the most important research areas in the free radical autoxidation of organic compounds, LPO has been attracting intensive research attention for the past several decades. The oxidation of PUFAs by the free radical mechanism has been well established based on the four elemental reactions mentioned earlier. These reactions have been applied to understand the oxidation of multiple PUFAs. We take linoleic acid and arachidonic acid as examples to illustrate the free radical oxidation of lipids, focusing on the reaction kinetics and the roles of vitamin E on the products' distributions.

9.3.1 Free Radical Oxidation of Linoleic Acid

As shown in Figure 9.3A, free radical oxidation of linoleic acid starts by abstraction of one of the bis-allylic hydrogen atoms at C11 to form a delocalised pentadienyl radical. The odd electron spin is principally distributed on the two terminal carbons (C9 or C13) and the central carbon C11. In the propagation step, the fast oxygen addition can occur at either of these three positions, but it took almost two decades to isolate and characterise the 11-HpODE (11-hydroperoxyoctadecadienoic acid).[30–33] Trapping the peroxyl radical at C11 was achieved in the presence of a good hydrogen atom donor, α-tocopherol (1 M), due to the fast β-fragmentation of the oxygen to form the pentadienyl radical (2.6×10^6 s^{-1}).[31] Trapping the peroxyl radical at C9 or C13 by hydrogen atoms gives rise to *Z*,*E*-9-HpODE and *Z*,*E*-13-HpODE, respectively. These three primary hydroperoxides are collectively designated as kinetic products.[34–36] As shown in Figure 9.3B, in the presence of 0.3 M of α-tocopherol, LA oxidation leads to equal distribution of these three kinetic products.

On the other hand, in the absence of good hydrogen atom donors, the fragmentation of a peroxyl radical due to the reversible oxygen addition can compete with trapping the peroxyl radical. As shown in Figure 9.3A, the peroxyl radical at C9 (or C13) can fragment to give rise to the pentadienyl radical with a rate constant of 6.9×10^2 s^{-1}. And the resulting delocalised radical rearranges before the subsequent oxygen addition, which may produce two *trans*,*trans*-HpODEs at C9 or C13 (*E*,*E*-13-HpODE, *E*,*E*-9-HpODE). These products are called the thermodynamic products of the lipid peroxidation. Similarly, the peroxyl radical at C13 will lead to a similar set of thermodynamic products: *E*,*E*-9-HpODE and *E*,*E*-13-HpODE. Thus, the distribution of the oxidation products from LA is strongly dependent on the concentration of the hydrogen atom donors, especially the presence of antioxidants such as

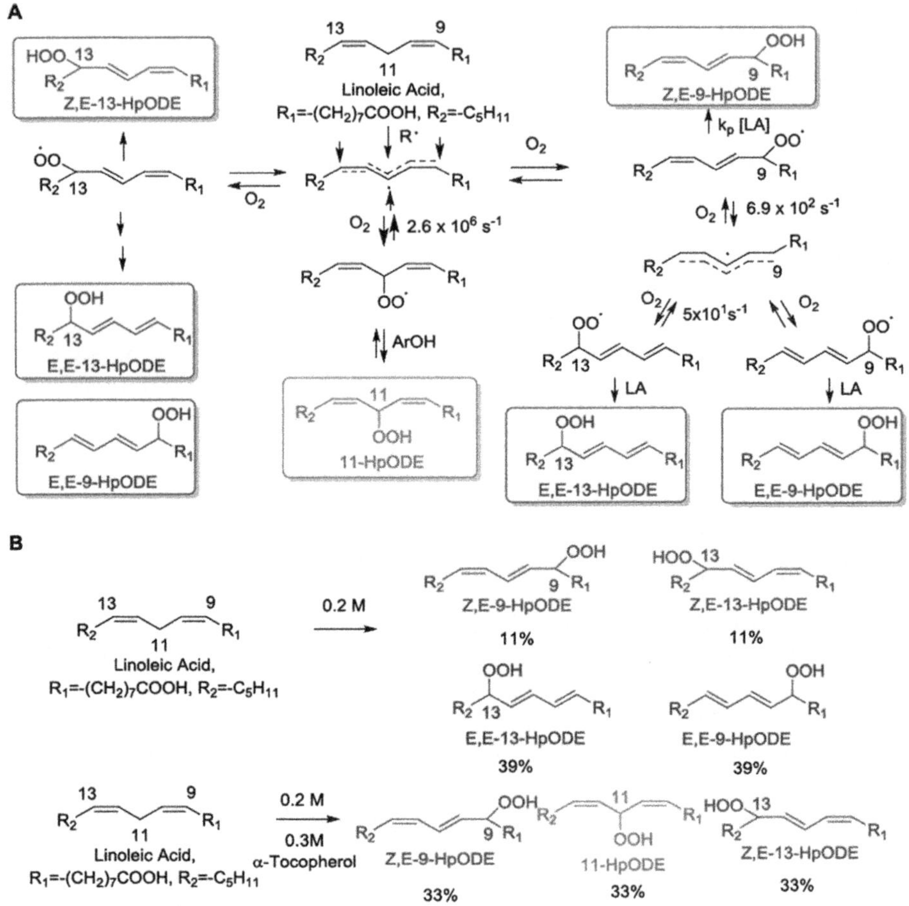

Figure 9.3 Free radical oxidation of linoleic acid. (A) Kinetic and thermodynamic products of LA oxidation. (B) Product distribution of LA oxidation in the presence and absence of α-tocopherol.

vitamin E. As shown in Figure 9.3B, in the absence of α-tocopherol, LA oxidation (0.2 M) gives rise to four primary oxidation products: kinetic products *Z*,*E*-9-HpODE (11%) and *Z*,*E*-13-HpODE (11%); thermodynamic products *E*,*E*-9-HpODE (39%) and *E*,*E*-13-HpODE.[1]

Because linoleic acid is one of the most abundant essential fatty acids in humans, efforts have been made to use the LA oxidation products, such as HpODE and their corresponding reducing products HODE (hydroxyoctadecadienoic acid), as a measure of oxidative stress *in vivo*.[37,38] Protocols to analyse the oxidation products of LA, primarily HODEs, have been reported, paying special attention to the artificial oxidation *ex vivo* during the work-up before analysis.[39–43] Furthermore, the ratio of *Z*,*E*–*E*,*E* products *in vivo* has been suggested as a means to measure the hydrogen atom donating characteristics of the membranes. The HODE *Z*,*E*–*E*,*E* product ratios in early reports in the literature varied from 0.5 to 2.0 in the liver and brain, whereas the plasma ratios were somewhat higher: 2.0 to 3.0. Later studies found that the sodium borohydride used in the previous protocol led to substantial *ex vivo* oxidation and triphenylphosphine is a better reducing agent in this case.[38] When triphenylphosphine was used in the sample preparation, the *Z,E*–*E,E* HODE product ratio was found to be 6.0 in human plasma and close to 20 in mouse liver. Furthermore, the HODE product ratio has also been used to evaluate the antioxidant capability *in vivo* in humans and animal models.[15,44] Removing vitamin E from the diet also increased the levels of total HODE and decreased the *Z,E*–*E,E*-HODE ratio, whereas natural and synthetic antioxidants, including vitamin E isoforms, chlorogenic acid, caffeic acid, coffee, and 2,3-dihyro-5-hydroxy-4,6-di-*tert*-butyl 2,2-dipentylbenzoylfuran (BO-653) increased the *Z,E*–*E,E*-HODE ratio.[45–47] Similarly, LA product distribution has been used to evaluate novel synthetic antioxidants, 6-amino-3-pyridinols, which reach an almost diffusion-controlled rate constant for breaking free radical chain reactions.[48–50] Moreover, LA oxidation products and their distribution has been used as a useful marker for low-density lipoprotein oxidation *in vitro*.[51,52]

9.3.2 Free Radical Oxidation of Arachidonic Acid

Free radical oxidation of arachidonic acid is more complex than linoleic acid due to the presence of two more double bonds in the molecule. All the elemental reactions for free radical chain reactions mentioned in the previous section can be applied to understand the oxidation of arachidonic acid. As shown in Figure 9.4, hydrogen atom abstraction by a free radical R$^•$ can occur at three bis-allylic positions, C7, C10 or C13. Abstraction of a hydrogen atom at C7 forms a delocalised pentadienyl radical among C5, C6, C7, C8 and C9. Oxygen addition to the terminal position gives rise to hydroperoxyl radical at C5 and C9, which forms 5-hydroperoxyeicosatetraenoic acid (5-HpETE) or 9-HpETE, respectively. Similarly, hydrogen atom abstraction at C10 will lead to the formation of 8-HpETE and 12-HpETE. Interestingly, linoleic acid oxidation may lead to the thermodynamic product *E,E*-HpODE in the absence of a good hydrogen atom donor, while these *E,E* products are not readily formed for arachidonic acid. Instead, the internal peroxyl radicals from arachidonic acid, including 8-, 9-, 11-, and 12-hydroperoxyl radicals, can undergo cyclisation to form cyclic peroxides. The rate constant for 5-*exo* cyclisation has not been rigorously established but is estimated to be in the range of 500–1000 s^{-1}, which allows cyclisation to compete with β-fragmentation to form the *E,E*-products.[53] For example, the 11-hydroperoxyl radical **4b** either traps a hydrogen to form 11-HpETE (**4c**), or undergoes *5-exo* cyclisation to form a cyclic peroxide (**4d**), which has a carbon radical at C8. Oxygen addition to this carbon radical to form a peroxyl radical (**4e**), which can further cyclise to form serial cyclic peroxides (**4g**), is followed by oxygen addition. Intermediate **4e** can form monocyclic peroxide **4f** after hydrogen atom abstraction.[34,54–56] Interestingly, the carbon radical **4d** may undergo a

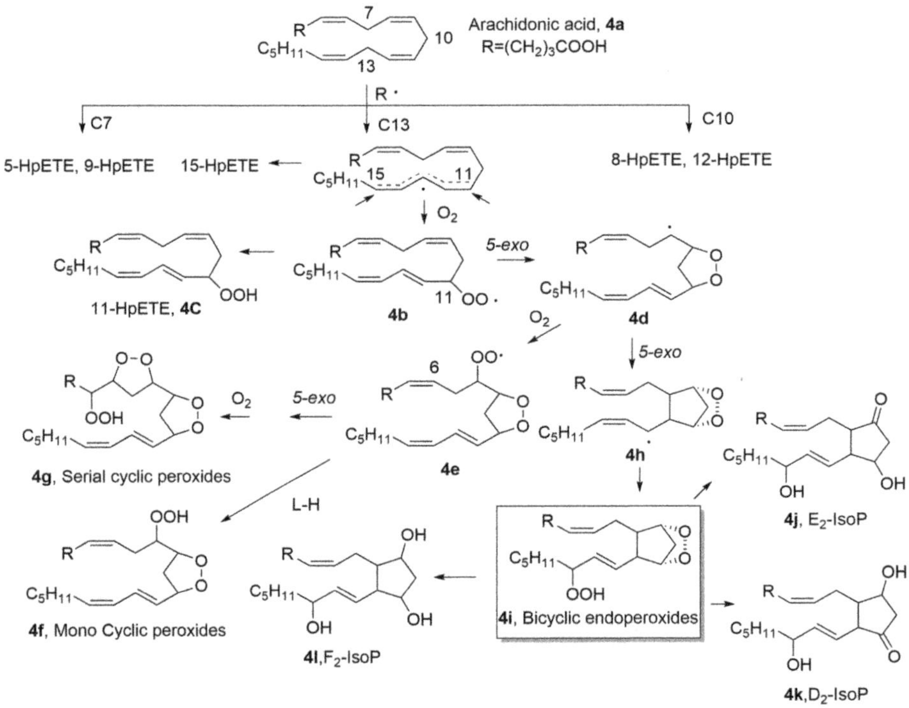

Figure 9.4 Free radical oxidation of arachidonic acid.

5-exo cyclisation to form a bicyclic peroxide (**4e**), which serves as an intermediate to form bicyclic endoperoxides (**4f**), followed by oxygen addition at C15 and hydrogen atom abstraction.[20] These types of bicyclic endoperoxides resemble the structure of prostaglandin G_2 (PGG_2) from cyclooxygenase pathways.[57] This unstable bicyclic endoperoxides can be reduced to form F-type isoprostanes (F_2-IsoP, **4l**), isomers of prostaglandins, or rearrange to form E_2-IsoP (**4g**) or D_2-IsoP (**4h**). Formation of these stereoisomers of prostaglandins by free radical pathways was predicted in the 1970s[58] but formation of these compounds *in vivo* was not discovered until 1990.[59,60] These compounds were termed isoprostanes because they are stereoisomers of prostaglandins. Serendipitous discovery of these compounds from LPO *in vivo* revolutionised the field of LPO and oxidative stress. Subsequent studies have identified the stereoisomers of IsoPs and the chemical mechanism of their formation.[1,2,21,61,62] For the past three decades, analysis of isoprostanes by mass spectrometry techniques has evolved as the gold standard to assess oxidative stress status *in vivo*.[63,64] The advantages of using F_2-IsoP as a biomarker for oxidative stress *in vivo* over other methods have been extensively investigated and the levels of F_2-IsoPs in biological fluids and tissues have been established in the context of numerous human diseases that are linked to lipid peroxidation and oxidative stress,[65–67] including cardiovascular diseases,[68–71] cancer,[72,73] and neurodegenerative diseases, such as Parkinson's disease and Alzheimer's disease.[74–79] Furthermore, overwhelming evidence supports the notion that LPO products can not only serve as biomarkers for oxidative stress but they also possess potent biological functions in the context of human diseases.[80–83]

9.4 Antioxidants and LPO: Vitamin E as an Antioxidant for LPO

Oxidation of arachidonic acid by free radical reactions is much more complicated than that of linoleic acids. Several competing routes leading to different sets of oxidation products are dependent on the hydrogen atom donation capability in the surroundings. Thus, analysis of the primary hydroperoxides, IsoPs, and monocyclic peroxides and serial cyclic peroxides may serve as biomarkers for different diseases as well as evaluating the potency of different chain breaking antioxidants, such as vitamin E.[6,15,84] The role of vitamin E and LPO *in vivo* has been extensively studied by using isoprostanes as a biomarker. A study was designed to define the dose-dependent effects of vitamin E (*RRR*-α-tocopherol) to suppress plasma concentrations, the marker for oxidative stress of F_2-IsoP in participants with polygenic hypercholesterolemia and a population at risk for cardiovascular events.[85] Participants were supplemented with 3200 IU (international unit) per day of vitamin E for 20 weeks and the study found that a maximum suppression of plasma F_2-isoprostane concentrations did not occur until 16 weeks of supplementation. Interestingly, this study also found a linear trend between the dosage of vitamin E and percentage reduction in plasma F_2-IsoP concentrations, which reached significance at doses of 1600 IU (35 ± 2%, $p < 0.035$) and 3200 IU (49 ± 10%, $p < 0.005$), which provides information on the dosage of vitamin E that decreases systemic oxidant stress *in vivo* in humans. In a separate study, 1000 mg per day vitamin C and 800 IU vitamin E, or a placebo, were administered to a cohort of non-obese, non-smokers, for two months. This study found that the effect of vitamins C and E on the reduction of F_2-IsoPs depended on the baseline levels: when baseline F_2-isoprostane was >50 μg mL^{-1}, vitamin C reduced F_2-IsoP by 22% ($p = 0.01$), whereas vitamin E reduced it by 9.8% ($p = 0.46$). Below that cut-off point, neither treatment produced further reductions.[86] Vitamin E (300 mg α-tocopherol) and C (1000 mg) supplementation prevents exercise-induced lipid peroxidation as indicated by the F_2-isoP levels in plasma in ultramarathon runners without reducing the inflammation.[87]

It remains controversial to use the antioxidants to prevent and treat human diseases associated with oxidative stress, such as atherosclerosis.[7] Some observational studies showed an inverse association between plasma vitamin E levels and incidences of cardiovascular disease (CVD).[88] But many interventional studies have failed to show benefits of vitamin E supplementation against CVD.[89,90] Although existing data appear to support vitamin E as an efficient lipophilic radical scavenger, vitamin E is not an efficient scavenger for other radicals, such as hydroxyl, alkoxyl, thiyl, or NO_2 radicals.[7,15] The rate constant for α-tocopherol reacting with the peroxyl radical is in the range of 10^5–10^6 M^{-1} s^{-1}, whereas the rate constants for α-tocopherol reacting with peroxynitrite, NO_2 radical, $CO_3^{\bullet -}$, HOCl and singlet oxygen are 1.8×10^5,[91] 1×10^5, 10^7–10^8, 1.3×10^3, and 3×10^8 M^{-1} s^{-1}, respectively.[15] Considering the rate constants k_p for PUFA and $k_{\alpha T}$ for α-tocopherol are 10^2 and 10^6 M^{-1} s^{-1}, respectively, the majority of peroxyl radicals (>90%) in the cellular membrane and lipoproteins can be scavenged by α-tocopherol to prevent LPO. However, the efficiency of vitamin E in the inhibition of LPO in atherosclerotic plaque was questioned based on the fact that vitamin E is not exhausted even in advanced atherosclerotic plaque.[92,93] Later studies carried out by Stocker and colleagues found that α-tocopherol functioned as a pro-oxidant to induce lipid peroxidation, a process which is termed tocopherol-mediated peroxidation (TMP).[94–96] TMP may play a limited role in the lipid peroxidation of the cellular bilayer membrane due to the absence of a lipid core.[97] It is conceivable that vitamin E acts as an effective radical scavenger to inhibit LPO in combination with vitamin C or other reductants. A lack of efficacy in interventional clinical studies against cardiovascular outcomes does not necessary contradict with the oxidative hypothesis of atherosclerosis because other pathophysiological processes are also involved, such as inflammation.[98,99]

9.5 Summary and Future Perspectives

Lipids are important membrane components that maintain proper cellular function by compartmentalisation, providing energy by fatty acid β-oxidation, and generating bioactive lipid mediators for signalling purposes. Fatty acids in membrane lipids can be taken up from diet fat or *de novo* synthesised. The chemical structures of fatty acids have a profound impact on the membrane structures as well as generating bioactive lipids through enzymatic processes or free radical lipid oxidation. The free radical chain reactions consist of initiation, propagation, and termination steps. The free radical chain reactions can be inhibited by hydrogen-atom-donating phenolic antioxidants, such as α-tocopherol, a major component in vitamin E.

The chemical mechanisms for free radical lipid peroxidation of PUFAs, such as linoleic acid and arachidonic acid, have been well established. The biological relevance of LPO has been extensively studied in the context of human diseases. A general consensus is that oxidation products from LPO can serve as a better marker for human diseases associated with LPO and oxidative stress, and the oxidation products are also actively involved in the pathogenesis of these diseases. The efficacy of free radical chain breaking antioxidants can be evaluated based on the product distribution of linoleic acid or arachidonic acid. Even though mounting *in vitro* data exist to support the inhibition of LPO by vitamin E and other antioxidants, the clinical efficacy of these antioxidants to treat human diseases associated with oxidative stress remains to be seen. The field of lipid oxidation and oxidative stress is transforming by incorporating state-of-the-art technologies, including genomics, lipidomics, and metabolomics. The pleotropic roles of α-tocopherols beyond chain breaking properties start to emerge, such as the modulation of gene expression. The complexity of lipid oxidation *in vivo* is increasingly recognised in different human diseases, such as atherosclerosis and cancer. Accumulating evidence appears to support an adaptive or even beneficial role of ROS and LPO in human pathogenesis. Thus, the timing of antioxidant interventions may be important for clinical applications, besides a better understanding of the roles of LPO and antioxidants. The field of lipid oxidation faces enormous challenges, but an exciting horizon is emerging.

Acknowledgements

This work was supported by National Natural Science Foundation of China (31470831, 91439103, 91539127, and 31401015) and the National Key R&D Program of China administered by Chinese Ministry of Science and Technology (MOST) (2016YFD0400205, 2016YFC0903403) and a grant from the Chinese Academy of Sciences (ZDBS-SSW-DQC-02).

References

1. H. Yin, L. Xu and N. A. Porter, *Chem. Rev.*, 2011, **111**, 5944.
2. H. Yin and N. A. Porter, *Antioxid. Redox Signaling*, 2005, **7**, 170.
3. N. A. Porter, *J. Org. Chem.*, 2013, **78**, 3511.
4. E. Niki, *Biochim. Biophys. Acta*, 2014, **1840**, 809.
5. E. Niki, *Free Radical Res.*, 2015, **49**, 827.
6. E. Niki, *Free Radical Biol. Med.*, 2009, **47**, 469.
7. E. Niki, *Free Radical Biol. Med.*, 2014, **66**, 3.
8. N. A. Porter, *J. Org. Chem.*, 2013, **78**, 3511–3524.
9. J. L. Bolland and G. Gee, *Trans. Faraday Soc.*, 1946, **42**, 236.
10. O. Boutaud and J. A. Oates, *Methods Mol. Biol.*, 2010, **644**, 67.
11. O. Boutaud, K. P. Moore and B. J. Reeder, *et al.*, *Proc. Natl. Acad. Sci. U. S. A.*, 2010, **107**, 2699.
12. S. A. Everett, M. F. Dennis, K. B. Patel, S. Maddix, S. C. Kundu and R. L. Willson, *J. Biol. Chem.*, 1996, **271**, 3988.
13. D. I. Pattison, C. L. Hawkins and M. J. Davies, *Chem. Res. Toxicol.*, 2003, **16**, 439.
14. F. H. Doleiden, S. R. Fahrenholtz, A. A. Lamola and A. M. Trozzolo, *Photochem. Photobiol.*, 1974, **20**, 519.
15. E. Niki, *Free Radical Biol. Med.*, 2018, **120**, 425.
16. K. Miyashita, N. Hara, K. Fujimoto and T. Kaneda, *Lipids*, 1985, **20**, 578.
17. C. Schneider, W. E. Boeglin, H. Yin, N. A. Porter and A. R. Brash, *Chem. Res. Toxicol.*, 2008, **21**, 895.

18. W. Liu, N. A. Porter, C. Schneider, A. R. Brash and H. Yin, *Free Radical Biol. Med.*, 2011, **50**, 166.
19. H. Yin and M. Zhu, *Free Radical Res.*, 2012, **46**, 959.
20. H. Yin, C. M. Havrilla, J. D. Morrow and N. A. Porter, *J. Am. Chem. Soc.*, 2002, **124**, 7745.
21. H. Yin, J. D. Morrow and N. A. Porter, *J. Biol. Chem.*, 2004, **279**, 3766.
22. H. Yin, C. M. Havrilla, L. Gao, J. D. Morrow and N. A. Porter, *J. Biol. Chem.*, 2003, **278**, 16720.
23. H. Yin, J. D. Brooks, L. Gao, N. A. Porter and J. D. Morrow, *J. Biol. Chem.*, 2007, **282**, 29890.
24. H. Yin, W. Liu, K. Goleniewska, N. A. Porter, J. D. Morrow and R. S. Peebles Jr, *Free Radical Biol. Med.*, 2009, **47**, 622.
25. H. Yin, E. S. Musiek, L. Gao, N. A. Porter and J. D. Morrow, *J. Biol. Chem.*, 2005, **280**, 26600.
26. G. A. Russell, *J. Am. Chem. Soc.*, 1957, **79**, 3871.
27. G. W. Burton and K. U. Ingold, *Acc. Chem. Res.*, 1986, **19**, 194.
28. E. Niki, T. Saito, A. Kawakami and Y. Kamiya, *J. Biol. Chem.*, 1984, **259**, 4177.
29. G. W. Burton and K. U. Ingold, *J. Am. Chem. Soc.*, 1981, **103**, 6472.
30. A. R. Brash, *Lipids*, 2000, **35**, 947.
31. K. A. Tallman, D. A. Pratt and N. A. Porter, *J. Am. Chem. Soc.*, 2001, **123**, 11827.
32. K. A. Tallman, C. L. Rector and N. A. Porter, *J. Am. Chem. Soc.*, 2009, **131**, 5635.
33. K. A. Tallman, B. Roschek and N. A. Porter, *J. Am. Chem. Soc.*, 2004, **126**, 9240.
34. N. A. Porter, S. E. Caldwell and K. A. Mills, *Lipids*, 1995, **30**, 277.
35. H. Yin and N. A. Porter, *Antioxid. Redox Signaling*, 2005, **7**, 170.
36. D. A. Pratt, K. A. Tallman and N. A. Porter, *Acc. Chem. Res.*, 2011, **44**, 458.
37. A. Umeno, M. Morita, Y. Yoshida, Y. Naito and E. Niki, *Arch. Biochem. Biophys.*, 2017, **635**, 96.
38. W. Liu, H. Yin, Y. O. Akazawa, Y. Yoshida, E. Niki and N. A. Porter, *Chem. Res. Toxicol.*, 2010, **23**, 986.
39. Y. Yoshida, S. Kodai, S. Takemura, Y. Minamiyama and E. Niki, *Anal. Biochem.*, 2008, **379**, 105.
40. Y. Yoshida, M. Hayakawa, Y. Habuchi, N. Itoh and E. Niki, *Lipids*, 2007, **42**, 463.
41. Y. Yoshida, N. Itoh and M. Hayakawa, *et al.*, *Nutrition*, 2006, **22**, 303.
42. Y. Yoshida and E. Niki, *BioFactors*, 2006, **27**, 195.
43. Y. Yoshida, N. Itoh, M. Hayakawa, R. Piga, O. Cynshi, K.-I. Jishage and E. Niki, *Toxicol. Appl. Pharmacol.*, 2005, **208**, 87.
44. Y. Yoshida, A. Umeno and M. Shichiri, *J. Clin. Biochem. Nutr.*, 2013, **52**, 9.
45. Y. Yoshida, Y. Saito, M. Hayakawa, Y. Habuchi, Y. Imai, Y. Sawai and E. Niki, *Lipids*, 2007, **42**, 439.
46. Y. Yoshida, M. Hayakawa and E. Niki, *J. Oleo Sci.*, 2008, **57**, 691.

47. Y. Yoshida, M. Hayakawa and N. Itoh, *et al.*, *Biochem. Pharmacol.*, 2007, **74**, 1010.
48. M. Wijtmans, D. A. Pratt, L. Valgimigli, G. A. DiLabio, G. F. Pedulli and N. A. Porter, *Angew. Chem. Int. Ed. Engl.*, 2003, **42**, 4370.
49. H. Y. Kim, D. A. Pratt, J. R. Seal, M. Wijtmans and N. A. Porter, *J. Med. Chem.*, 2005, **48**, 6787.
50. T. G. Nam, C. L. Rector and H. Y. Kim, *J. Am. Chem. Soc.*, 2007, **129**, 10211.
51. J. A. Kenar, C. M. Havrilla, N. A. Porter, J. R. Guyton, S. A. Brown, K. R. Klemp and E. Selinger, *Chem. Res. Toxicol.*, 1996, **9**, 737.
52. G. L. Milne and N. A. Porter, *Lipids*, 2001, **36**, 1265.
53. N. A. Porter, L. S. Lehman, B. A. Weber and K. J. Smith, *J. Am. Chem. Soc.*, 1981, **103**, 6447.
54. N. A. Porter, A. N. Roe and A. T. McPhail, *J. Am. Chem. Soc.*, 1980, **102**, 7574.
55. N. A. Porter, B. A. Weber, H. Weenen and J. A. Khan, *J. Am. Chem. Soc.*, 1980, **102**, 5597.
56. C. M. Havrilla, D. L. Hachey and N. A. Porter, *J. Am. Chem. Soc.*, 2000, **122**, 8042.
57. C. D. Funk, *Science*, 2001, **294**(5548), 1871.
58. N. A. Porter and M. O. Funk, *J. Org. Chem.*, 1975, **40**, 3614.
59. J. D. Morrow, T. M. Harris and L. J. Roberts, 2nd, *Anal. Biochem.*, 1990, **184**, 1.
60. J. D. Morrow, E. Hill, R. F. Burk, T. M. Nammour, K. F. Badr and L. J. Roberts, Jr, *Proc. Natl. Acad. Sci. U. S. A.*, 1990, **87**, 9383.
61. H. Yin, N. A. Porter and J. D. Morrow, *J. Chromatogr. B*, 2005, **827**, 157.
62. H. Yin, L. Gao, H.-H. Tai, L. J. Murphey, N. A. Porter and J. D. Morrow, *J. Biol. Chem.*, 2007, **282**, 329.
63. M. B. Kadiiska, B. C. Gladen and D. D. Baird, *et al.*, *Free Radical Biol. Med.*, 2005, **38**, 698.
64. M. B. Kadiiska, B. C. Gladen and D. D. Baird, *et al.*, *Free Radical Biol. Med.*, 2005, **38**, 711.
65. G. L. Milne, E. S. Musiek and J. D. Morrow, *Biomarkers*, 2005, **10**(Suppl 1), S10.
66. J. L. Watters, J. A. Satia and K. A. da Costa, *et al.*, *Biomarkers*, 2009, **14**, 587.
67. G. L. Milne, H. Yin, J. D. Brooks, S. Sanchez, L. Jackson Roberts, 2nd and J. D. Morrow, *Methods Enzymol.*, 2007, **433**, 113.
68. J. Lu, B. Chen and T. Chen, *et al.*, *Redox Biol.*, 2017, **12**, 899.
69. L. J. Roberts and J. D. Morrow, *Isoprostanes a Markers of Lipid Peroxidation in Atheriosclerosis*, Humana Press, Totowa, NJ, 1999.
70. M. P. Stojiljkovic, H. F. Lopes, D. Zhang, J. D. Morrow, T. L. Goodfriend and B. M. Egan, *J. Hypertens.*, 2002, **20**, 1215.
71. S. Voutilainen, J. D. Morrow, L. J. Roberts, II, G. Alfthan, H. Alho, K. Nyyssonen and J. T. Salonen, *Arterioscler., Thromb. Vasc. Biol.*, 1999, **19**, 1263.

72. T. Daikoku, D. Wang, S. Tranguch, J. D. Morrow, S. Orsulic, R. N. DuBois and S. K. Dey, *Cancer Res.*, 2005, **65**, 3735.
73. D. A. Barocas, S. Motley and M. S. Cookson, *et al.*, *J. Urol.*, 2011, **185**, 2102.
74. T. J. Montine and J. D. Morrow, *Am. J. Pathol.*, 2005, **166**, 1283.
75. T. J. Montine, J. F. Quinn and D. Milatovic, *et al.*, *Ann. Neurol.*, 2002, **52**, 175.
76. T. J. Montine, W. R. Markesberry, J. D. Morrow and L. J. Roberts, Jr, *Ann. Neurol.*, 1998, **44**, 410.
77. T. J. Montine, K. S. Montine, W. McMahan, W. R. Markesbery, J. F. Quinn and J. D. Morrow, *Antioxid. Redox Signaling*, 2005, **7**, 269.
78. J. F. Quinn, K. S. Montine, M. Moore, J. D. Morrow, J. A. Kaye and T. J. Montine, *J. Alzheimer's Dis.*, 2004, **6**, 93.
79. T. J. Montine, M. F. Beal and M. E. Cudkowicz, *et al.*, *Neurology*, 1999, **52**, 562.
80. L. J. Roberts, 2nd and J. D. Morrow, *Cell. Mol. Life Sci.*, 2002, **59**, 808.
81. L. J. Roberts, Jr and J. D. Morrow, *Free Radical Biol. Med.*, 2000, **28**, 505.
82. G. L. Milne, H. Yin and J. D. Morrow, *J. Biol. Chem.*, 2008, **283**, 15533.
83. J. Lu, S. Guo and X. Xue, *et al.*, *J. Biol. Chem.*, 2017, **292**, 5378.
84. E. Niki, Y. Yoshida, Y. Saito and N. Noguchi, *Biochem. Biophys. Res. Commun.*, 2005, **338**, 668.
85. L. J. Roberts, 2nd, J. A. Oates and M. F. Linton, *et al.*, *Free Radical Biol. Med.*, 2007, **43**, 1388.
86. G. Block, C. D. Jensen and J. D. Morrow, *et al.*, *Free Radical Biol. Med.*, 2008, **45**, 377.
87. A. Mastaloudis, J. D. Morrow, D. W. Hopkins, S. Devaraj and M. G. Traber, *Free Radical Biol. Med.*, 2004, **36**, 1329.
88. K. F. Gey, P. Puska, P. Jordan and U. K. Moser, *Am. J. Clin. Nutr.*, 1991, **53**, 326S.
89. L. Schwingshackl, H. Boeing, M. Stelmach-Mardas, M. Gottschald, S. Dietrich, G. Hoffmann and A. Chaimani, *Adv. Nutr.*, 2017, **8**, 27.
90. F. Galli, A. Azzi and M. Birringer, *et al.*, *Free Radical Biol. Med.*, 2017, **102**, 16.
91. K. I. Priyadarsini, S. Kapoor and D. B. Naik, *Chem. Res. Toxicol.*, 2001, **14**(5), 567.
92. C. Suarna, R. T. Dean, J. May and R. Stocker, *Arterioscler., Thromb. Vasc. Biol.*, 1995, **15**, 1616.
93. J. M. Upston, A. C. Terentis, K. Morris, J. F. Keaney Jr and R. Stocker, *Biochem. J.*, 2002, **363**, 753.
94. V. W. Bowry and K. U. Ingold, *Acc. Chem. Res.*, 1999, **32**, 27.
95. V. W. Bowry and R. Stocker, *J. Am. Chem. Soc.*, 1993, **115**, 6029.
96. S. R. Thomas and R. Stocker, *Free Radical Biol. Med.*, 2000, **28**, 1795.
97. M. Alessi, T. Paul, J. C. Scaiano and K. U. Ingold, *J. Am. Chem. Soc.*, 2002, **124**, 6957.
98. C. Z. Han, I. J. Juncadella and J. M. Kinchen, *et al.*, *Nature*, 2016, **539**, 570.

99. K. J. Moore and I. Tabas, *Cell*, 2011, **145**, 341.
100. F. Antunes, R. E. Pinto, L. R. C. Barclay and M. R. Vinqvist, *Int. J. Chem. Kinet.*, 1998, **30**(10), 753–767.
101. L. Xu, T. A. Davis and N. A. Porter, *J. Am. Chem. Soc.*, 2009, **131**(36), 13037–13044.
102. J. A. Howard and K. U. Ingold, *Can. J. Chem.*, 1967, **45**(8), 793–802.

CHAPTER 10

Antioxidant Defense Network and Vitamin E

ETSUO NIKI*

Research Center for Advanced Science and Technology, University of Tokyo, 153-8904, Japan
*E-mail: etsuo-niki@emer.rcast.u-tokyo.ac.jp

10.1 Introduction: Antioxidant Defense Network

Oxygen is a double-edged sword. Foods, beverages, oils, plastics, and rubber are oxidatively damaged and deteriorated in air. The same is true of humans and other living organisms. Aerobic organisms are always under threat from oxidative stress and have had to develop an efficient defense network against oxidative stress during the course of evolution over a period of a billion years. On the one hand, specific reactive oxygen species are generated *in vivo* in a regulated manner as a host defense mechanism or as a cell signaling mechanism, but on the other hand, random and excessive production of reactive oxygen species causes oxidative modification of biological molecules, which may result in various disorders and diseases. Probably, humans are protected from oxidative stress by the most sophisticated antioxidant defense network and, thanks to this, humans can live as long as one hundred years in air at ambient temperature, despite foods consisting of similar molecules being oxidatively deteriorated over several days, even in a refrigerator.[1]

As described in Chapter 8, various oxidants with different reactivities and selectivities are involved in oxidative stress. It is therefore necessary to

Food Chemistry, Function and Analysis No.11
Vitamin E: Chemistry and Nutritional Benefits
Edited by Etsuo Niki

Published by the Royal Society of Chemistry, www.rsc.org

construct multiple defense lines to protect us from detrimental oxidative stress. Various antioxidants with different functions play their respective roles to cope with oxidative damage mediated by different reactive oxidants. An antioxidant is a substance that, when present at low concentrations compared to those of the oxidizable substrate, significantly delays, or inhibits, oxidation of that substrate.[1] Some antioxidants are enzymes and proteins, while other antioxidants are small molecules such as vitamins E and C.[1,2]

In the first defense line, the antioxidants prevent the production of reactive oxidants by reducing hydrogen peroxide and hydroperoxides to water and hydroxides, respectively, or by sequestering metal ions such as iron and copper.[1,3] Catalase reduces hydrogen peroxide to water, while glutathione peroxidases reduce hydrogen peroxide and lipid hydroperoxides to water and lipid alcohols, respectively. Peroxiredoxins and thioredoxins are also important antioxidant enzymes. Superoxide dismutase (SOD) catalyzes dismutation of superoxide to inhibit the production of peroxynitrite by the reaction of superoxide with nitric oxide and also to inhibit the reduction of transition metal ions by superoxide to the lower valency state, an active form of metal ions for the decomposition of lipid and protein hydroperoxides. These are important since the decomposition of peroxides by redox active metal ions is an important reaction to produce reactive oxygen radicals.

In the second defense line, antioxidants scavenge reactive oxidants before the oxidants react with substrates. Vitamins E and C are major physiological, lipophilic and hydrophilic radical scavenging antioxidants, respectively. Polyphenolic compounds found in plants, including flavonoids and phenolic acids, also act as radical scavenging antioxidants. Carotenoids are potent scavengers of singlet oxygen and inhibit photooxidation. In the third defense line, various enzymes function by repairing damage, clearing waste, and reconstituting lost functions.

In addition, the adaptation mechanism functions as the fourth defense line, in which appropriate antioxidants are generated at the right time and transferred to the right position in the right concentration. Furthermore, there is now increasing evidence showing that some antioxidants act as a cellular signaling messenger to regulate the level of antioxidant compounds and enzymes.[4,5]

As described, foods and industrial materials are also oxidatively damaged in air, which can be protected by natural and synthetic antioxidants. However, the antioxidant functions in the third and fourth defense lines mentioned are specific for the aerobic organisms and cannot be supplied to inanimate materials.

10.2 Role of Vitamin E in the Antioxidant Defense Network

Vitamin E acts as an important antioxidant both *in vitro* and *in vivo* by scavenging reactive oxidants, especially lipid peroxyl radicals, which play a key role in lipid peroxidation as chain carrying species.[6–9] The outline of lipid peroxidation and its inhibition by vitamin E *in vivo* is described later. The oxidative damage of foods and organic materials and its inhibition *in vitro* proceeds largely by the same mechanisms as *in vivo*. The detailed mechanisms of lipid peroxidation are described in Chapter 9.

Lipid peroxidation involves three steps: chain initiation, chain propagation, and chain termination, as shown in Figure 10.1. The oxidation of lipids (LH) begins with the formation of lipid radical $L^•$ from the lipids by the attack of reactive species X (reaction 1). Reactive species X may be formed as a byproduct of normal physiological processes, by the decomposition of hydrogen peroxide and hydroperoxides by metal ions, or by other thermal or photochemical reactions. Importantly, the subsequent chain propagation proceeds by the same mechanisms independent of the type of initiating radical X.

There are two chain propagating steps; one is the very fast reaction of the carbon-centered radical derived from the lipid with molecular oxygen to produce lipid peroxyl radial $LOO^•$ (reaction 2) and the other is hydrogen atom abstraction from the lipids by the peroxyl radical to produce lipid hydroperoxide (LOOH) and a new lipid radical (reaction 3), which attacks the lipid and continues chain propagation by the sequence of reactions 2 and 3.

The chain reaction is terminated by bimolecular reactions of lipid peroxyl radicals to give stable non-radical products (reaction 4). Since the rate of reaction 2 is much faster than that of reaction 3, the contribution of the lipid radical in the chain termination is generally very small. In the presence of potent antioxidant IH, the lipid peroxyl radical is scavenged by the antioxidant to produce, in general, lipid hydroperoxide and antioxidant-derived radical $I^•$ (reaction 5), which reacts with another lipid peroxyl radical to produce stable products. The products of the chain termination reaction by vitamin E are described in Chapter 13.

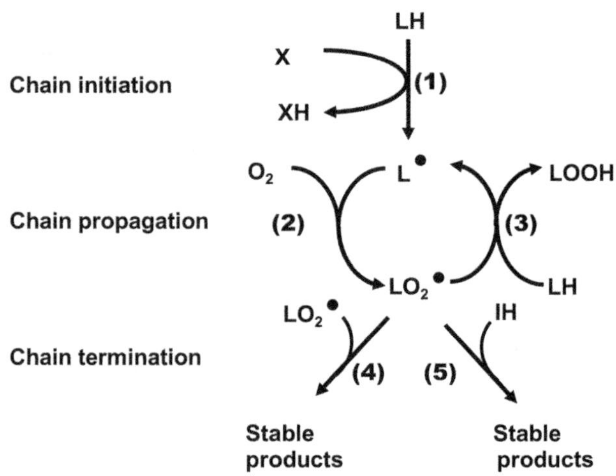

Figure 10.1 Free radical mediated lipid peroxidation by chain mechanisms.

The efficacy of scavenging oxidants is determined by the reactivity and concentration of both the antioxidant and substrate. Vitamin E and other radical scavenging antioxidants trap the lipid peroxyl radical to break the chain propagation and terminate the chain reaction (reaction 5). The efficacy of the antioxidant is determined by the competition between the chain propagation (reaction 3) and chain inhibition (reaction 5) (Figure 10.2).[10] The ratio of the rates of chain propagation to that of chain inhibition by the antioxidant is given by $kp[LOO^{\bullet}][LH]/k_{inh}[LOO^{\bullet}][IH]$, where kp and k_{inh} are the rate constants for chain propagation and chain inhibition, and LH and IH are the lipid and antioxidant, respectively. Since the molar ratio [LH]–[IH] *in vivo* is generally 10^2–10^3 : 1, the ratio k_{inh}/kp should be larger than 10^4 in order for the antioxidant to scavenge more than 90% of the peroxyl radicals efficiently before they attack lipids. It is noteworthy that the efficacy of chain inhibition is dependent on kp as well as k_{inh}.

10.3 Factors that Determine the Antioxidant Efficacy of Vitamin E

The efficacy of antioxidants for inhibition of lipid peroxidation is determined by the following factors:

(1) Chemical reactivity of antioxidant and lipid toward free radicals.
(2) Fate of antioxidant-derived radical.
(3) Interaction with other antioxidants.
(4) Concentration and mobility in the environment.
(5) With regard to the efficacy *in vivo*, absorption, distribution, retention, metabolism, and excretion.

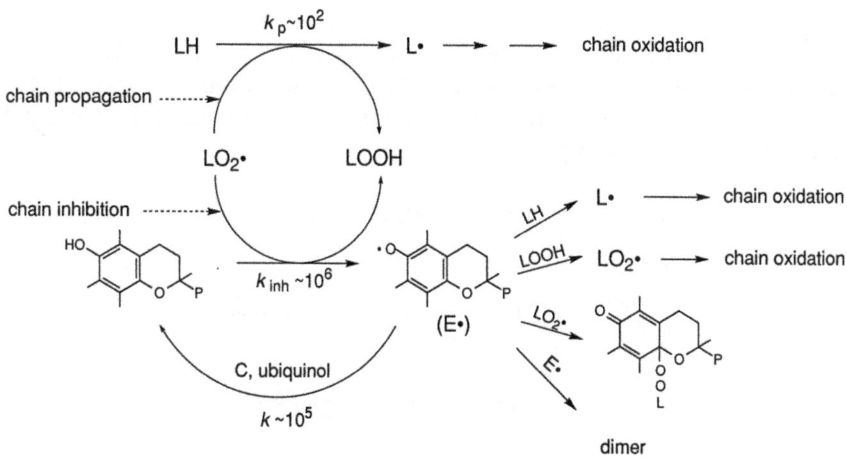

Figure 10.2 Inhibition of lipid peroxidation by vitamin E. LH, lipid; C, vitamin C; P, phytyl side chain. k is the rate constant in $M^{-1}\ s^{-1}$. Adapted from ref. 10 with permission from Portland Press.

10.3.1 Chemical Reactivity toward Oxidants

The antioxidant must have high reactivity toward the peroxyl radical. Structure–reactivity relationships for phenolic antioxidants have been the subject of many studies and have been discussed in detail (Chapter 11 and ref. 7–9). It has been shown that electron-donating substituents on the aromatic ring enhance radical scavenging capacity. It was found, in the oxidation of styrene inhibited by *meta*- and *para*-substituted phenols, that the styrylperoxyl radical scavenging rate constants by 4-methoxy, 4-methyl, 4-chloro-phenol, and unsubstituted phenol were 47.8, 9.17, 2.15, and 2.91 × 10^3 M^{-1} s^{-1} at 65 °C, respectively.[11] It may be noted that some strong electron donor groups may lower the ionization potential to the point that there is a direct transfer to dioxygen to produce superoxide.

Another important factor is the resonance stabilization of the phenoxyl radical produced when the antioxidant scavenges the peroxyl radical. The high reactivity of α-tocopherol is due to the efficient resonance stabilization of the chromanoxyl radical by delocalization of the unpaired electron to the p-type lone pair of the O-atom in the chroman ring, which reduces the O–H bond dissociation energy and hence increases the value of k_{inh}. This makes tocopherols much more reactive toward peroxyl radicals than hindered phenols such as 2,6-di-*tert*-butyl-4-methylphenol, a representative synthetic antioxidant used commercially, k_{inh} being 3.2×10^6 and 1.4×10^4 M^{-1} s^{-1} for α-tocopherol and 2,6-di-*tert*-butyl-4-methylphenol, respectively.[9]

The absolute rate constants (k_{inh}) for vitamin E isoforms and related antioxidants have been measured under different conditions. Many methods for determining the absolute rate constants of radical reactions have been applied including classical methods based on the rotating sector, flash photolysis, pulse radiolysis, and chemiluminescence. Later, the rate constants (k_{inh}) for α-tocopherol and related antioxidants were measured from the rate of regulated inhibited oxidation of hydrocarbons or lipids in the presence of α-tocopherol induced by peroxyl radicals produced from a radical initiator at a constant rate.[12–19] Under such conditions, the oxidation proceeds by the reaction scheme represented in the following equations and the steady state kinetic analysis can be applied.

$$\text{Chain initiation: production of } L^{\bullet} \text{ or } LOO^{\bullet} \tag{10.1}$$

$$\text{Chain propagation: } L^{\bullet} + O_2 \rightarrow LOO^{\bullet} \tag{10.2}$$

$$LOO^{\bullet} + LH \rightarrow LOOH + L^{\bullet} \tag{10.3}$$

$$\text{Chain termination: } LOO^{\bullet} + LOO^{\bullet} \rightarrow \text{non-radical products} \tag{10.4}$$

In the presence of potent radical scavenging antioxidant IH, chain termination occurs by the reactions shown in eqn (10.5) and (10.6).

$$\text{Chain termination: LOO}^{\bullet} + \text{IH} \rightarrow \text{LOOH} + \text{I}^{\bullet} \quad (10.5)$$

$$\text{LOO}^{\bullet} + \text{A}^{\bullet} \rightarrow \text{non-radical products} \quad (10.6)$$

The rates of uninhibited and inhibited oxidation, $V(-\text{IH})$ and $V(+\text{IH})$, are represented by eqn (10.7) and (10.8), respectively:

$$V(-\text{IH}) = k\text{p}[\text{LH}]V\text{i}^{1/2}/(2k_{\text{t}})^{1/2} \quad (10.7)$$

$$V(+\text{IH}) = k\text{p}[\text{LH}]V\text{i}/2k_{\text{inh}}[\text{IH}] \quad (10.8)$$

where kp, kt, and k_{inh} are the rate constants for chain propagation, chain termination (eqn (10.4)), and chain inhibition (eqn (10.5)), and Vi is the rate of chain initiation, respectively. A distinct lag phase, t(lag), is observed under these conditions, which is given by $t(\text{lag}) = n[\text{IH}]/V\text{i}$, where n is a stoichiometric number, that is the number of radical molecules scavenged by each molecule of antioxidant. The n is known to be 2.0 for α-tocopherol and related hindered phenols. The rate of oxidation is monitored by measuring oxygen absorption or conjugated diene formation. Since $V(-\text{IH})$, $V(+\text{IH})$, t(lag), and Vi are obtained experimentally and [LH] and [IH] are known, the ratio of the rate constants k_{inh}–kp can be calculated and, further, the absolute rate constant k_{inh} can be obtained if kp is known.

It may be noteworthy that, under the controlled oxidation of methyl linoleate in homogeneous solution initiated by an azo compound, the oxidation proceeds by a chain mechanism with a long kinetic chain length and the amounts of substrate reacted, oxygen absorbed, conjugated diene and hydroperoxides produced all agree well.[20]

A peroxyl radical clock method has been developed by Porter and his colleagues based on the competition between the unimolecular rearrangement of the peroxyl radical and its bimolecular reaction with a hydrogen atom donor.[8,21] This methodology has been successfully applied to determine the inhibition and propagation rate constants ranging from 10^0 to 10^7 $\text{M}^{-1}\,\text{s}^{-1}$.[21]

It was observed that the corresponding tocopherols and tocotrienols exerted the same reactivities toward peroxyl radicals and the same inhibitory effects against peroxidation of methyl linoleate. The reactivity of tocopherol and tocotrienol homologues toward peroxyl radicals and antioxidant effects have been observed to decrease in the order of $\alpha > \beta = \gamma > \delta$.[12,14,22–24]

Many kinds of reactive oxidants with different reactivities and selectivities are involved in oxidative stress *in vivo*. Among them, hydroxyl, alkoxyl and peroxyl radicals, peroxynitrite, nitrogen dioxide radical, hypochlorite, singlet oxygen, and ozone are important mediators of oxidative damage. In fact, the oxidation products produced specifically by these oxidants have been found in human samples including plasma, urine, and tissues.[25] Interestingly, it has been shown that all of the biomarkers produced specifically by peroxyl radicals, nitrogen dioxide radical, hypochlorite, lipoxygenase, and singlet oxygen are elevated in human atherosclerotic lesions,[25] suggesting that all these oxidants may contribute to the pathogenesis of atherosclerosis in humans and that multiple antioxidants effective against these oxidants are required, since no single antioxidant can effectively inhibit the oxidation mediated by these multiple oxidants.

Importantly, the efficacy of antioxidants depends on the nature of oxidants. The rate constants for the reactions of biologically important oxidants with lipids including oleic acid, linoleic acid, and cholesterol and with several antioxidants including α-tocopherol, ascorbic acid, β-carotene, uric acid, and glutathione are summarized in Table 10.1. It shows that the rate constants vary markedly depending on the nature of the oxidants and also on the target substrates.

With respect to the free-radical-mediated lipid peroxidation, the rate constants for the reaction of the peroxyl radical with oleic acid, linoleic acid, arachidonic acid, eicosapentaenoic acid (EPA), docosahexaenoic acid (DHA), and cholesterol are 0.9, 62, 197, 249, 334, and 11 $M^{-1}\ s^{-1}$ at 37 °C in solution, respectively.[26] The rate constant for scavenging the peroxyl radical by α-tocopherol depends on the reaction milieu. It is 5×10^5–$3 \times 10^6\ M^{-1}\ s^{-1}$ in solution, so that the ratio kp–$k_{inh} > 10^3$, suggesting that α-tocopherol can scavenge the peroxyl radical efficiently in competition with the substrates. However, hydroxyl and alkoxyl radicals react with polyunsaturated fatty acids (PUFA) rapidly with rate constants larger than 10^8 and $10^6\ M^{-1}\ s^{-1}$, respectively, making it difficult for any antioxidant to scavenge these two radicals efficiently before they react with PUFA and other biological molecules. Further, the data

Table 10.1 Rate constants for the reactions of selected oxidants with linoleate and antioxidants (in $M^{-1}\ s^{-1}$).[a]

Oxidant:	Peroxyl radical	ONOO(H)	NO_2 radical	$CO_3^{\bullet-}$	HOCl	$GS^{\bullet}$	$^1O_2$
Substrate							
Linoleate (18:2)	62		2×10^5	$>2 \times 10^5$	18	8×10^6	1.3×10^5
α-Tocopherol	10^5–10^6[c]	1.8×10^5[b]	1×10^5	10^7–10^8	1.3×10^3[b]		3×10^8
Ascorbic acid	2.2×10^6	2.4×10^2	3.5×10^7	1.1×10^9	6×10^6	3.6×10^8	1.9×10^6
β-Carotene	10^5		1.1×10^8		2.3×10^4	2.2×10^8	$>10^{10}$
Urate	$<1.4 \times 10^6$	5×10^2	2×10^7	2.9×10^8	2.0×10^5		3.6×10^8

[a]Modified from refs. 6 and 25. References for k are shown in ref. 25.
[b]Data for Trolox.
[c]See text.

in Table 10.1 show that vitamin E is not a potent scavenger of several oxidants including peroxynitrite, hypochlorite, and lipoxygenase. This suggests that other antioxidants are required to cope with the oxidative stress induced by multiple oxidants having different selectivities. Importantly, this also suggests that human intervention studies using a single antioxidant or even a mixture of limited antioxidants may not give positive results.

10.3.2 Fate of Antioxidant-derived Radicals

When an antioxidant scavenges a radical, the antioxidant is generally converted to a free radical, which may undergo multiple reactions. The fate of this antioxidant-derived radical is important in determining the effect of the antioxidant. When α-tocopherol scavenges a peroxyl radical, it is converted to α-tocopheroxyl radical, which may scavenge another peroxyl radical, react with another vitamin E radical, attack a lipid or lipid hydroperoxide to produce a lipid or lipid peroxyl radical, or react with a reductant such as ascorbate and ubiquinol to regenerate α-tocopherol (see Figure 10.2).[10]

The production and reaction of tocopheroxyl radicals can be monitored from their characteristic electron paramagnetic resonance (EPR) spectra.[27–30] The characteristic EPR spectra of α-, β-, and γ-tocopheroxyl radicals observed in benzene under a high vacuum are shown in Figure 10.3.[30] The EPR spectrum of α-tocopherol showed seven main lines with an additional hyperfine structure with a g-value of 2.0046.[28] In the presence of oxygen, a broad singlet line spectrum without hyperfine splitting is observed due to the dipole–dipole coupling between oxygen and the tocopheroxyl radical. The α-tocopheroxyl radical is stable due to resonance stabilization and reacts with methyl linoleate slowly with a rate constant of $2.7 \times 10^{-2}\ M^{-1}\ s^{-1}$,[31] but it is readily reduced by ascorbate with a rate constant of $10^5\ M^{-1}\ s^{-1}$.[29,32]

Hydroquinones such as ubiquinol and α-tocopherylhydroquinone scavenge the peroxyl radical rapidly to produce a semiquinone radical (eqn (10.9)). They may also reduce the α-tocopheroxyl radical to regenerate α-tocopherol and produce the semiquinone radical. The fate of the semiquinone radical derived from them determines the antioxidant effect. The semiquinone radical may scavenge another radical by hydrogen atom donation to produce quinone or react with oxygen to give quinone and a hydroperoxyl radical (eqn (10.10)), which may attack substrates and induce oxidation. Thus, although ubiquinol and α-tocopherylhydroquinone scavenge the peroxyl radical faster than α-tocopherol, the apparent antioxidant efficacy of ubiquinol and α-tocopherylhydroquinone against lipid peroxidation may become less than that of α-tocopherol.[33]

$$X^{\bullet} + QH_2 \rightarrow XH + QH^{\bullet} \tag{10.9}$$

$$QH^{\bullet} + O_2 \rightarrow Q + HO_2^{\bullet} \tag{10.10}$$

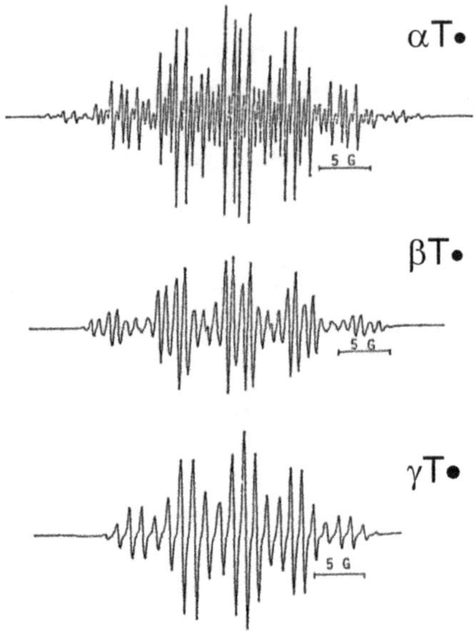

Figure 10.3 EPR spectra of α-, β-, and γ-tocopheroxyl radicals in benzene under vacuum. Adapted from ref. 30 with permission from The Chemical Society of Japan.

Carotenoids also act as potent scavengers of free radicals and singlet oxygen. Carotenoids scavenge radicals predominantly by addition reaction to the polyene structure to produce a resonance stabilized radical.[34] This carbon-centered radical reacts with oxygen rapidly to give a peroxyl radical that may attack substrates and continue chain oxidation. Therefore, the antioxidant efficacy of carotenoids is diminished with increasing oxygen concentration; the lower the oxygen concentration the higher the antioxidant effect.[34] On the other hand, carotenoids scavenge singlet oxygen primarily by physical quenching without a chemical reaction to inhibit photooxidation.[35]

Thiols (RSH) may scavenge a reactive oxygen radical to produce a thiyl radial ($RS^•$), which may undergo several secondary reactions.[36] Thiyl radicals react with PUFA with rate constants as large as 10^7 M^{-1} s^{-1} by hydrogen abstraction or addition to a double bond.[37] The fate of the thiyl radical determines the efficacy of thiol compounds as radical scavenging antioxidants.

10.3.3 Localization of Antioxidant and Oxidant

The prominent features of biological systems are heterogeneity and compartmentalization, which make the dynamics of antioxidant action *in vivo* much more complex than in a homogeneous solution. Hydrophilic antioxidants are present in the aqueous phase, while lipophilic antioxidants such as vitamin E are localized in lipophilic domains in membranes and lipoproteins. Some lipophilic antioxidants are localized at the surface of cellular membranes and lipoproteins, while others are within the membranes or lipoprotein core. Likewise, some oxidants are localized mainly in the aqueous phase, while other oxidants are localized in the lipophilic environment. The oxidants in the aqueous phase are scavenged primarily by aqueous antioxidants such as vitamin C, uric acid and protein thiols, whereas the oxidants present within the membranes or lipoproteins are scavenged by lipophilic antioxidants. Although uric acid does not reduce the α-tocopheroxyl radical, it can spare vitamin E during the oxidation of membranes or lipoproteins induced by aqueous radicals.[38] This suggests that the sparing of vitamin E by some antioxidant does not always mean they interact with each other or exert a synergistic antioxidant action.

The phenolic O–H bond of vitamin E is assumed to be located at the membrane surface (see Chapter 3), which makes it possible to scavenge aqueous radicals attacking membrane lipids from the outside. On the other hand, carotenes are localized within the interior of membranes. β-Carotene is chemically less reactive toward the peroxyl radical than α-tocopherol, but β-carotene scavenges peroxyl radicals at the interior of the membrane faster than α-tocopherol, while α-tocopherol scavenges aqueous radicals faster than β-carotene.[39]

It has been assumed that vitamin E is localized in the non-raft domain of the membranes together with phosphatidylcholines containing polyunsaturated fatty acids, which is convenient for vitamin E to inhibit lipid peroxidation (see Chapter 3). Thus, the efficacy of scavenging oxidants by antioxidants is dependent on their localization.

10.3.4 Interaction Between Antioxidants

Antioxidants function *in vivo* not only independently but also cooperatively or sometimes synergistically with other antioxidants. For example, vitamin E scavenges lipid peroxyl radicals to inhibit production of lipid hydroperoxides, while selenium-containing glutathione peroxidases reduce lipid hydroperoxides to the stable hydroxides. Thus, the deficiency of both vitamin E and selenium gives rise to serious oxidative damage, while the deficiency of either of them may be complemented by the other. Depletion of glutathione peroxidase 4 (phospholipid hydroperoxide glutathione peroxidase, PHGPx) induces lipid peroxidation-dependent cell death termed ferroptosis,[40] which is rescued, at least in part, by vitamin E.[41]

It has been shown that the combination of vitamins E and C exerts synergistic antioxidant effects, greater than simply an additive effect.[4,6,42] Vitamin E is present in the lipophilic domain of membranes and lipoproteins, while vitamin C is present in the aqueous phase. Vitamin E scavenges free radicals localized within the membrane and lipoproteins and also the radicals attacking from the aqueous phase. Vitamin E is oxidized to produce a vitamin E radical. Vitamin C scavenges reactive radicals in the aqueous phase and also reduces the vitamin E radical to regenerate vitamin E at the water–membrane interface.[43,44] Vitamin C is capable of reducing the vitamin E radical in the membrane, low density lipoprotein, plasma, and whole blood, although the efficacy of reducing the radical decreases as the radical goes deeper into the interior of the membrane or lipoprotein.[45,46] Thus, in the oxidation of LDL, plasma and cells, it is observed that only vitamin C is consumed initially, while vitamin E is spared and begins to be consumed after most of the vitamin C is depleted. Importantly, little lipid peroxidation proceeds during the presence of vitamin C.

It was confirmed that the α-tocopheroxyl radical was reduced to regenerate α-tocopherol by phenolic compounds as well as ascorbate. A pulse radiolysis study showed that the phenoxyl radical derived from Trolox, a water-soluble vitamin E analogue, was readily reduced by ascorbate ($k = 8.3 \times 10^6\ M^{-1}\ s^{-1}$) and certain thiols ($k < 10^5\ M^{-1}\ s^{-1}$) but not by urate, NADH or propyl gallate.[47] Ubiquinol and α-tocopherylhydroquinone reduce the α-tocopheroxyl radical faster than ascorbate.[33] Flavonoids, a group of natural phenolic compounds found in fruits, vegetables, tea and wine, are also capable of reducing the α-tocopheroxyl radical. It was reported that the rate constants for the reduction of the 2,6-di-*tert*-butyl-4-phenoxyl radical by flavonoids increased in the order of epicatechin < epicatechin gallate < epigallocatechin < epigallocatechin gallate (EGCG).[48] It may be noteworthy that a significant steric effect is observed in the reduction of the phenoxyl radical[31] and that the relative importance of the reductant depends also on the localization of the phenoxyl radical and the reductant.[4]

Uric acid also acts as a major hydrophilic radical scavenging antioxidant.[49–51] Uric acid scavenges radicals in the aqueous phase and may spare vitamin E, despite the fact that uric acid does not reduce the vitamin E radical. Uric acid and vitamin E function independently.

It was found that the regeneration of α-tocopherol by much less reactive phenolic co-antioxidants such as 2,6-di-*tert*-butyl-4-methoxyphenol ($k_{inh} = 1.1 \times 10^5\ M^{-1}\ s^{-1}$) was possible in a homogeneous hydrocarbon solution, which makes the oxidation kinetics indistinguishable from those of an oxidation inhibited by highly reactive α-tocopherol alone but with stoichiometries reflected by the total amount of α-tocopherol and phenolic co-antioxidants.[52]

Under certain circumstances, vitamin E acts as a prooxidant, since the vitamin E radical may attack polyunsaturated fatty acids, although slowly, to initiate lipid oxidation. It may be noted that the bond dissociation energies of the O–H bond of α-tocopherol and the bis-allylic C–H bond are 77.1 and 75 kcal mol^{-1},[7] suggesting that the hydrogen atom abstraction reaction from polyunsaturated fatty acids by α-tocopherol is exothermic. Stocker and his colleagues showed that this reaction is particularly important in the oxidation of LDL and that the reduction of the α-tocopheroxyl radical by ascorbate and other reductants inhibits this tocopherol-mediated peroxidation (TMP) and at the same time spares vitamin E.[53,54]

10.3.5 Concentration and Mobility in the Environment

As mentioned, the oxidation and anti-oxidation *in vivo* proceed in complex heterogeneous systems. Obviously, distribution and local concentrations of antioxidants are important factors that determine the antioxidant effect. The distribution of vitamin E in the raft and non-raft domains of the membranes is also an important factor.

In contrast to homogeneous solutions, the mobility of vitamin E is restricted in the membranes and lipoproteins and the apparent rate of scavenging radicals is smaller than that in solution. In fact, the reported rate constants for scavenging the peroxyl radical by α-tocopherol vary from $3.2–3.8 \times 10^6$ in chlorobenzene,[12,17,21,22] 6.6×10^5 in acetonitrile,[17] 5.1×10^5 in *tert*-butyl alcohol/methanol,[13] 5.9×10^5 in LDL,[55] 3.7×10^4 in micelles,[23] to $2.8–5.8 \times 10^3$ M^{-1} s^{-1} in PC liposomal membranes.[56,57]

The ratio k_{inh}–kp, that is, competition between chain breaking and chain propagation, is an important factor that determines the antioxidant efficacy. Both k_{inh} and kp are dependent on the reaction milieu. It was found that the ratio k_{inh}–kp in the oxidation of methyl linoleate in *tert*-butyl alcohol solution, 5.2×10^3, was about two orders larger than that (58) in the oxidation of soybean phosphatidylcholine liposomal membranes composed primarily of the same linoleate as the oxidizable substrate.[58] The dynamics of lipid peroxidation and its inhibition by antioxidants in homogeneous solutions have been studied in detail and are understood fairly well,[7–9,59] while those in the membranes and lipoproteins have to be addressed in future studies.

As described, the reactivity of tocopherols decreases in the order $\alpha > \beta = \gamma > \delta$ and α-tocopherol reduces β, γ, and δ-tocopheroxyl radicals.[60] It was found that, when liposomal membranes containing both α-tocopherol and other tocopherol isoforms were oxidized, only α-tocopherol was consumed initially and other forms began to be consumed after α-tocopherol was depleted, whereas α-tocopherol and other tocopherols were consumed at the same rate when they were incorporated into different liposomal membranes, suggesting the mobility of tocopherols between the membranes is slow.[58]

The phytyl side chain of α-tocopherol is necessary for incorporation and retainment in the membrane and lipoprotein,[61] but it reduces the mobility of α-tocopherol within and between the membranes.[62,63] The length of the side chain at the 2-position of 2,5,7,8-tetramethyl-6-chromanol does not affect the chemical reactivity toward free radicals. Further, identical EPR signals were observed, independent of the side chain length,[45] but the apparent antioxidant effect decreases with increasing side chain length. The inhibitory effect against erythrocyte hemolysis induced by peroxyl radicals decreased with increasing length of the side chain length of 2-alkyl-2,5,7,8-tetramethyl-6-chromanol incorporated into liposomal membranes.[63] Interestingly, distinct effects were observed between α-tocopherol and 2,2,5,7,8-pentamethyl-6-chromanol (PMC) on the disappearance of *N*-oxyl-4,4′-dimethyloxazolidine derivatives of stearic acid (NS) incorporated into LDL. As shown in Figure 10.4, 5-NS or 16-NS was consumed during the oxidation induced by a water-soluble azo initiator. PMC spared both 5-NS and 16-NS efficiently, whereas α-tocopherol spared only 5-NS, but not 16-NS, implying that PMC scavenges peroxyl radicals efficiently but

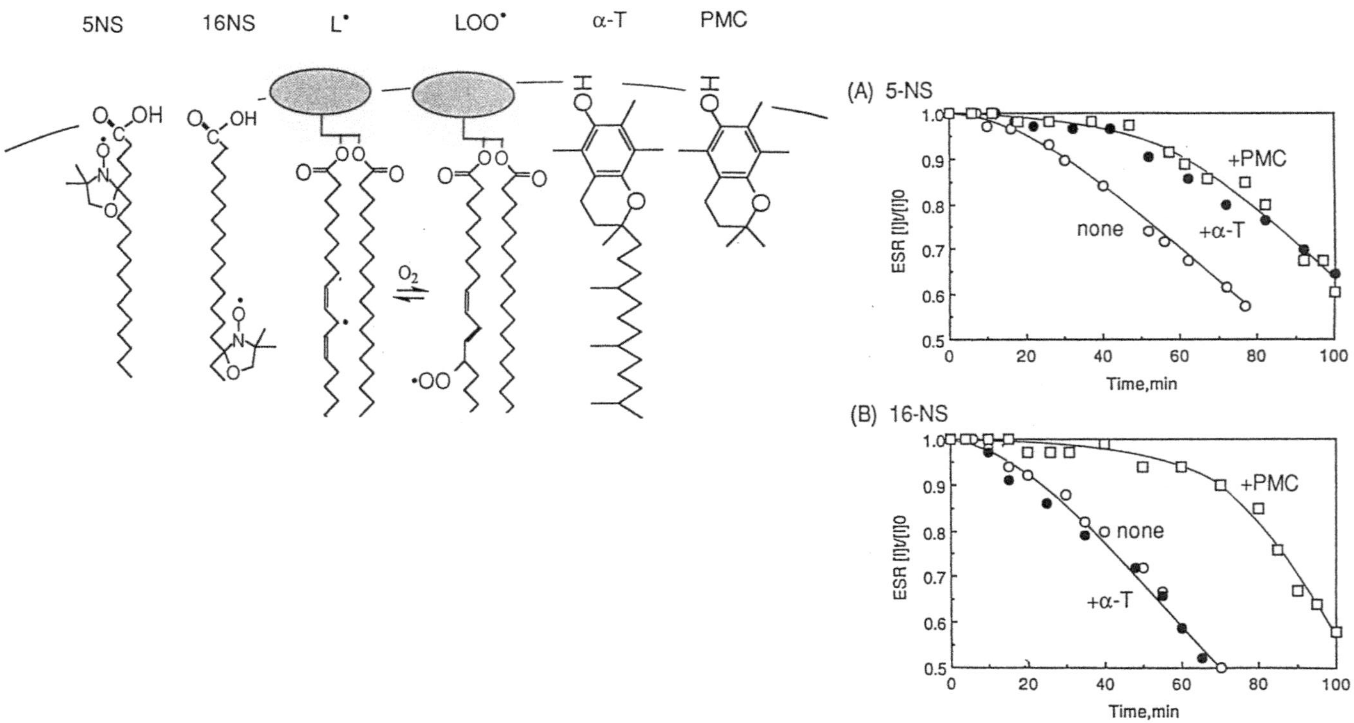

Figure 10.4 Effects of α-tocopherol and PMC on the consumption of (A) 5-NS and (B) 16-NS during the oxidation of LDL (1.69 mg protein per mL) AAPH. Adapted from ref. 46 with permission from Taylor & Francis, Copyright 1995.

that the efficacy of scavenging peroxyl radicals by α-tocopherol decreases as the peroxyl radicals go deeper into the interior of LDL particles.[46] Although PMC acts as a potent antioxidant against oxidation in the membranes, LDL suspensions, and cultured cell systems, it is a poor antioxidant *in vivo* because it is not retained in the membranes and lipoproteins to maintain a sufficient physiological concentration.

The nitroxide group of NS is reduced rapidly by ascorbate in homogeneous solutions, but the rate of the reaction decreases as the nitroxide group goes deeper into the interior of the LDL. Furthermore, the nitroxide group (16-NS) attached to cholesteryl stearate was not reduced by ascorbate in LDL, but it was reduced by 6-*O*-palmitoyl ascorbic acid.[46]

Ubiquinol, a reduced form of coenzyme Q, is also a potent peroxyl radical scavenger with a longer side chain than tocopherols.[64] It was observed that ubiquinols with short isoprenoid chains exerted a more potent inhibition of lipid peroxidation than the longer chain homologues.[65–67]

These data show that, although the side chain of vitamin E and ubiquinol is necessary for incorporation and retainment in the membranes and lipoproteins, it reduces the mobility and apparent antioxidant efficacy in the membranes and lipoproteins.

10.3.6 Absorption, Distribution, Retention, Metabolism, and Excretion

One important factor for humans and other organisms is bioavailability, which is determined by absorption, distribution, retention, metabolism, and excretion. The physiological concentrations of vitamin E and other antioxidants are tightly controlled, and, for example, the plasma concentration of vitamin E does not increase in proportion to the increasing intake.[68] Polyphenolic compounds are metabolized *in vivo* to glucuronidate, sulfate, and *O*-methyl derivatives, which are not active as free radical scavenging antioxidants. Another important issue is the safety of or potential harm from antioxidant supplementation. The details of these issues are described in other chapters.

References

1. B. Halliwell and J. M. C. Gutteridge, *Free Radicals in Biology & Medicine*, Oxford University Press, 5th edn, 2015.
2. E. Niki, N. Noguchi, H. Tsuchihashi and N. Gotoh, *Am. J. Clin. Nutr.*, 1995, **62**, 1322S.
3. S. Toppo, L. Flohé, F. Ursini, S. Vanin and M. Maiorino, *Biochim. Biophys. Acta*, 2009, **1790**, 1486.
4. E. Niki, *Free Radical Biol. Med.*, 2010, **49**, 503.
5. H. J. Forman, K. J. Davies and F. Ursini, *Free Radical Biol. Med.*, 2014, **66**, 24.

6. E. Niki, *Free Radical Biol. Med.*, 2014, **66**, 3.
7. M. Lucarini and G. F. Pedulli, *Chem. Soc. Rev.*, 2010, **39**, 2106.
8. D. A. Pratt, K. A. Tallman and N. A. Porter, *Acc. Chem. Res.*, 2011, **44**, 458.
9. K. U. Ingold and D. A. Pratt, *Chem. Rev.*, 2014, **114**, 9022.
10. E. Niki, N. Noguchi and N. Gotoh, *Biochem. Soc. Trans.*, 1993, **21**, 313.
11. J. A. Howard and K. U. Ingold, *Can. J. Chem.*, 1963, **41**, 1744.
12. G. W. Burton and K. U. Ingold, *J. Am. Chem. Soc.*, 1981, **103**, 6472.
13. E. Niki, T. Saito, A. Kawakami and Y. Kamiya, *J. Biol. Chem.*, 1984, **259**, 4177.
14. W. A. Pryor, J. A. Cornicelli, L. J. Devall, B. Tait, B. K. Trivedi, B. T. Witiak and M. Wu, *J. Org. Chem.*, 1993, **58**, 3521.
15. L. Valgimigli, D. Bartolomei and R. Amorati, *et al.*, *Beilstein J. Org. Chem.*, 2013, **9**, 2781.
16. B. Li and D. A. Pratt, *Free Radical Biol. Med.*, 2015, **82**, 187.
17. E. A. Haidasz, A. T. Van Kessel and D. A. Pratt, *J. Org. Chem.*, 2016, **81**, 737.
18. E. Niki, *Food Funct.*, 2016, **7**, 2156.
19. R. Amorati and L. Valgimigli, *J. Agric. Food Chem.*, 2018, **66**, 3324.
20. Y. Yamamoto, E. Niki and Y. Kaimya, *Lipids*, 1982, **17**, 870.
21. B. Roschek Jr, K. A. Tallman, C. L. Rector, J. G. Gillmore, D. A. Pratt, C. Punta and N. A. Porter, *J. Org. Chem.*, 2006, **271**, 3527.
22. G. W. Burton, T. Doba, E. Gabe, L. Hughes, F. L. Lee, L. Prasad and K. U. Ingold, *J. Am. Chem. Soc.*, 1985, **107**, 7053.
23. W. A. Pryor, T. Strickland and D. F. Church, *J. Am. Chem. Soc.*, 1988, **110**, 2224.
24. Y. Yoshida, E. Niki and N. Noguchi, *Chem. Phys. Lipids*, 2003, **123**, 63.
25. E. Niki, *Free Radical Biol. Med.*, 2018, **120**, 425.
26. L. Xu, T. A. Davis and N. A. Porter, *J. Am. Chem. Soc.*, 2009, **131**, 13037.
27. W. Boguth and H. Niemann, *Biochim. Biophys. Acta*, 1971, **248**, 121.
28. T. Ozawa, A. Hanaki, S. Matsumoto and M. Matsuo, *Biochim. Biophys. Acta*, 1978, **531**, 72.
29. E. Niki, J. Tsuchiya, R. Tanimura and Y. Kamiya, *Chem. Lett.*, 1982, **11**, 789.
30. J. Tsuchiya, E. Niki and Y. Kamiya, *Bull. Chem. Soc. Jpn.*, 1983, **56**, 229.
31. A. Watanabe, N. Noguchi, A. Fujisawa, T. Kodama, K. Tamura, O. Cynshi and E. Niki, *J. Am. Chem. Soc.*, 2000, **122**, 5438.
32. J. E. Packer, T. F. Slater and R. L. Willson, *Nature*, 1979, **278**, 737.
33. H. Shi, N. Noguchi and E. Niki, *Free Radical Biol. Med.*, 1999, **27**, 334.
34. G. W. Burton and K. U. Ingold, *Science*, 1984, **224**, 569.
35. W. Stahl and H. Sies, *Mol. Aspects Med.*, 2003, **24**, 345.
36. C. C. Winterbourn and M. B. Hampton, *Free Radical Biol. Med.*, 2008, **45**, 549.
37. C. Schöneich, U. Dillinger, F. von Bruchhausen and K. D. Asmus, *Arch. Biochem. Biophys.*, 1992, **292**, 456.

38. K. Sato, E. Niki and H. Shimasaki, *Arch. Biochem. Biophys.*, 1990, **279**, 402.
39. H. Tsuchihashi, M. Kigoshi, M. Iwatsuki and E. Niki, *Arch. Biochem. Biophys.*, 1995, **323**, 137.
40. M. Maiorino, M. Conrad and F. Ursini, *Antioxid. Redox Signaling*, 2018, **29**, 61.
41. H. Imai, M. Matsuoka, T. Kumagai, T. Sakamoto and T. Koumura, *Curr. Top. Microbiol. Immunol.*, 2017, **403**, 143.
42. C. Golumbic and H. A. Mattill, *J. Am. Chem. Soc.*, 1941, **63**, 1279.
43. T. Doba, G. W. Burton and K. U. Ingold, *Biochim. Biophys. Acta*, 1985, **835**, 298.
44. E. Niki, A. Kawakami, Y. Yamamoto and Y. Kamiya, *Bull. Chem. Soc. Jpn.*, 1985, **58**, 1971.
45. M. Takahashi, J. Tsuchiya, E. Niki and S. Urano, *J. Nutr. Sci. Vitaminol.*, 1988, **34**, 25.
46. N. Gotoh, N. Noguchi, J. Tsuchiya, K. Morita, H. Sakai, H. Shimasaki and E. Niki, *Free Radical Res.*, 1996, **24**, 123.
47. M. J. Davies, L. G. Forni and R. L. Willson, *Biochem. J.*, 1988, **255**, 513.
48. K. Mukai, S. Mitani, K. Ohara and S. Nagaoka, *Free Radical Biol. Med.*, 2005, **38**, 1243.
49. B. N. Ames, R. Cathcart, E. Schwiers and P. Hochstein, *Proc. Natl. Acad. Sci. U. S. A.*, 1981, **78**, 6858.
50. E. Niki, M. Saito, Y. Yoshikawa, Y. Yamamoto and Y. Kamiya, *Bull. Chem. Soc. Jpn.*, 1986, **59**, 471.
51. M. G. Simic and S. V. Jovanovic, *J. Am. Chem. Soc.*, 1989, **111**, 5778.
52. R. Amorati, F. Ferroni, M. Lucarini, G. F. Pedulli and L. Valgimigli, *J. Org. Chem.*, 2002, **67**, 9295.
53. V. W. Bowry, K. U. Ingold and R. Stocker, *Biochem. J.*, 1992, **288**, 34.
54. P. K. Witting, C. Westerlund and R. Stocker, *J. Lipid Res.*, 1996, **37**, 853.
55. S. M. Culbertson, F. Antunes, C. M. Havrilla, G. L. Milne and N. A. Porter, *Chem. Res. Toxicol.*, 2002, **15**, 870.
56. L. R. C. Barclay, K. A. Baskin, K. A. Dakin, S. J. Locke and M. R. Vinqvist, *Can. J. Chem.*, 1990, **68**, 2258.
57. L. R. Barclay, F. Antunes, Y. Egawa, K. L. McAllister, K. Mukai, T. Nishi and M. R. Vinqvist, *Biochim. Biophys. Acta*, 1997, **1328**, 1.
58. E. Niki, M. Takahashi and E. Komuro, *Chem. Lett.*, 1986, **15**, 1573.
59. N. A. Porter, *J. Org. Chem.*, 2013, **78**, 3511.
60. E. Niki, J. Tsuchiya, Y. Yoshikawa, Y. Yamamoto and Y. Kamiya, *Bull. Chem. Soc. Jpn.*, 1986, **59**, 497.
61. B. Maggio, A. T. Diplock and J. A. Lucy, *Biochem. J.*, 1977, **161**, 111.
62. E. Niki, A. Kawakami, M. Saito, Y. Yamamoto, J. Tsuchiya and Y. Kamiya, *J. Biol. Chem.*, 1985, **260**, 2191.
63. E. Niki, E. Komuro, M. Takahashi, S. Urano, E. Ito and K. Terao, *J. Biol. Chem.*, 1988, **263**, 19809.
64. B. Frei, M. C. Kim and B. N. Ames, *Proc. Natl. Acad. Sci. U. S. A.*, 1990, **87**, 4879.

65. V. E. Kagan, E. A. Serbinova, R. A. Bakalova, T. S. Stoytchev, A. N. Erin, L. L. Prilipko and R. P. Evstigneeva, *Biochem. Pharmacol.*, 1990, **40**, 2403.
66. V. E. Kagan, E. A. Serbinova and G. M. Koynova, *et al.*, *Free Radical Biol. Med.*, 1990, **9**, 117.
67. E. Niki, *Mol. Aspects Med.*, 1997, **18**, S63.
68. M. G. Traber, *J. Lipid Res.*, 2013, **54**, 2295.

CHAPTER 11

Vitamin E Inspired Synthetic Antioxidants

LUCA VALGIMIGLI* AND RICCARDO AMORATI

Department of Chemistry "G. Ciamician", University of Bologna, Via S. Giacomo 11, 40126 Bologna, Italy
*E-mail: luca.valgimigli@unibo.it

11.1 Introduction

The identification of the correct relative reactivity of different tocopherols (Scheme 11.1 and Table 11.1) with alkylperoxyl radicals (ROO•),[1] highlighting the parallelism between vitamin E activity *in vivo* and antioxidant activity *in vitro*, allowed an understanding of the structure–activity relationship governing the properties of the most active α-tocopherol,[2] and opened the way to the rational design of synthetic analogues. Indeed, the search for vitamin-E-inspired synthetic antioxidants - which is currently ongoing[3] - has stolen the scene for over three decades[4] and could benefit from the previous three decades of research in synthetic antioxidants.[5] This study has served, and continues to serve, different scopes, namely: (i) to fully rationalize the structure–activity relationships for both the antioxidant and biological properties of vitamin E; (ii) to afford easily available and economically viable antioxidants that would perform as well as Nature's premiere lipid-soluble antioxidant for the protection of manufactured goods; and (iii) to better

Food Chemistry, Function and Analysis No.11
Vitamin E: Chemistry and Nutritional Benefits
Edited by Etsuo Niki

Published by the Royal Society of Chemistry, www.rsc.org

vitamin E by developing more active compounds for pharmaceutical or other high-end applications. This last goal, which is arguably the most challenging, will be more specifically addressed.

11.2 Influence of Simple Structural Modifications on the Antioxidant Activity of Vitamin E

Electing α-tocopherol as the "lead structure", the main molecular regions that were considered for structural modification are identified in Scheme 11.1. The transfer of the phenolic hydrogen atom (HAT) to chain-carrying peroxyl radicals (eqn (11.1)) is the key reaction involved in the antioxidant

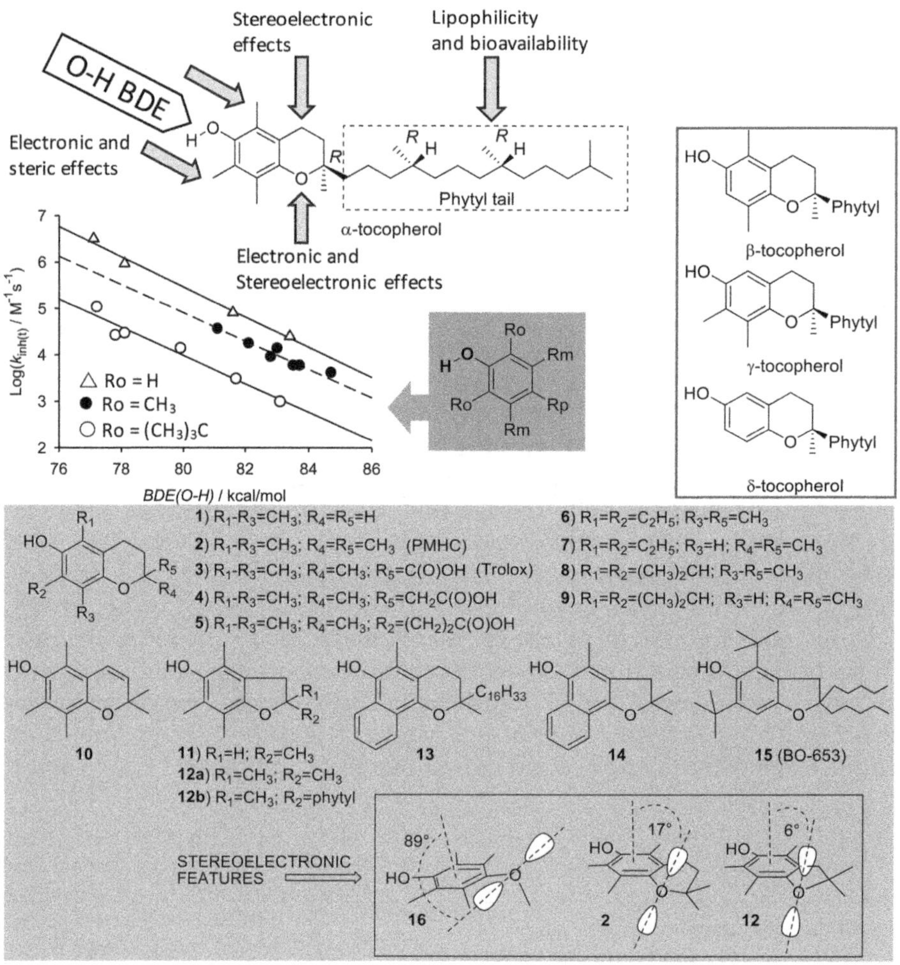

Scheme 11.1 Structure and modification of natural tocopherols: simpler phenolic analogues.

Table 11.1 Reactivity toward alkylperoxyl radicals (k_{inh}, M^{-1} s^{-1}) and phenolic O–H bond dissociation enthalpies (BDE_{OH}, kcal mol^{-1}) of selected chalcogen-containing tocopherol derivatives.

Phenol	$k_{inh}/M^{-1}s^{-1}$ [a]	n	BDE_{OH}/kcal mol^{-1} [b]	Reference
α-Tocopherol	3.2×10^6	2.0	77.1	1,4
β-Tocopherol	1.3×10^6	2.1	—	1,4
γ-Tocopherol	1.4×10^6	1.8	—	1,4
δ-Tocopherol	4.4×10^5	2.0	—	1,4
1	2.7×10^6	—	—	4
2	3.8×10^6	1.9	—	1,4
3	1.1×10^6	—	—	4
4	1.9×10^6	—	—	4
5	3.7×10^6	—	—	4
6	2.1×10^6	2.1	—	10
7	1.6×10^6	2.0	—	10
8	8.8×10^5	1.6	—	10
9	1.1×10^6	1.9	—	10
10	2.5×10^6	—	—	4
11	5.4×10^6	—	—	4
12a	5.7×10^6	—	—	4
13	1.1×10^7	1.5	—	10
14	2.9×10^7	1.6	—	10
15	1.1×10^6 [c]	—	76 [c]	13
17	1.4×10^6	1.8	—	15
18	1.3×10^6	1.9	78.8	18
19	7.6×10^5	1.9	80.6	18
20	5.8×10^5	1.9	80.7	18
21	3.0×10^5	1.9	81.8	18
23	1.3×10^6	1.8	78.9	20
25	3.4×10^6	1.8	77.2	21
26	1.5×10^6	1.9		22
28	5.0×10^6	1.9	—	23
30	7.4×10^6	1.6	—	23
31	1.2×10^6	1.9	78.1	24
32	3.8×10^5	2.0	81.6	25
33	1.5×10^6	1.7	77.6	25
35	1.0×10^7	0.5	—	27
37	7.5×10^5	0.3	—	27
42	2.0×10^6	—	—	4
44a	8.8×10^7	—	76.3	33
45	2.8×10^8	—	75.2	33
46	5.2×10^7 [d]	—	—	42

[a]30 °C, solvent chlorobenzene.
[b]Solvent benzene. All BDEs measured before 2005 have been downscaled by 1.1 kcal mol^{-1} to account for the revised value of phenol (ref. 35).
[c]Based on the relative reactivity with α-TOH.
[d]Value for the analogue with the phytyl chain replaced by a methyl group.

activity of α-tocopherol; therefore, rate constant k_{inh} is the key parameter to meter the performance. The second relevant parameter is the "stoichiometric factor" n, *i.e.* the number of peroxyl radicals trapped by one molecule of antioxidant, which is $n = 2$ for all tocopherols (in homogeneous solution and in the absence of co-antioxidants), as a second peroxyl radical is trapped by the tocopheroxyl radical at a diffusion-controlled rate (eqn (11.2)).[2,6]

$$\alpha\text{-TOH} + \text{ROO}^{\bullet} \xrightarrow{k_{inh}} \alpha\text{-TO}^{\bullet} + \text{ROOH} \tag{11.1}$$

$$\alpha\text{-TO}^{\bullet} + \text{ROO}^{\bullet} \xrightarrow{diff.} \text{non-radical adduct} \tag{11.2}$$

Modification of substituents in the aromatic ring of chromane affords results identical to those observed with simpler phenolic antioxidants.[6] Substituents influence the reactivity with peroxyl radicals mainly according to their electronic properties: electron donating groups (EDGs) decrease the bond dissociation enthalpy (BDE_{OH}) of the phenolic O–H, making the HAT to peroxyl radicals faster, while electron withdrawing groups (EWG) have the opposite effect.[7]

The effect is more marked when the substituent is in the *ortho* (or *para*) position, conjugated with the –OH group, while it is less important in the *meta* position. For instance, the BDE_{OH} of a phenol bearing two methyl substituents in *ortho* is lowered by 3.5 kcal mol^{-1} as compared to bare phenol, while two methyls in *meta* only lower the BDE_{OH} by 1.0 kcal mol^{-1}.[6,7] There exists a linear free energy correlation between the BDE_{OH} and Brown and Okamoto's σ^+ substituent values.[6,8,9] However, for EDGs of the type –XR with X bearing at least a lone pair (*e.g.* –OR, –SR, *etc.*) in the *ortho* position, the occurrence of intramolecular H-bonding to the –OH group is to be considered.[9] This stabilizes the phenol with respect to the phenoxyl radical, increasing the BDE_{OH}, thereby partly compensating the lowering effect due to the ED character.[9] The effect of the same substituents in the *para* position will instead be influenced by stereoelectronic features, as discussed in the following section. Although less prominent than electronic effects, substituents exert also steric effects, particularly in the *ortho* position. Scheme 11.1 shows the linear free energy relationship between BDE_{OH} and $\text{Log}\,k_{inh}$ for several substituted phenols including α-TOH. Interestingly, data points fall in three parallel lines depending on the size of the substituents in the *ortho* position: phenols having the same BDE_{OH} react more slowly when they have in *ortho* two *tert*-butyls compared to those with two methyls, which are faster than those with no substituents. A combination of the two effects – electronic and steric – fully explains the relative reactivity of α-, β-, δ- and γ-tocopherol (Table 11.1).[6] It also explains the reactivity of synthetic derivatives **6–9** compared to **2**.[10]

Extension of the conjugated system stabilizes the phenoxyl radical similar to EDGs in the ring. Therefore, naphthol **13** was over three-fold more reactive than α-TOH.[10] Truncation of the phytyl tail to afford 2,2,5,7,8-pentamethylchromanol (PMHC, **2**) does not influence significantly the reactivity with peroxyl radicals.[4] Replacement with a carboxylic group as in Trolox (**3**) or with aliphatic carboxylic acids (**4**, **5**), affords water-soluble variants of vitamin E.

In an aprotic solution, these molecules have antioxidant activity similar to that of α-tocopherol;[4] when the EW carboxylic group is directly attached to the chromanol (*e.g.* Trolox) or is separated only by a $-CH_2-$, there is a minor negative influence on the reactivity due to polar effects; however, these functions cause enhanced reactivity in water solution at pH > pK_a upon conversion of $-COOH$ into the mild ED $-COO^-$.[11]

11.2.1 Manipulation of Stereoelectronic Effects to Alter the Reactivity of Tocopherol

Comparison of α-TOH and 4-methoxy-2,3,5,6-tetramethylphenol (**16**) – so to have the same set of substituents in the aromatic ring – in terms of BDE_{OH} (77.1 kcal mol^{-1} *vs.* 80.8 kcal mol^{-1}) and reactivity toward peroxyl radicals (k_{inh} at 303 K: 3.2×10^6 M^{-1} s^{-1} *vs.* 3.9×10^5 M^{-1} s^{-1}) illustrates the importance of stereoelectronic effects.[4,11] In order to fully express its ED contribution by conjugation, one lone pair of the RO– substituent should align parallel to the axis of the aromatic π-system. In **16**, this is prevented by the steric hindrance of the methyls in the 3- and 5-position, and both EPR and the crystal structure indicate that the dihedral angle is 89°. On the other hand, in α-TOH or PMHC (**2**), the dihedral angle is forced to about 17° by the geometrical constrains provided by the fused oxane ring, which causes higher conjugation.[3,4,11]

Synthetic analogues were prepared by altering the geometry of the fused heterocyclic ring so as to alter the stereoelectronic effect on the heteroatom. Unsaturation in the oxane ring conjugated with the aromatic π-system (**10**) caused the dihedral angle to open to 38°, worsening the conjugation of the oxygen lone pair.[4] This compensated the favorable extension of the C–C π-system: as a result, k_{inh} decreased slightly (2.5×10^6 M^{-1} s^{-1}).[4] On the other hand, shrinking the fused heterocycle to a five-membered (furanoid) ring, as in compounds **11** and **12**, decreased the dihedral angle to 6°, improving the conjugation of the oxygen lone pair, which boosted the reactivity by nearly two-fold (Table 11.1). Interestingly, compound **12b**, bearing the phytyl tail, had about twice the vitamin E activity *in vivo*.[12] Similarly, **14** was almost three-fold more reactive than **13**.[10] In BO-653 (**15**), both the *tert*-butyl groups *ortho* to the –OH and the benzofuranoid structure contribute to lowering the BDE_{OH}; however, the steric hindrance of *tert*-butyls works in the opposite direction concerning the reactivity with ROO$^•$, which is half that of α-TOH.[13]

Alkylamino is a stronger EDG than alkoxy due to the lower electronegativity of nitrogen: this would suggest an improved performance by analogues **42** and **43** (see Section 11.4). When tested, compound **42** gave instead only $k_{inh} = 2.0 \times 10^6$ M^{-1} s^{-1} due to unfavorable stereoelectronics: steric interaction with the methyl in the 8-position forced the *N*-ethyl to adopt an axial position, leaving the nitrogen lone pair nearly aligned with the plane of the aromatic ring.[4] The steric interaction was removed in compound **43**; however, the phenol became too electron-rich to be stable in air (even in the crystalline form) and was completely useless as an antioxidant.[4] A strategy to overcome this limitation will be discussed in Section 11.4.

11.3 Vitamin-E-inspired Antioxidants Containing Chalcogens

Introduction of chalcogen (S, Se and Te) atoms into the tocopherol scaffold has been a major strategy to develop improved antioxidants. Stepping down the 16th group of the periodic table, electronegativity decreases. Thus, electrons should be more available for stabilizing the phenoxyl radical. In addition, organochalcogens would behave as preventive antioxidants too, because they are able to reduce organic hydroperoxides to the corresponding alcohols. The selenium-containing enzyme glutathione peroxidase catalyzes the reduction of hydroperoxides and hydrogen peroxide using glutathione and other thiols as sacrificial reducing agents. It is therefore expected that this approach would lead to better antioxidants, endowed with additional enzyme-like activity. Although the synthesis of thiatocopherol dates back to 1986,[14] only recently have these promises been reached, together with a better comprehension of the radical chemistry of chalcogen-containing phenols.

11.3.1 Sulfur-containing Compounds

Replacement of endocyclic oxygen with sulfur was achieved by a multistep low overall yield synthesis of all-racemic **17** (Scheme 11.2).[14] Disappointingly, reactivity of **17** toward ROO• was 2–3 times lower than that of α-tocopherol.[15] Such a low reactivity was also reflected in the three-fold lower activity in a curative myopathy bioassay in vitamin-E-deficient rats.[16]

Interest in sulfur-containing analogues was then revived about 10 years later, with the discovery of a straightforward route to S-containing heterocycles resembling the chromanol core. The reaction is based on the transient generation of an *ortho*-thioquinone, which reacts with electron-rich dienes giving rise to aryl-substituted benzoxathiin cycloadducts in high yields (see Scheme 11.2).[17] This strategy afforded a variety of antioxidants: α-, β-, γ-, δ-thia-tocopherols **18–21**,[18] and the tocopherol-flavonoids conjugates with the S-atom in the *meta* (**22** and **23**)[19,20] or *ortho* position (**24** and **25**) relative to the reactive OH group.[21]

The reactivity of analogues **18–23** is smaller than the natural counterpart, because the S-atom causes a distortion of the molecular geometry that reduces the conjugation of the O-atom with the phenolic ring (the Ar–O–R dihedral angle in **23** is 30°).[20] Compounds **24** and **25** are instead as reactive as α-TOH. The conformation of the six-membered heterocyclic ring prevents the formation of an intramolecular hydrogen bond between the OH group and the S atom, while ensuring a good stabilization by electron donation to

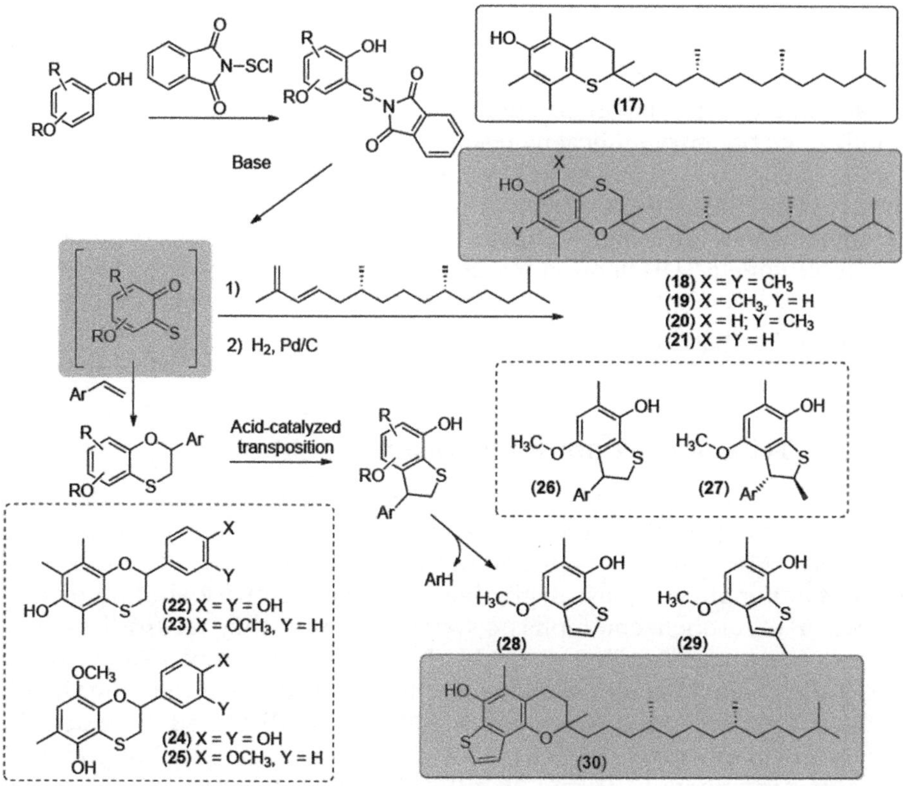

Scheme 11.2 Sulfur-containing tocopherol derivatives.

the phenoxyl radical formed upon reaction with ROO• radicals.[21] This unexpected result led to the synthesis of compounds **26** and **27**, which have a slightly lower reactivity than α-TOH.[22] Interestingly, the latter compounds are easily converted into the corresponding benzothiophenes **28** and **29**, which are more reactive than α-TOH due to the onset of an unusual sulfur–oxygen non-covalent interaction in the phenoxyl radical. This peculiar feature was used to build the tocopherol-like antioxidant **30**, which was about three-fold more reactive than α-TOH (Table 11.1).[23] Additionally, binding experiments showed that **30** binds to the tocopherol transfer protein (TTP) with an affinity very similar to that of all-rac-α-tocopherol (see Section 11.5).[23]

11.3.2 Selenium-containing Tocopherols

Selenium is seen as an appealing element to be included in the tocopherol scaffold because it may afford additional antioxidant mechanisms, such as the ability to participate in a catalase-like redox cycle. All rac α-selenotocopherol **31** was synthesized in 2006 in 11 steps in a 6.6% total yield. The compound, however, showed a lower reactivity with ROO• than α-tocopherol, due to a 1 kcal mol^{-1} larger BDE_{OH} (Table 11.1).[24] Selenium-containing analogues **32** and **33**, lacking the phytyl tail and featuring a five-membered fused ring, were also less reactive than α-TOH.[25] These results were explained by considering that the larger size of Se (compared to O) reduces its ability to stabilize the phenoxyl radical by resonance.[26] Interestingly, when studied in a water–chlorobenzene biphasic system containing *N*-acetylcysteine (NAC) in the aqueous phase, the duration of the inhibition due to **32** was six-fold longer than without NAC, suggesting the regeneration of the phenoxyl radical by the thiol, while α-TOH showed no regeneration. On the other hand, **31** and **33** were not regenerated under the same conditions, presumably because the methyl substituents, that reduce their BDE_{OH}, also make the reduction to the parent phenols more difficult.[25]

11.3.3 Tellurium-containing Tocopherol Mimics

The discovery of a thiol-mediated regeneration of the Se-containing compound **32** prompted the exploration of the effect of tellurium on tocopherol analogues. Mechanistic studies on the simple phenols **34–37** showed that Te-containing phenols have negligible antioxidant activity when studied in the water–chlorobenzene biphasic system, because they are oxidized irreversibly to unreactive telluroxides by the traces of hydroperoxides always present in the oxidizable substrate (methyl linoleate). On the other hand, in the presence of NAC, Te-containing **35** and **37** displayed an extraordinary antioxidant reactivity (Table 11.1) and a longer inhibited period than Se-containing phenols **34** and **36**.[27] The key for the high k_{inh} of **35** is the presence of the Te-alkyl group *ortho* to the phenolic OH, which changes the mechanism of ROO˙ quenching to a sequential O-atom plus H-atom transfers (Scheme 11.3).[3,28]

These exciting results led to the development of a straightforward synthesis of the Te-containing tocopherols **38** and **41**,[29,30] which also afforded the S- and Se-containing counterparts (Scheme 11.3).[29]

As expected, autoxidation experiments in a biphasic system with NAC in the aqueous phase showed that the reactivity with ROO˙ and the regenerability in the presence of thiols increases on going from S to Te (**38:** R_{inh} = 5, T_{inh} = 591; **39:** R_{inh} = 104, T_{inh} = 25; **40:** R_{inh} = 94, T_{inh} = 33; **41:** R_{inh} = 2, T_{inh} = 653; **α-TOH:** R_{inh} = 25, T_{inh} = 97; where R_{inh}, μM h^{-1}, is the rate of oxidation in the presence of the antioxidant and T_{inh}, min, is the length of the inhibition period).[29,30]

11.4 Insertion of Nitrogen in the Aromatic Ring: from Phenols to 3-Pyridinols

The prevailing strategy to improve the antioxidant performance of phenols is the insertion of stronger ED substituents on the aromatic ring to lower the BDE_{OH}. However, pushing it below the value of 77 kcal mol^{-1} of α-tocopherol

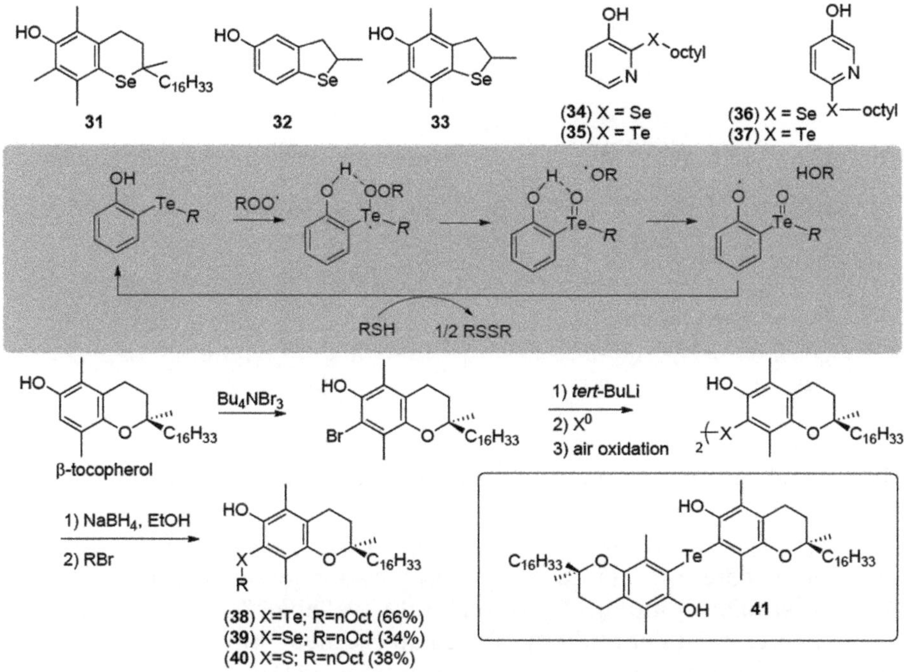

Scheme 11.3 Selenium- and tellurium-containing tocopherol derivatives.

also dramatically impairs air stability, since the ionization potential (IP) will also be lowered to the point that the compound can react directly with O_2. It has been observed with amino-derivative **43** (see Section 11.2.1).[4] This makes the compound useless as an antioxidant, and it turns it into a pro-oxidant, since the product of the reaction with oxygen is superoxide ($O_2^{-\bullet}$) or the hydroperoxyl radical ($HOO^\bullet$) (Scheme 11.4).

A way to overcome this limitation had proven effective with simpler phenolic antioxidants: to insert heterocyclic nitrogens in the phenolic ring in place of =CH– functions, turning the phenol into a 5-pyrimidinol (Scheme 11.4). Being more electronegative than carbon, nitrogen makes the aromatic ring electron-poorer thereby causing an enormous increase in IP and a more modest increase in BDE_{OH}.[31] Calculations and experiments showed that a series of 5-pyrimidinols had BDE_{OH} higher by about 2.5 kcal mol^{-1} compared to equivalently substituted phenols; however, the IP had raised by 24.3 kcal mol^{-1} (almost 10-fold as much) and the oxidation potential by about 0.5 V.[31,32] While the increase in IP and oxidation potential largely increased air stability, the unfavorable increase in BDE_{OH} could easily be compensated by ring substitution with stronger ED groups.[32] Transfer of the same strategy to the vitamin E structure was possible by inserting only one heterocyclic nitrogen, so to allow closure of the aliphatic ring.[33] Compound **44a**, obtained *via* a six-step synthesis,[34] had a BDE_{OH} lower by 1.9 kcal mol^{-1} compared to α-TOH (75.2 *vs.* 77.1 kcal mol^{-1})[33,35] and was 28-fold more reactive toward peroxyl

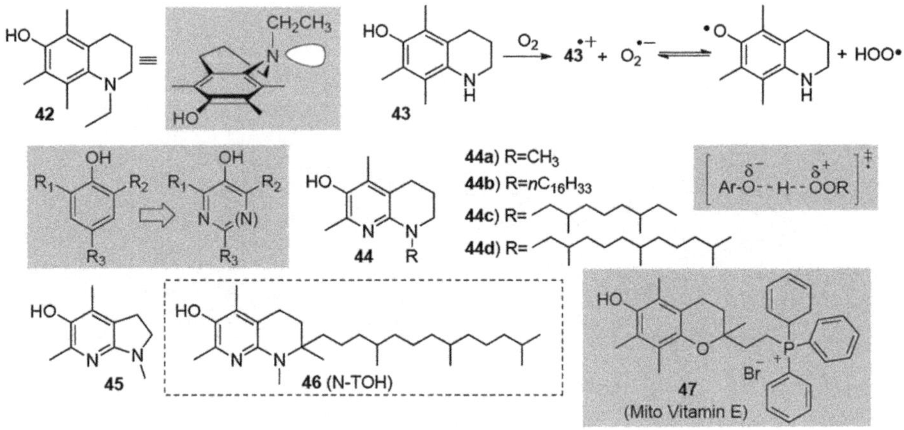

Scheme 11.4 Nitrogen-containing vitamin E analogues and other structures.

radicals (k_{inh} = 8.8 × 10^7 M^{-1} s^{-1} at 303 K),[33] being nonetheless more stable toward air oxidation.[34] An improved synthetic strategy starting from vitamin B6 made similar compounds conveniently available.[36] Homologous **44b,** having an increased lipophilicity, proved an excellent antioxidant in human LDL, where it was even able to spare endogenous α-TOH.[37]

The extraordinary reactivity of **44a** is not solely attributed to the lower BDE_{OH}, as it also benefits from polar effects. The electron-poorer pyridinol moiety can better accommodate a partial negative charge arising from the contribution of ionic configuration in the transition state wavefunction, lowering the activation energy for HAT (Scheme 11.4).[32,38] Pyridinols are more reactive than phenols of comparable BDE_{OH}; therefore, they are able to spare them, or to be more easily regenerated by co-antioxidants affording synergic behavior.[38,39]

Shrinking the fused aliphatic ring from six-membered to five-membered as in compound **45** allowed the exploitation of more favorable stereoelectronics. As a result, **45** has a BDE_{OH} of only 74.3 kcal mol^{-1} and is able to trap peroxyl radicals with a k_{inh} = 2.8 × 10^8 M^{-1} s^{-1} (at 303 K), being 88-fold faster than α-TOH: the fastest phenol-type antioxidant ever reported![33] It should be noted that such an extraordinary rate has reached the limit for diffusion control.[33] Although **45** pushed the reactivity to the limit, its moderate stability and somewhat difficult synthesis[34] suggested that **44a** would be a better "lead" for further development.

11.5 Activity in Biological Systems: The Role of the Lipophilic Tail

The phytyl tail does not influence the antioxidant activity of α-TOH analogues in organic solution; however, it is important for the biological function of vitamin E. It conveys a large lipophilicity, confining the molecule in the lipid domain of LDL or biomembranes, where it is supposed to play its antioxidant role. It is well established that α-TOH does not protect human LDL in the absence of ubiquinol in the particle or of ascorbate in the surrounding medium. Upon reaction with ROO•, the resulting tocopheroxyl radical is also too lipophilic to escape the particle: only if co-antioxidants (*e.g.* ascorbate) reduce it back is α-TOH is available for further protection.[40] In the absence of co-antioxidants, under typical settings, the tocopheroxyl radical will react with polyunsaturated lipids, propagating the autoxidation: a phenomenon known as tocopherol-mediated peroxidation (TMP).[41] Another important role of the phytyl tail is to regulate tocopherol's bioavailability. To avoid oxidation by P450 enzymes, upon absorption, vitamin E binds to tocopherol transport protein (TTP), which is apparently optimized to accommodate the phytyl tail. TTP also mediates the movement of vitamin E between lipid regions, *i.e.*, between lipoproteins and cells, and between subcellular regions. Therefore, the binding to TPP appears critical for bioactivity.

Among synthetic analogues, compound **30** bearing the phytyl chain was shown to bind to TTP with an affinity very similar to that of all-rac-α-tocopherol, indicating that the benzothiophene moiety is well tolerated.[40]

Insertion of a methyl and a phytyl tail in the 2-position in the structure of **44a** afforded *N*-TOH (**46**) a close copy of α-TOH.[42] This compound spares endogenous α-TOH in LDL and has an improved mobility profile in model membrane systems.[42] Interestingly *N*-TOH binds to human TTP with a higher affinity than α-TOH itself, indicating the potential for *in vivo* activity.[42] Despite the interest, *N*-TOH required a 17-step synthesis, which prompted further investigation of the less onerous **44** to evaluate the impact of the N1 alkyl side chain length and branching.[43] Derivatives with eight or more carbons had affinities for TTP similar to α-TOH (*e.g.* **44b** and **44c**), while the *N*-phytyl derivative, **44d**, had an affinity 10-fold better than α-TOH.[43] All the compounds protected human LDL about 30-fold better than α-TOH and did so in the absence of ascorbic acid, without giving TMP. This is because they have a lower BDE_{OH} (by about 2 kcal mol^{-1}) than α-TOH, which impairs the reaction of the phenoxyl radical with lipids. Additionally, all naphthyridinols **44** were more effectively regenerated than α-TOH by ascorbate, urate and *N*-acetylcysteine.[43]

As an alternative approach, in BO-653 (**15**), sufficient lipophilicity is guaranteed by combination of the two *n*-pentyl tails and the two *tert*-butyls. BO-653 was shown to accumulate in LDL and localize deeper inside than was observed for α-TOH, which caused lower regeneration by ascorbic acid at the interface.[44] It protects human LDL from autoxidation sparing α-TOH or regenerating it from the tocopheroxyl radical.[12,44] Indeed, EPR studies[12] indicate that BO-653 has a lower BDE_{OH} than α-TOH by at least 1 kcal mol^{-1}. As a consequence of the higher stability of the phenoxyl radical, BO-653 does not give TMP. In rat plasma, BO-653, but not its analogue bearing two methyls in place of *tert*-butyls, was shown to inhibit lipid peroxidation better than α-TOH in the absence of ascorbic acid.[45] Most notably, it was shown to protect lipids in human plasma following oral ingestion,[46] and to have antiatherogenic effects in three animal models, which granted BO-653 admission to clinical trials.[47]

Worth mentioning is the so-called "Mito Vitamin E" (**47**), a mimic of α-TOH in which the side tail was modified to maintain lipophilicity while accommodating a phosphonium cation to selectively target mitochondria.[48] The compound was accumulated hundreds-fold in the mitochondrial matrix, due to the organelle's large membrane potential, protecting it from oxidative injury better than α-TOH.[48] It also protected bovine aorta endothelial cells from oxidative stress – a model in which α-TOH was not effective.[49]

11.6 Future Perspectives

The search for vitamin-E-inspired novel antioxidants has significantly contributed to rationalization of the chemical–physical basis of vitamin E bioactivity. A number of very promising molecules with improved antioxidant activity have been investigated and a few of them have also shown improved bioactivity compared to the natural model. However, with the exception of troglitazone, which was approved as an anti-diabetic drug by the FDA in 1997 and withdrawn from the market in 2000 due to hepatotoxicity,[50] none of these promising molecules have so far entered medical practice. Clearly more research is needed in this fascinating area. Meanwhile, the frontier in antioxidants research has moved to nanomaterials and vitamin E continues to offer much inspiration for innovation.[51]

References

1. G. W. Burton and K. U. Ingold, *J. Am. Chem. Soc.*, 1981, **103**, 6472.
2. G. W. Burton and K. U. Ingold, *Acc. Chem. Res.*, 1986, **19**, 194.
3. K. U. Ingold and D. A. Pratt, *Chem. Rev.*, 2014, **114**, 9022.
4. G. W. Burton, T. Doba, E. Gabe, L. Hughes, F. L. Lee, L. Prasad and K. U. Ingold, *J. Am. Chem. Soc.*, 1985, **107**, 7053.
5. J. A. Howard and K. U. Ingold, *Can. J. Chem.*, 1963, **41**, 2800.
6. L. Valgimigli and D. A. Pratt, in *Encyclopedia of Radicals in Chemistry, Biology and Materials*, ed. C. Chatgilialoglu and A. Studer, John Wiley & Sons, Ltd, Chichester, UK, 2012, p. 1623.
7. M. Lucarini and G. F. Pedulli, *Chem. Soc. Rev.*, 2010, **39**, 2106.
8. H. C. Brown and Y. Okamoto, *J. Am. Chem. Soc.*, 1958, **80**, 4979.
9. R. Amorati and L. Valgimigli, *Org. Biomol. Chem.*, 2012, **10**, 4147.
10. L. R. C. Barclay, M. R. Vinqvist, K. Mukai, S. Itoh and H. Morimoto, *J. Org. Chem.*, 1993, **58**, 7416.
11. R. Amorati, A. Baschieri, G. Morroni, R. Gambino and L. Valgimigli, *Chem. Eur. J.*, 2016, **22**, 7924.
12. K. U. Ingold, G. W. Burton and D. O. Foster, *et al.*, *FEBS J.*, 1986, **205**, 117.
13. N. Noguchi, Y. Iwaki and M. Takahashi, *et al.*, *Arch. Biochem. Biophys.*, 1997, **342**, 236.
14. B. Robillard, L. Hughes, M. Slaby, D. A. Lindsay and K. U. Ingold, *J. Org. Chem.*, 1986, **51**, 1700.

15. H. A. Zahalka, B. Robillard and L. Hughes, *et al.*, *J. Org. Chem.*, 1988, **53**, 3739.
16. K. U. Ingold, G. W. Burton, L. Hughes, D. O. Foster and B. Robillard, *Free Radical Res. Commun.*, 1990, **11**, 207.
17. G. Capozzi, P. Lo Nostro, S. Menichetti, C. Nativia and P. Sarri, *Chem. Commun.*, 2001, 551.
18. S. Menichetti, R. Amorati, M. G. Bartolozzi, G. F. Pedulli, A. Salvini and C. Viglianisi, *Eur. J. Org. Chem.*, 2010, 2218.
19. R. Amorati, M. G. Fumo, G. F. Pedulli, S. Menichetti, C. Pagliuca and C. Viglianisi, *Helv. Chim. Acta*, 2006, **89**, 2462.
20. R. Amorati, A. Cavalli and M. G. Fumo, *et al.*, *Chem. Eur. J.*, 2007, **13**, 8223.
21. C. Viglianisi, M. G. Bartolozzi, G. F. Pedulli, R. Amorati and S. Menichetti, *Chem. Eur. J.*, 2011, **17**, 12396.
22. C. Viglianisi, R. Amorati, L. Di Pietro and S. Menichetti, *Chem. Eur. J.*, 2015, **21**, 16639.
23. S. Menichetti, R. Amorati, V. Meoni, L. Tofani, G. Caminati and C. Viglianisi, *Org. Lett.*, 2016, **18**, 5464.
24. D. Shanks, R. Amorati, M. G. Fumo, G. F. Pedulli, L. Valgimigli and L. Engman, *J. Org. Chem.*, 2006, **71**, 1033.
25. S. Kumar, H. Johansson, L. Engman, L. Valgimigli, R. Amorati, M. G. Fumo and G. F. Pedulli, *J. Org. Chem.*, 2007, **72**, 2583.
26. R. Amorati, G. F. Pedulli, L. Valgimigli, H. Johansson and L. Engman, *Org. Lett.*, 2010, **12**, 2326.
27. S. Kumar, H. Johansson and T. Kanda, *et al.*, *J. Org. Chem.*, 2010, **75**, 716.
28. R. Amorati, L. Valgimigli, P. Dinér, K. Bakhtiari, M. Saeedi and L. Engman, *Chem. Eur. J.*, 2013, **19**, 7510.
29. J.-F. Poon, V. P. Singh, J. Yan and L. Engman, *Chem. Eur. J.*, 2015, **21**, 2447.
30. J.-F. Poon, J. Yan, V. P. Singh, P. J. Gates and L. Engman, *J. Org. Chem.*, 2016, **81**, 12540.
31. D. A. Pratt, G. A. DiLabio, G. Brigati, G. F. Pedulli and L. Valgimigli, *J. Am. Chem. Soc.*, 2001, **123**, 4625.
32. L. Valgimigli, G. Brigati, G. F. Pedulli, G. A. DiLabio, M. Mastragostino, C. Arbizzani and D. A. Pratt, *Chem. Eur. J.*, 2003, **9**, 4997.
33. M. Wijtmans, D. A. Pratt, L. Valgimigli, G. A. DiLabio, G. F. Pedulli and N. A. Porter, *Angew. Chem., Int. Ed.*, 2003, **42**, 4370.
34. M. Wijtmans, D. A. Pratt, J. Brinkhorst, R. Serwa, L. Valgimigli, G. F. Pedulli and N. A. Porter, *J. Org. Chem.*, 2004, **69**, 9215.
35. P. Mulder, H.-G. Korth, D. A. Pratt, G. A. DiLabio, L. Valgimigli, G. F. Pedulli and K. U. Ingold, *J. Phys. Chem. A*, 2005, **109**, 2647.
36. R. Serwa, T.-G. Nam and L. Valgimigli, *et al.*, *Chem. Eur. J.*, 2010, **16**, 14106.
37. H.-Y. Kim, D. A. Pratt, J. R. Seal, M. Wijtmans and N. A. Porter, *J. Med. Chem.*, 2005, **48**, 6787.
38. L. Valgimigli and D. A. Pratt, *Acc. Chem. Res.*, 2015, **48**, 966.
39. L. Valgimigli, D. Bartolomei and R. Amorati, *et al.*, *Beilstein J. Org. Chem.*, 2013, **9**, 2781.

40. K. U. Ingold, V. W. Bowry, C. Walling and R. Stocker, *Proc. Natl. Acad. Sci. U. S. A.*, 1993, **90**, 45.
41. V. W. Bowry and K. U. Ingold, *Acc. Chem. Res.*, 1999, **32**, 27.
42. T.-G. Nam, C. L. Rector and H.-Y. Kim, *et al.*, *J. Am. Chem. Soc.*, 2007, **129**, 10211.
43. B. Li, J. R. Harjani, N. S. Cormier, H. Madarati, J. Atkinson, G. Cosa and D. A. Pratt, *J. Am. Chem. Soc.*, 2013, **135**, 1394.
44. N. Noguchi, Y. Okimoto, J. Tsuchiya, O. Cynshi, T. Kodama and E. Niki, *Arch. Biochem. Biophys.*, 1997, **347**, 141.
45. N. Itoh, Y. Yoshida, M. Hayakawa, N. Noguchi, T. Kodama, O. Cynshi and E. Niki, *Biochem. Pharmacol.*, 2004, **68**, 813.
46. R. Yamauchi, P. Southwell-Keely, C. Suarna, S. Ray, M. Raftery, O. Cynshi and R. Stocker, *Free Radical Biol. Med.*, 2005, **38**, 32.
47. K. Tamura, Y. Kato and A. Ishikawa, *et al.*, *J. Med. Chem.*, 2003, **46**, 3083.
48. R. A. Smith, C. M. Porteous, C. V. Coulter and M. P. Murphy, *Eur. J. Biochem.*, 1999, **263**, 709.
49. A. Dhanasekaran, S. Kotamraju and S. V. Kalivendi, *et al.*, *J. Biol. Chem.*, 2004, **279**, 37575.
50. M. T. Smith, *Chem. Res. Toxicol.*, 2003, **16**, 679.
51. M. Massaro, S. Riela and S. Guernelli, *et al.*, *J. Mater. Chem. B*, 2016, **4**, 2229.

CHAPTER 12

Action of Vitamin E Against Lipid Peroxidation and Cell Death

NORIKO NOGUCHI*

Systems Life Sciences Laboratory, Department of Medical Life Systems, Faculty of Life and Medical Sciences, Doshisha University, 1-3 Miyakodani, Tatara, Kyotanabe, Kyoto 610-0394, Japan
*E-mail: nnoguchi@mail.doshisha.ac.jp

12.1 Introduction

Many large-scale epidemiological studies have supported the beneficial effects of vitamin E against cardiovascular diseases[1–5] and neurodegenerative diseases such as Alzheimer's disease (AD)[6] and Parkinson's disease (PD).[7] Vitamin E is the most abundant lipid-soluble radical-scavenging antioxidant in our body and is expected to protect us from diseases in which lipid peroxidation is involved. In this chapter, the action of vitamin E against lipid peroxidation in different systems is described and, in addition, novel mechanisms rather than radical scavenging activity by which vitamin E inhibits different types of cell death are discussed.

Food Chemistry, Function and Analysis No.11
Vitamin E: Chemistry and Nutritional Benefits
Edited by Etsuo Niki

Published by the Royal Society of Chemistry, www.rsc.org

12.2 Inhibition of Lipid Peroxidation in Homogeneous Solution

Alpha-tocopherol (α-Toc) is the most abundant vitamin E and its antioxidant activity has been studied extensively compared to other Tocs and tocotrienols (Toc3s). The reactivity of vitamin E toward peroxyl radicals can be measured using spectrometry. The actual rate of free radical flux from azo compounds such as 2,2′-azobis(2,4-dimethylvaleronitrile) (AMVN) is measured by using *N*,*N*′-diphenyl-1,4-phenylenediamine (DPPD) as a radical scavenger in homogeneous solution (Figure 12.1).[8,9] One molecule of DPPD scavenges two radicals rapidly to give *N*,*N*′-diphenyl-*p*-benzoquinone diimine (DPBQ), which has a strong absorption at 440 nm. In the absence of antioxidant, the absorption at 440 nm increases with time without any induction period and then reaches a plateau when all the DPPD is converted to DPBQ. The plateau absorbance is directly proportional to the initial DPPD concentration, independent of radical initiator concentration. The addition of α-Toc, γ-Toc, α-Toc3 or γ-Toc3 to this mixture suppresses the formation of DPBQ due to competition with DPPD in scavenging peroxyl radicals. The same inhibitory effect was observed for α-Toc and α-Toc3 and for γ-Toc and γ-Toc3. For both types of Toc and Toc3, α forms showed more effective inhibition than γ forms. The reactivity toward peroxyl radicals can also be measured by following a decrease in the maximum absorption at 479 nm of pyrogallol red. The reactivities of Toc3 against peroxyl radicals in homogeneous solution are substantially the same as the corresponding Toc and the relative reactivities decrease in the order of α > β = γ > δ (Figure 12.1).[10]

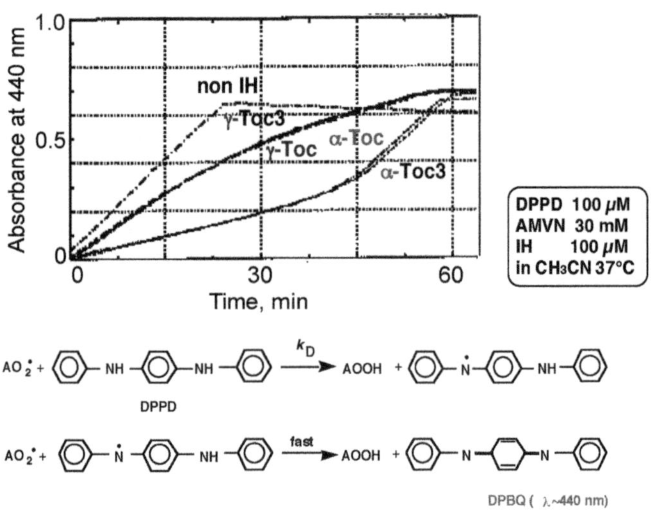

Figure 12.1 Radical scavenging antioxidant activity of vitamin E homologues. DPPD (100 μM) was reacted with AMVN (30 mM) in the absence or presence of vitamin E homologue (100 μM) in acetonitrile at 37 °C.

The antioxidant activity against lipid peroxidation has been evaluated by several parameters and methods such as oxygen consumption, conjugated diene, lipid hydroperoxides, and induction period (lag phase). It has been observed that the corresponding Toc and Toc3 isoforms exert the same antioxidant activities against lipid peroxidation in homogeneous solution.[9,10] α-Toc and α-Toc3 exert the same antioxidant activities against lipid peroxidation in acetonitrile solution and γ-Toc and γ-Toc3 are also the same but less potent than α forms. The antioxidant efficacy decreases in the order $\alpha > \beta = \gamma > \delta$, in agreement with the relative reactivities toward peroxyl radicals. Similar antioxidant activities were reported for α-Toc and α-Toc3, and also for γ-Toc and γ-Toc3 against oxidation of cholesterol in aqueous dispersions containing the polyoxyethylene sorbitol ester Tween 20 initiated by a water-soluble radical initiator, 2,2′-azobis(2-amindinopropane) dihydrochloride (AAPH).[11]

12.3 Inhibition of Lipid Peroxidation in Liposomal Membranes

The antioxidant activities in the membranes can be assessed by using liposome systems. When β-linoleoyl-γ-palmitoyl phosphatidylcholine (PLPC) or soybean PC liposomal membranes are oxidized by lipophilic radical initiator AMVN, α-Toc and α-Toc3 incorporated into the membranes show the same inhibitory effect.[10] When the competition between α-Toc and α-Toc3 for scavenging peroxyl radicals in PLPC liposomes oxidized by AMVN is studied, α-Toc and α-Toc3 are consumed at the same rate. The study on the action of T and T3 isoforms in the liposomal membranes by fluorescence quenching experiments using 2,3-diazabicyclo[2,2,2]-oct-2-ene and fulorazophone L showed that the rate decreases in the order $\alpha > \beta = \gamma > \delta$ and the rate constant for the corresponding Toc and Toc3 is the same in liposomal membranes.[12] These results suggest that the antioxidant action of Toc and Toc3 in the liposomal membranes is almost the same.

Reduction of α-tocopheroxyl radical by ubiquinol-10 (UQ) or vitamin C (VC) to regenerate α-Toc has been well demonstrated using liposomal membranes in which oxidation is initiated by peroxyl radicals produced in either an aqueous phase or lipid phase (Figure 12.2A and B).[13] UQ and VC compete to react with α-tocopheroxyl radical. α-Toc and UQ are incorporated in the liposomal membranes, while VC exists in the aqueous phase. When the oxidation of liposomal membranes is induced by hydrophilic AAPH, VC is consumed first, followed by UQ, and α-Toc is spared (Figure 12.2A). When the oxidation of liposomal membranes is induced by lipophilic AMVN, UQ is consumed faster than VC, and α-Toc is spared (Figure 12.2B). In both oxidation systems, the production of PC hydroperoxide (PCOOH) is suppressed during the presence of α-Toc and the rapid production of PCOOH is observed after depletion of α-Toc.

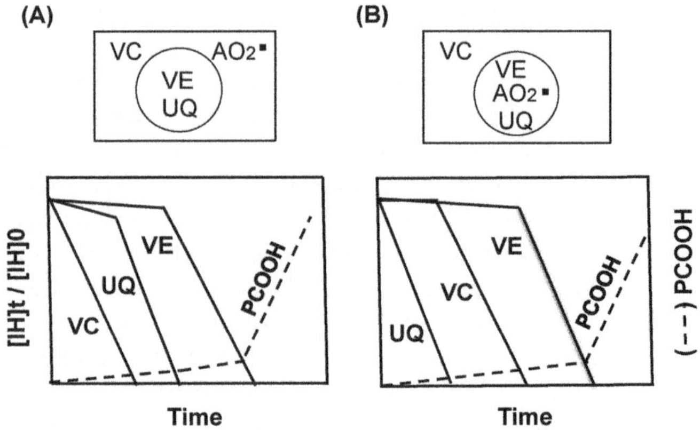

Figure 12.2 Patterns of consumption of antioxidants and formation of hydroperoxide in the oxidation of phosphatidylcholine (PC) liposomal membranes. Oxidation of PC liposomal membranes was induced by a hydrophilic (A) or lipophilic (B) radical initiator at 37 °C in air in the presence of vitamin C (VC), α-Toc (VE) and ubiquinol (UQ). Formation of PC hydroperoxide (PCOOH) is shown by a broken line.

12.4 Inhibition of Lipid Peroxidation in Lipoproteins

With increasing evidence of biochemical, cell biological, clinical and epidemiological studies showing the important role of the oxidative modification of low density lipoprotein (LDL) in atherogenesis, the actions of antioxidants have received much attention.[14,15] LDL is composed of a core containing cholesteryl esters and triglyceride and an outer monolayer of phospholipids. Vitamin E (α-Toc and γ-Toc), carotenoid, and UQ are the major lipophilic antioxidants in the LDL particle (Table 12.1).[16] α-Toc is the most abundant lipophilic antioxidant in LDL.[17–19] The sequence of radical-induced consumption of antioxidants in freshly isolated, *in vitro* and *in vivo* Toc3-enriched LDL is UQ > α-Toc = α-Toc3 > carotenoids > γ-Toc = γ-Toc3.[20]

Although it has not been clearly shown what initiates LDL oxidation *in vivo*, LDL oxidation *in vitro* can be initiated by a copper ion, radical initiators, lipoxygenase, myeloperoxidase, and singlet oxygen. Regardless of how oxidation is initiated, the free-radical-mediated lipid peroxidation proceeds by a chain mechanism in which lipid peroxyl radicals play an important role as chain carriers giving rise to lipid hydroperoxide accumulation. The oxidation of LDL can be followed by measuring oxygen uptake, formation of lipid hydroperoxides (cholesterol ester hydroperoxides, CEOOH and phosphatidylcholine hydroperoxide, PCOOH).[19] α-Toc is spared by UQ and ascorbate and the formation of CEOOH and PCOOH is suppressed during the presence of UQ and ascorbate. After depletion of UQ and ascorbate, rapid formation of CEOOH and PCOOH occurs even in the presence of α-Toc, in which the α-tocopheroxyl radical attacks polyunsaturated lipids to induce oxidation in

Table 12.1 Antioxidants in human LDL and plasma.

Antioxidants	LDL (mol per particle)[16]	Plasma (μM)
Vitamin C		30–150
Uric acid		160–450
Bilirubin		5–20
Vitamin E	8–10	15–40
Ubiquinol-10	0.5–1.0	0.4–1.0
β-Carotene	0.4	0.3–0.6
Lycopene	0.7	0.5–1.0

LDL particles. This oxidation is termed as "tocopherol-mediated peroxidation (TMP)".[20] Stocker suggested that UQ and other reducing coantioxidants capable of reducing the α-tocopheroxyl radical inhibit lipid peroxidation.[21] If the oxidation of LDL is initiated by either AAPH or AMVN in the presence of ascorbate, similar oxidation patterns are observed as shown in the oxidation of PC liposomal membranes (Figure 12.2A and B). Both ascorbate and uric acid suppress the oxidation of LDL induced by AAPH and spare α-Toc by scavenging radicals in the aqueous phase. When the LDL oxidation is induced by lipophilic AMVN, α-Toc is spared by ascorbate but not by uric acid since uric acid, unlike ascorbate, does not reduce the α-tocopheroxyl radical to regenerate α-Toc and inhibit TMP.[22,23]

When the oxidation is induced by an azo radical initiator in plasma[24] or whole blood,[25] we can follow the consumption of antioxidants and accumulation of lipid hydroperoxides as in LDL oxidation. Given that hydrophilic antioxidants such as vitamin C and uric acid are present in the aqueous phase (Table 12.1), these antioxidants are consumed first in the oxidation induced by the hydrophilic radical initiator AAPH. When the oxidation is induced by the lipophilic radical initiator AMVN, UQ is consumed first, followed by VC, and after complete consumption of VC, the level of α-Toc starts to decrease as observed in the oxidation of PC liposomal membranes (Figure 12.2A and B).[26]

12.5 Inhibition of Cell Death by Vitamin E

Cell death is caused by oxidative stress such as exposure to hydrogen peroxide, cumene hydroperoxide (CumOOH), and aldehydes. Regardless of the type of cell death, for example, apoptosis or necrosis, lipid peroxidation proceeds and lipid hydroperoxides are accumulated in cells.[27,28]

Glutathione peroxidase 4 (GPx4) is known to reduce lipid hydroperoxides to the corresponding lipid hydroxides and protect cells against oxidative stress.[29] It has been shown that knockdown of GPx4 induces accumulation of lipid hydroperoxides, resulting in apoptotic cell death.[30] Depletion of GPx4 in mice induces lipid-peroxide-dependent cell death in several tissues, which is significantly inhibited by α-Toc both *in vitro* and *in vivo*. Recently, it has been reported that an iron-dependent and non-apoptotic form of cell death, ferroptosis, can be induced by inhibition of GPx4 or glutathione biosynthesis using chemical inhibitors, respectively,[31] and that vitamin E is effective in the suppression of ferroptosis.[32] More recently, Kagan *et al.* reported that the oxidation of phosphatidylethanolamines (PE) with arachidonoyl and adrenoyl by 15-lipoxygenase (15-LOX) is important to induce ferroptosis, which is suppressed by α-Toc and α-Toc3; with a higher effectiveness of α-Toc3 than α-Toc.[33] They suggest that radical scavenging can contribute to the inhibitory effects of α-Toc and α-Toc3 along with their ability to compete for the substrate binding site of LOX (so-called "corking"). Their computational modeling analysis indicates that the affinity of α-Toc3 with LOX is higher than that of α-Toc. When we evaluate the effectiveness of Toc and Toc3 in cells, we should consider the different rates of their uptake into the cells.[9,34] It has been shown that, regardless of the concentration of vitamin E homologues added to the medium, a good reverse correlation is observed between the intracellular concentration of vitamin E and the number of adhesion monocytes to endothelial cells (Figure 12.3).[9]

We have shown that 24*S*-hydroxycholesterol (24*S*-OHC), a neuron-specific oxysterol that is known to increase in the plasma of AD patients, induces non-apoptotic programmed cell death.[35–37] We have reported that 24*S*-OHC is esterified with polyunsaturated fatty acid by acyl-CoA : cholesterol acyltransferase 1 (ACAT1), resulting in the formation of a lipid droplet-like structure along with endoplasmic reticulum (ER) and that 24*S*-OHC-induced

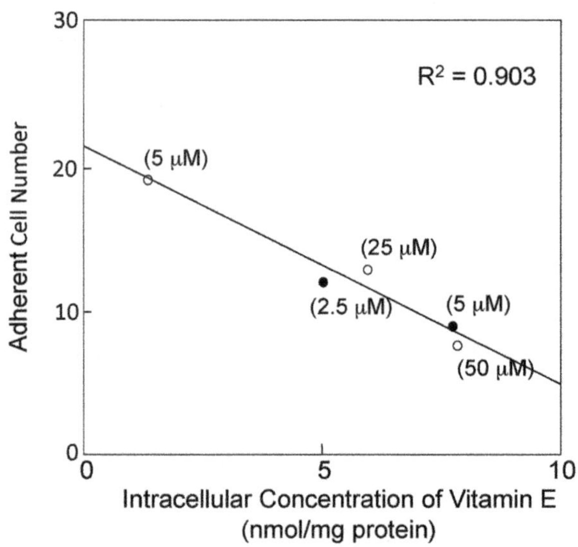

Figure 12.3 Correlation between intracellular concentrations of vitamin E and adhesion numbers of monocytes. Adhesion numbers of monocytes to human umbilical vein endothelial cells (HUVECs) were plotted against the intracellular concentration of α-Toc (open) and α-Toc3 (solid). The numbers shown in parentheses are the concentrations of vitamin E with which the HUVECs were treated. Reproduced from ref. 9 with permission from Elsevier, Copyright 2003.

cell death is suppressed by α-Toc and γ-Toc but not by α-Toc3 nor γ-Toc3 in human neuroblastoma SH-SY5Y cells. In contrast, CumOOH-induced cell death is inhibited by both Tocs and Toc3s, with an even higher effectiveness of Toc3s than Tocs (Scheme 12.1).[38,39] Again, the intracellular concentration of vitamin E may affect their effectiveness in cells (Figure 12.3).[9,34] We have concluded that the radical-scavenging activity of Toc is not responsible for inhibition of 24*S*-OHC-induced cell death and that the number of double bonds in the side chain of vitamin E is more important than the position of a phenol group at the chroman ring in the inhibition of 24*S*-OHC-induced cell death, in agreement with our observation that 24*S*-OHC treatments caused little reactive oxygen species (ROS) generation.[38,40]

Further studies are needed in order to elucidate more details of the underlying mechanisms; however, it may be noted that several reports on the physical effects of the side chain in liposomal membranes may give us a clue to understand the different effects of Toc and Toc3 on the inhibition of 24*S*-OHC-induced cell death. It has been shown that the phytyl side chain in α-Toc causes an increase in the rigidity of the phospholipid liposomal membrane but that the isoprenoid side chain in α-Toc3 does not.[10,41,42] Since the formation of the lipid droplet-like structure induced by 24*S*-OHC is not affected by either Toc or Toc3,[38,39] the effects of Toc and Toc3 on the integrity of ER membranes may be different (Scheme 12.1). Although the consequence of the formation of the lipid droplet-like structure at ER induced by 24*S*-OHC

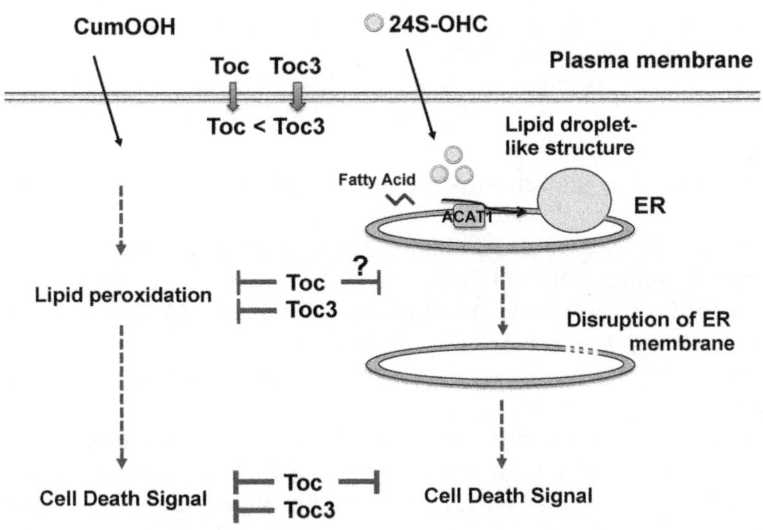

Scheme 12.1 Lipid peroxidation proceeds and activates cell death signaling in CumOOH-treated cells. Both Toc and Toc3 suppress CumOOH-induced cell death. In 24*S*-OHC-treated cells, 24*S*-OHC is esterified with fatty acid by action of ACAT1 and a lipid droplet-like structure is formed at the ER. Disruption of ER membranes may occur, resulting in the onset of cell death signaling. 24*S*-OHC-induced cell death is suppressed by Toc but not by Toc3 even if Toc3 is taken up by cells more efficiently.

is under investigation, disruption of ER is also observed, which may have an important role in the activation of cell death signaling.

12.6 Conclusion

The efficacy of vitamin E homologues in the inhibition of lipid peroxidation is determined by the methyl substituents on the chroman ring. The side chain affects their incorporation, retention, and mobility within and between the membranes. Toc3 usually affects cell function with a higher intracellular concentration because it has a higher uptake by cells than Toc. However, our recent studies have shown that 24*S*-OHC-induced cell death is inhibited by Toc but not by Toc3, whereas free-radical-induced cell death is inhibited by all of the vitamin E homologues. The critical effect of the side chain of vitamin E in the inhibition of 24*S*-OHC-induced cell death is indicated. The mechanism of the different effects of Toc and Toc3 on the integrity of ER membranes and cell death signaling should be explored further.

References

1. M. Sano, C. Ernesto and R. G. Thomas, *et al.*, *N. Engl. J. Med.*, 1997, **336**, 1216.
2. M. C. Morris, L. A. Beckett, P. A. Scherr, L. E. Hebert, D. A. Bennett, T. S. Field and D. A. Evans, *Alzheimer Dis. Assoc. Disord.*, 1998, **12**, 121.
3. M. C. Morris, D. A. Evans, J. L. Bienias, C. C. Tangney and R. S. Wilson, *Arch. Neurol.*, 2002, **59**, 1125.
4. M. W. Dysken, M. Sano and S. Asthana, *et al.*, *JAMA*, 2014, **311**, 33.
5. K. G. Coupland, G. D. Mellick and P. A. Silburn, *et al.*, *Mov. Disord.*, 2014, **29**, 1606.
6. P. Gamba, M. Guglielmotto and G. Testa, *et al.*, *Aging Cell*, 2014, **13**, 561.
7. C. Y. Lee, R. C. Seet, S. H. Huang, L. H. Long and B. Halliwell, *Antioxid. Redox Signaling*, 2009, **11**, 407.
8. N. Yoshida, T. Yoshikawa, H. Manabe, Y. Terasawa, M. Kondo, N. Noguchi and E. Niki, *J. Leukocyte Biol.*, 1999, **65**, 757.
9. N. Noguchi, R. Hanyu, A. Nonaka, Y. Okimoto and T. Kodama, *Free Radical Biol. Med.*, 2003, **34**, 1614.
10. Y. Yoshida, E. Niki and N. Noguchi, *Chem. Phys. Lipids*, 2003, **123**, 64.
11. Z. Xu, N. Hua and J. S. Godber, *J. Agric. Food Chem.*, 2001, **49**, 2007.
12. A. F. Sonnen, H. Bakirci, T. Netscher and W. M. Nau, *J. Am. Chem. Soc.*, 2005, **127**, 15575.
13. Y. Yamamoto, E. Komuro and E. Niki, *J. Nutr. Sci. Vitaminol.*, 1990, **36**, 505.
14. H. Esterbauer, J. Gebicki, H. Puhl and G. Jurgens, *Free Radical Biol. Med.*, 1992, **13**, 341.
15. R. Amarowicz and R. B. Pegg, *Curr. Pharm. Des.*, 2017, **23**, 2754.

16. R. Stocker, *Asia Pac. J. Clin. Nutr.*, 1993, **2**(Suppl 1), 15.
17. H. Esterbauer, G. Striegl, H. Puhl and M. Rotheneder, *Free Radical Res. Commun.*, 1989, **6**, 67.
18. Y. Yamamoto, K. Kawamura, K. Tatsuno, S. Yamashita, E. Niki and C. Naito, *Oxidative Damage and Repair*, ed. K. J. A. Davies, Pergamon Press, NY, 1990, pp. 287–291.
19. N. Noguchi and E. Niki, *Free Radical Res.*, 1998, **28**, 561.
20. C. Suarna, R. L. Hood, R. T. Dean and R. Stocker, *Biochim. Biophys. Acta*, 1993, **24**, 163.
21. V. W. Bowry, D. Mohr, J. Cleary and R. Stocker, *J. Biol. Chem.*, 1995, **270**, 5756.
22. K. Sato, E. Niki and H. Shimasaki, *Arch. Biochem. Biophys.*, 1990, **279**, 402.
23. N. Gotoh, N. Noguchi, J. Tsuchiya and E. Niki, *Free Radical Res.*, 1996, **24**, 123.
24. E. Niki, Y. Yamamoto and M. Takahashi, *J. Nutr. Sci. Vitaminol.*, 1988, **34**, 507.
25. *Natural Antioxidants in Human Health and Disease*, ed. B. Frei, Academic Press, San Diego, 1994.
26. E. Niki and N. Noguchi, *Vitamin C in Health and Disease*, ed. L. Packer and E. Cadenas, Marcel Dekker Inc., New York, 1997, pp. 183–192.
27. R. Gundamaraju, R. Vemuri, W. C. Chong, D. P. Geraphty and R. Eri, *Curr. Pharm. Des.*, 2018, DOI: 10.2174/1381612824666180711122753.
28. P. Vandenabeele, L. Galluzzi, T. Vanden Berghe and G. Kroemer, *Nat. Rev. Mol. Cell Biol.*, 2010, **11**, 700.
29. F. Ursini, M. Maiorino and C. Gregolin, *Biochim. Biophys. Acta*, 1985, **839**, 62.
30. H. Imai, M. Matsuoka, T. Kumagi, T. Sakamoto and T. Koumura, *Curr. Top. Microbiol. Immunol.*, 2017, **403**, 143.
31. S. J. Dixon, K. M. Lemberg and M. R. Lamprecht, *et al.*, *Cell*, 2012, **149**, 1060.
32. M. Conrad, V. E. Kagan, H. Bayir, G. C. Pagnussat, B. Head, M. G. Traber and B. R. Stockwell, *Genes Dev.*, 2018, **32**, 602.
33. V. E. Kagan, G. Mao and F. Qu, *et al.*, *Nat. Chem. Biol.*, 2017, **13**, 81.
34. Y. Saito, Y. Yoshida, K. Nishio, M. Hayakawa and E. Niki, *Ann. N. Y. Acad. Sci.*, 2004, **1031**, 368.
35. K. Yamanaka, Y. Saito, T. Yamamori, Y. Urano and N. Noguchi, *J. Biol. Chem.*, 2011, **286**, 24666.
36. N. Noguchi, Y. Urano, W. Takabe and Y. Saito, *Free Radical Biol. Med.*, 2015, **87**, 366.
37. W. Takabe, Y. Urano and D. K. Vo, *et al.*, *J. Lipid Res.*, 2016, **57**, 2005.
38. T. Nakazawa, Y. Miyanoki, Y. Urano, M. Uehara, Y. Saito and N. Noguchi, *J. Steroid Biochem. Mol. Biol.*, 2017, **169**, 69.
39. Y. Kimura, M. Asa, Y. Urano, Y. Saito, K. Nishikawa and N. Noguchi, *Biochimie*, 2018, **153**, 203.

40. K. Yamanaka, Y. Saito, J. Sakiyama, Y. Ohuchi, F. Oseto and N. Noguchi, *RSC Adv.*, 2012, **2**, 7894.
41. M. Takahashi, J. Tsuchiya, E. Niki and S. Urano, *J. Nutr. Sci. Vitaminol.*, 1988, **34**, 25.
42. Y. Yoshida and E. Niki, in *Tocotrienols: Vitamin E Beyond Tocopherols*, ed. R. R. Watson and V. R. Preedy, CRC Press, New York, 2008, pp. 109–117.

CHAPTER 13

Oxidation Products of Vitamin E with Lipid-derived Free Radicals

RYO YAMAUCHI*

Gifu University, Department of Applied Life Science, Faculty of Applied Biological Sciences, 1-1 Yanagido, Gifu City, Gifu 501-1193, Japan
*E-mail: yamautir@gifu-u.ac.jp

13.1 Introduction

Vitamin E compounds (tocopherols and tocotrienols) are well recognized for their effective inhibition of lipid peroxidation in food and living cells.[1–4] Since the major vitamin E forms in dietary sources are α- and γ-tocopherols (αTH and γTH), these THs are more abundant than other vitamin E compounds in human and animal tissues.[5,6] The antioxidative activity of TH is related to scavenging free radicals to inhibit the propagation steps of lipid peroxidation.[1,2] Figure 13.1 shows the peroxidation (or autoxidation) process of unsaturated lipids and reactions with TH. The reactive substructure of unsaturated lipids is the unconjugated *cis*,*cis*-1,4-pentadiene unit common to all diene or polyene fatty acids and esters.[7,8] For example, linoleate is an 18-carbon diene lipid that undergoes free-radical oxidation. Lipid peroxidation is a chain reaction that proceeds in three stages: initiation, propagation, and termination.[7] In the initiation reaction, a carbon-centered lipid radical

Food Chemistry, Function and Analysis No.11
Vitamin E: Chemistry and Nutritional Benefits
Edited by Etsuo Niki

Published by the Royal Society of Chemistry, www.rsc.org

(an alkyl radical, L•) is produced by the abstraction of a hydrogen atom from an unsaturated fatty acyl moiety of lipid (LH). The produced L• immediately reacts with oxygen leading to a lipid-peroxyl radical (LOO•). The resulting LOO• propagates the radical chain reaction to accumulate a lipid hydroperoxide (LOOH) until the chain-propagating LOO• meets and combines with another radical to form inactive products. TH is a chain-breaking antioxidant to inhibit the propagation step of lipid peroxidation.[1,2] TH efficiently transfers a hydrogen atom to LOO• giving LOOH and TH radical (T•). TH can also react with L• rapidly, giving LH and T•.[9] The T• thus formed is sufficiently stable to be unable to continue the chain reaction and, instead, is removed by the radical–radical coupling reaction with a second free radical to form a non-radical product (T-OOL or T-L), or they react with each other to form TH dimer (T-T).[3,10,11]

Lipid hydroperoxides, the primary products of lipid peroxidation, are relatively stable, but they are decomposed by transition metal ions, heat, and light to produce a wide range of secondary products. Transition metal ions catalyze hemolysis of LOOH that is cleaved to a lipid-alkoxyl radical (LO•) by the reduced state of metal ions, such as a ferrous ion, whereas the oxidized state of metal ions, such as a ferric ion, leads to LOO•.[12,13] The free radicals produced in these redox processes stimulate the chain-propagating reaction of lipid peroxidation, while the LO• further cleaves to yield secondary

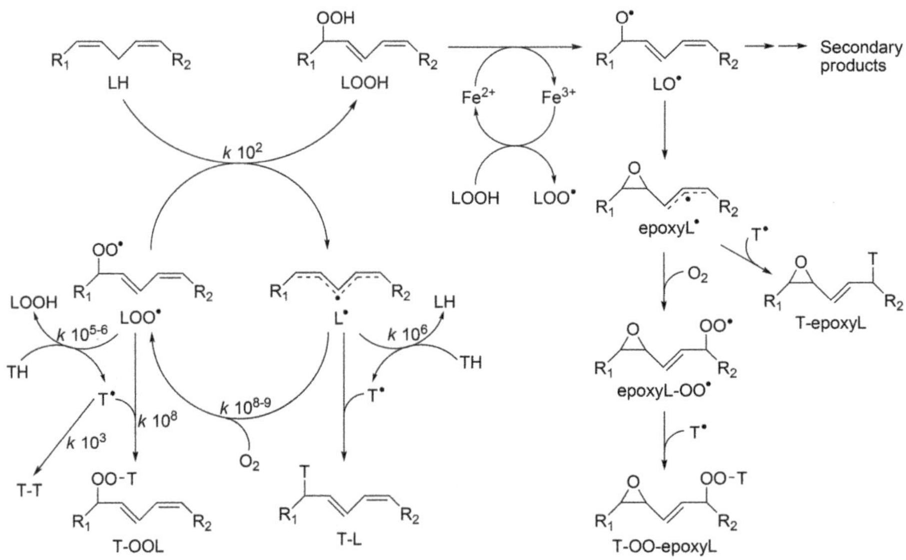

Figure 13.1 Peroxidation process of unsaturated lipids and reactions with tocopherol. LH, polyunsaturated lipid (the *cis*,*cis*-1,4-pentadiene structure of fatty acid moiety); L•, lipid-alkyl radical (carbon-centered pentadienyl radical); LO•, lipid-alkoxyl radical; LOOH, lipid hydroperoxide; LOO•, lipid-peroxyl radical; epoxyL•, epoxy-lipid allylic radical; epoxyLOO•, epoxy-lipid peroxyl radical; TH, tocopherol; T•, tocopheroxyl radical; T-T, tocopherol dimer; *k*, rate constant in M^{-1} s^{-1}.

oxidation products such as aldehydes.[14,15] In another pathway, the LO• rearranges intra-molecularly by addition to the adjacent double bond and forms a carbon-centered radical of epoxy-lipid (epoxyL•).[16] The epoxyL• couples with molecular oxygen at a diffusion-controlled rate and forms a peroxyl radical of epoxy-lipid (epoxyL-OO•). TH can trap these free radicals and might retard the secondary process of lipid peroxidation by forming non-radical addition products (T-epoxyL and T-OO-epoxyL).[17] Therefore, the formation of these TH products would provide us with more information on the antioxidant action of TH in food and biological systems.

This chapter describes the oxidation products of αTH and γTH during the peroxidation of unsaturated lipids, especially the degradation process of lipid hydroperoxides (the secondary oxidation process).

13.2 Oxidation Products of αTH with Lipid-derived Free Radicals

13.2.1 Products of αTH during the Peroxidation of Unsaturated Lipids

αTH is the most efficient chain-breaking antioxidant among the vitamin E compounds and inhibits the propagation step of lipid peroxidation by trapping peroxyl radicals.[1,2,18–20] Figure 13.2 shows the possible pathways for the reaction of αTH with lipid-derived free radicals. The first step in the oxidation of αTH (**1**) is a formation of a resonance-stabilized αTH radical, due to the donation of the phenolic hydrogen atom to a lipid-peroxyl or a lipid-alkyl radical. The delocalization of the unpaired electron also induces a radical site on the *ortho* (the C-5 and C-7) and *para* (the C-8a) positions of αTH. The αTH radical undergoes radical–radical coupling with another radical to form an inactive product. The αTH radical reacts differently with peroxyl and carbon-centered radicals. The lipid-peroxyl radical adds to the *para*-position to form 8a-(lipid-dioxy)-α-tocopherone (8a-(lipid-dioxy)-αT, **2**).[10,11] The peroxyl radical may also add to the *ortho*-position, forming 5- or 7-(lipid-dioxy)-αT followed by elimination of an alkoxyl radical and addition of oxygen to form epoxy-8a-hydroperoxy-αT (**3** or **4**).[10] These addition products (**2**–**4**) are easily hydrolyzed in aqueous solvents and in liposomes to form α-tocopherylquinone (αTQ, **5**), 2,3-epoxy-αTQ (**6**), and 5,6-epoxy-αTQ (**7**), respectively.[10] On the other hand, the lipid-alkyl radical generally added to the phenoxyl radical (the C-6 position of αTH) to form 6-*O*-lipid-αTH (**10**).[11,21] If the concentration of αTH in a reaction mixture is high, the αTH radical can undergo a self-coupling reaction to form the dimer (**8**) and trimer (**9**) of αTH.[22–24]

The reaction products of αTH with lipid-derived free radicals have been well-characterized using methyl linoleate,[25–27] cholesteryl linoleate,[28] and some molecular species of phosphatidylcholine[29,30] as the lipid sources. Furthermore, the highly-sensitive and specific methodologies have been developed for the analysis of αTH products in biological samples.[31–33]

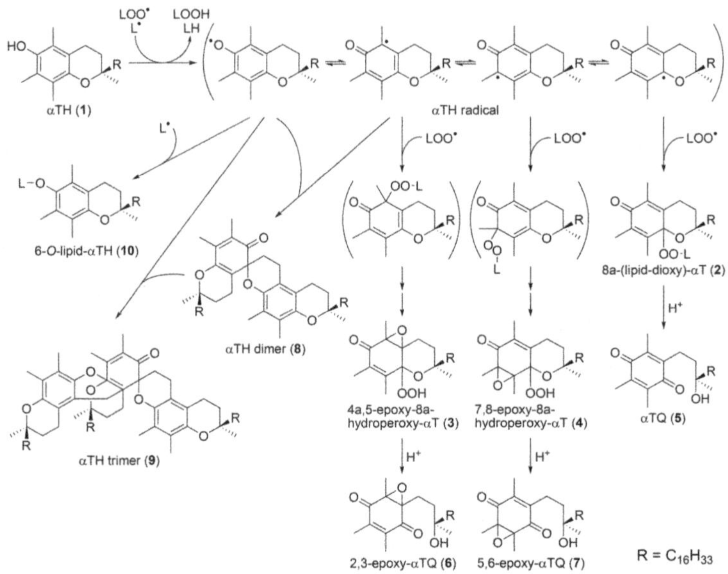

Figure 13.2 Reaction products of αTH during the peroxidation of unsaturated lipids. For abbreviations, see Figure 13.1.

13.2.2 Products of αTH on the Secondary Process of Lipid Peroxidation in Micelles and Liposomes

Lipid hydroperoxides are further cleaved oxidatively to yield peroxyl radicals or reductively to yield alkoxyl radicals by transition metal ions.[16] Both the peroxyl and alkoxyl radicals initiate new chain reactions, while the alkoxyl radicals further cleave to yield a wide range of secondary aldehydic products, such as alkanals, alkenals, 4-hydroxy-2-alkenals, and 4-oxo-2-alkenals.[14,15,34] These secondary products are more stable than hydroperoxides, and they may exert toxicological effects in biological systems.[35–37] Since such aldehydes are produced from lipid hydroperoxides *via* free radical reaction in hydrophobic circumstances, some lipid-soluble antioxidants may be expected to suppress the formation of aldehydes by trapping free radical intermediates. It has been reported that αTH could trap alkyl and peroxyl radicals derived from the iron-catalyzed decomposition of lipid hydroperoxides to form some addition products in organic solvents.[38–41]

Hemoproteins and hemin have been widely used to study initiation and propagation steps of lipid peroxidation and the decomposition of lipid hydroperoxides.[42–44] Hayashi *et al.*[43] reported that the decomposition of 1-palmitoyl-2-linoleoyl-3-*sn*-phosphatidylcholine 13-hydroperoxide (PLPC-OOH) in micellar solutions proceeded very rapidly by hemoproteins and hemin to generate secondary aldehydic products. Since αTH and other antioxidants inhibited the formation of such aldehydic products, αTH might trap free radicals derived from the heme-compound-catalyzed decomposition of PLPC-OOH. Therefore, the inhibitory effect of αTH on the formation of aldehydic products was assessed using the hemin-catalyzed reaction of PLPC-OOH in micellar and liposomal systems.[45] Hexanal was measured as the secondary product from PLPC-OOH because this aldehyde was one of the major aldehydes produced from linoleic acid hydroperoxides[46,47] and from the autoxidation of human low-density lipoprotein.[48] On the hemin-catalyzed decomposition of PLPC-OOH in micelles, αTH could trap the PLPC-OOH-derived free radicals and suppress the formation of hexanal.[45] This reaction brought the formation of an addition product (αT-OO-epoxyPLPC, **2a**) and its decomposed product, αTQ (**5**),[49] In addition, another addition product (αT-epoxyPLPC, **10a**) and αTH dimer (αTD, **8**) were detected. When the hemin-catalyzed reaction was performed in 1-palmitoyl-1-oleoyl-3-*sn*-phosphatidylcholine (POPC) liposomes, αTH could partly suppress the formation of hexanal, and the main products were αT-OO-epoxyPLPC (**2a**) and αTD (**8**).[45]

Figure 13.3 shows the possible mechanisms of the hemin-catalyzed reaction of PLPC-OOH with αTH in micellar and liposomal systems. The first step might be the reduction of PLPC-OOH by the reduced form of hemin (Hem_{red}) to produce the alkoxyl radical (PLPC-O$^•$) and the oxidized form of hemin (Hem_{ox}), or the oxidation of PLPC-OOH by Hem_{ox} to produce the corresponding peroxyl radical and Hem_{red}.[44,50] In the presence of αTH, Hem_{ox} is reduced by αTH immediately and forms Hem_{red} and the αTH radical (αT$^•$).[51] Consequently, the hemin-catalyzed reaction preferentially produces PLPC-O$^•$ and αT$^•$. The PLPC-O$^•$ then rearranges intra-molecularly by addition to the adjacent double bond and forms the carbon-centered radical of 12,13-epoxy-PLPC (epoxyPLPC$^•$).[16,52] The epoxyPLPC$^•$ couples with oxygen to form the peroxyl radical of 12,13-epoxy-PLPC (epoxyPLPC-OO$^•$) in the presence of oxygen.[53] The produced epoxyPLPC-OO$^•$ could be trapped by αT$^•$ to form a stable product, αT-OO-epoxyPLPC (**2a**), which was then easily hydrolyzed to αTQ (**5**) in micellar systems.[41,45] In addition, some epoxyPLPC$^•$ could react with αT$^•$ before the reaction with oxygen to produce αT-epoxyPLPC (**10a**) under the αTH-rich conditions in micelles. However, this reaction might be rare in biological systems due to the low production of αT-epoxyPLPC in POPC liposomes. The αTH radical also resulted in the formation of αTD (**8**) by a bimolecular self-reaction under the αTH-rich conditions. The overall hemin-catalyzed reaction proceeded very rapidly and terminated within several minutes.[43,45]

The addition products of αTH with epoxylipid-peroxyl radicals have been detected in the peroxidized human plasma.[17,32] Therefore, αTH may affect the secondary process of lipid peroxidation in biological systems.

13.3 Oxidation Products of γTH

13.3.1 Products of γTH During the Peroxidation of Unsaturated Lipids

γTH is one of the predominant homologs of vitamin E in dietary sources and acts as a chain-breaking antioxidant.[5,6,54] γTH is structurally different from αTH only by the absence of a methyl group in the 5-position of the chroman

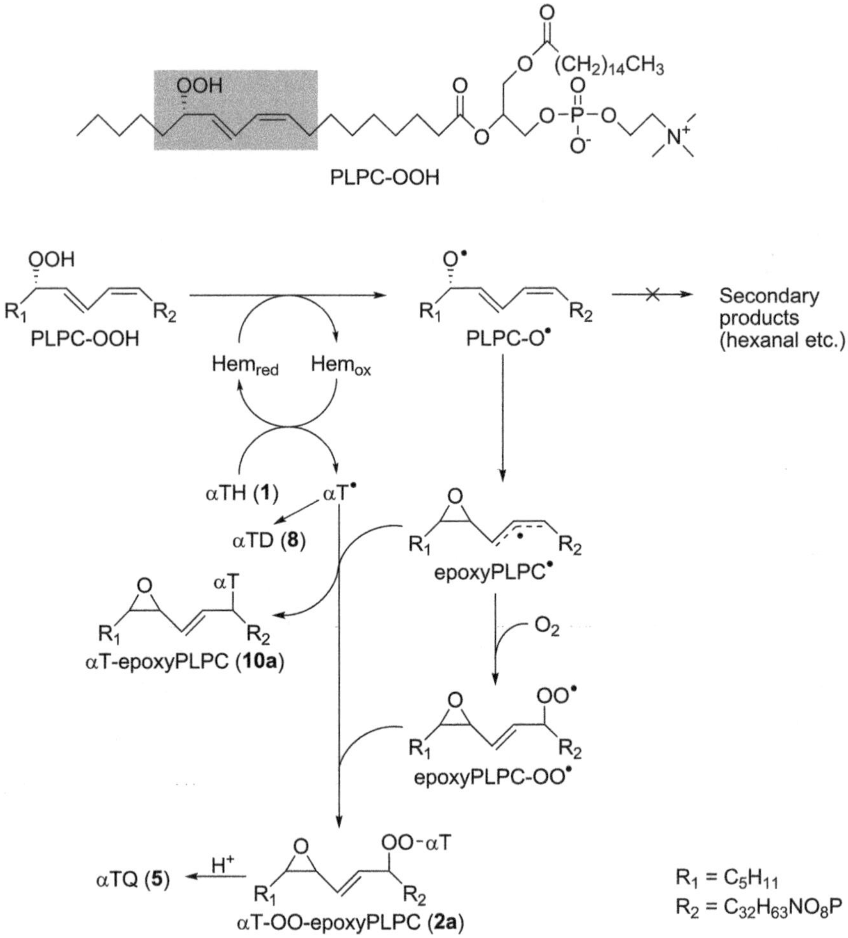

Figure 13.3 Hemin-catalyzed reaction of PLPC-OOH with αTH in micelles and liposomes. Hem$_{ox}$, the oxidized form of hemin; Hem$_{red}$, the reduced form of hemin.

ring. The unsubstituted C-5 position in γTH traps lipophilic electrophiles such as reactive nitrogen oxide species and thus counters the oxidative stress.[55–57] Although the reaction of γTH with lipid-derived free radicals would proceed in a manner similar to that of αTH, its oxidation products have not been satisfactorily characterized. There have only been reports of the products of γTH with alkylperoxyl radicals produced from thermal decomposition of an azo compound, 2,2′-azobis(2,4-dimethylvaleronitrile), in benzene.[58]

Figure 13.4 shows the possible pathways for the reaction of γTH with free radicals in lipids or lipophilic solvents.[3,11,58–60] γTH (**11**) transfers a hydrogen atom to a lipid-peroxyl or a lipid-alkyl radical. The resulting γTH radical is then resonance-stabilized and reacts irreversibly with another free radical or with another γTH radical. The reaction between the C-8a position of the γT radical and a lipid-peroxyl radical yields 8a-(lipid-dioxy)-γT (**12**), which

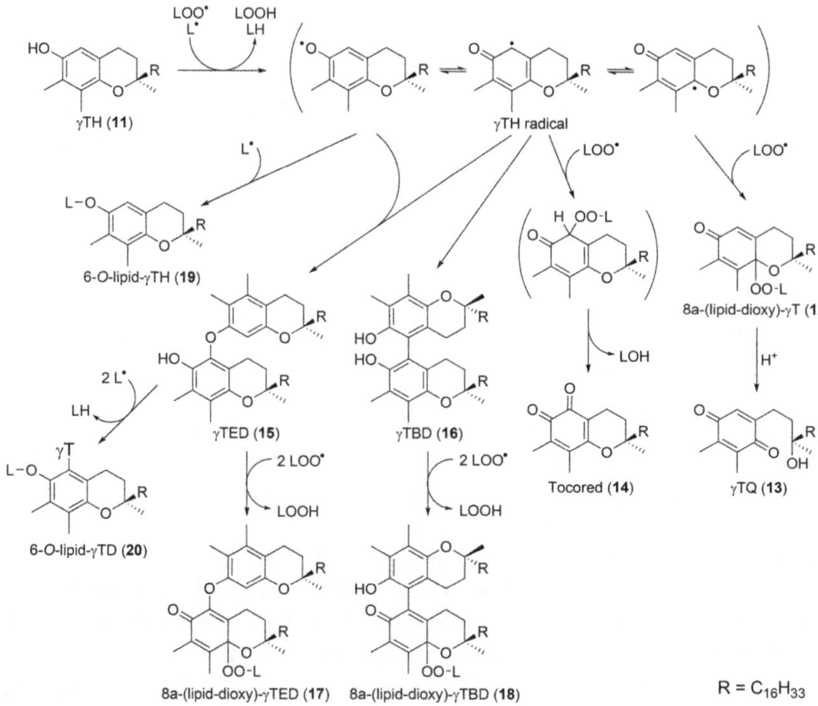

Figure 13.4 Reaction products of γTH during the peroxidation of unsaturated lipids. For abbreviations, see Figure 13.1.

then hydrolyzes to γTQ (**13**).[58,59] The formation of γTH-5,6-quinone (tocored, **14**) is thought to proceed *via* initial addition of a lipid-peroxyl radical to the C-5 position of the γT radical, followed by elimination of alcohol.[55] The rate constant for the bimolecular self-reaction of the γTH radical has been determined to be 10 times higher than that of the αTH radical,[61] and so the γTH radical easily reacts with another γTH radical and forms two types of γTH dimer: diphenylether dimer (γTED, **15**) and biphenyl dimer (γTBD, **16**).[58,62] These γTH dimers can further react with lipid-peroxyl radicals because of the presence of one or two phenolic hydroxyl groups in each molecule, and form addition products, 8a-(lipid-dioxy)-γTED (**17**) and 8a-(lipid-dioxy)-γTBD (**18**).[57,59] If the supply of oxygen is poor, γTH and its dimers may trap lipid-alkyl radicals to form 6-*O*-lipid-γTH (**19**) and 6-*O*-lipid-γTH dimers (6-*O*-lipid-γTD, **20**), respectively.[59]

13.3.2 Iron-catalyzed Reaction of Methyl Linoleate Hydroperoxides with γTH in Aprotic and Protic Solvents

The oxidation products of γTH with lipid-peroxyl radicals were characterized using the iron-catalyzed decomposition of methyl linoleate hydroperoxides (MeL-OOH).[39,40] γTH was reacted with MeL-OOH in the presence of a lipid-soluble iron chelate, Fe(III)-acetylacetonate, in toluene under air-sufficient conditions.[60] The reaction products of γTH were characterized as methyl (8a-dioxy-γ-tocopherone)-octadecadienoates (γT-OO-MeL, **12a**), methyl (8a-dioxy-γ-tocopherone)-epoxyoctadecenoates (γT-OO-epoxyMeL, **12b**), tocored (**14**), γTED (**15**), γTBD (**16**), and adducts of γTH dimers with the MeL-OOH-derived peroxyl radicals, γTED-OO-MeL (**17a**), γTED-OO-epoxyMeL (**17b**), and γTBD-OO-MeL (**18b**).

Figure 13.5 shows the possible mechanisms of the iron-catalyzed reaction of MeL-OOH with γTH in toluene solution. The iron-dependent decomposition of MeL-OOH in the aprotic solvent produces a peroxyl radical (MeL-OO˙) by the oxidized form of iron ion, ferric ion, and an alkoxyl radical (MeL-O˙) by the reduced form of iron ion, ferrous ion.[40] Ferric ion is much less reactive toward lipid hydroperoxides than ferrous ion.[51,63,64] Thus, ferric ion is a rate-limiting factor for the redox cycling between ferric and ferrous ions, and this is the reason why the rate of iron-catalyzed reaction in toluene solution has been slow.[60] The MeL-O˙ rapidly adds to the adjacent double bond to form the carbon-centered epoxy-MeL radical (epoxyMeL˙) followed by the reaction of oxygen to form epoxy-MeL-peroxyl radical (epoxyMeL-OO˙) under air-sufficient conditions.[40] γTH traps these MeL-OOH-derived peroxyl radicals by hydrogen-atom transfer, giving hydroperoxides (MeL-OOH and epoxyMeL-OOH) and the γTH radical (γT˙). The γT˙ reacts rapidly with the second peroxyl radical to form addition products, γT-OO-MeL (**12a**) and γT-OO-epoxyMeL (**12b**).

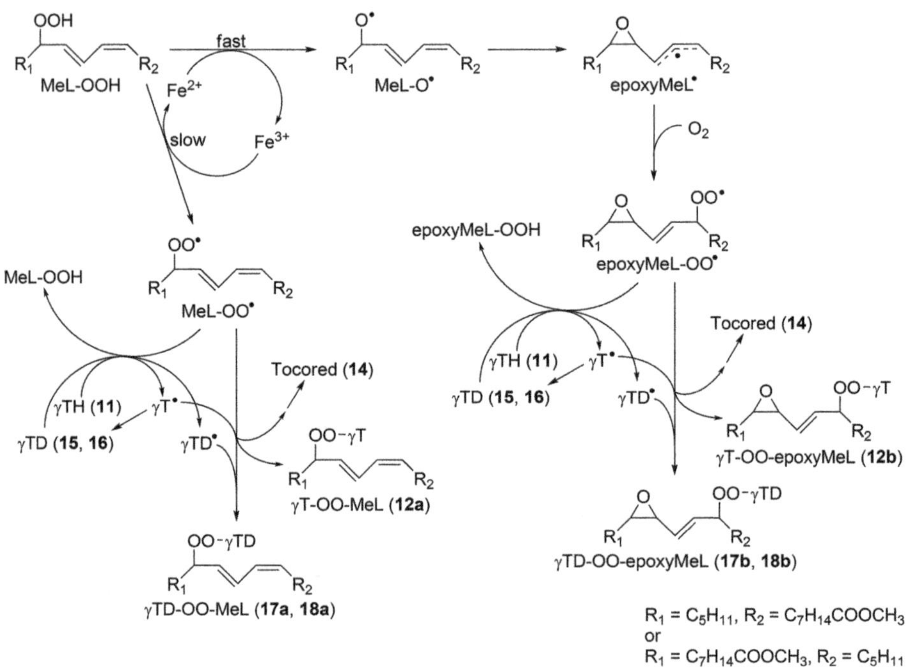

Figure 13.5 Iron-catalyzed reaction of MeL-OOH with γTH in toluene.

The bimolecular self-reaction of γT˙ yields γTH dimers (γTD), that is, γTED (**15**) and γTBD (**16**). The formed γTD can trap additional peroxyl radicals. Consequently, γTD-OO-MeL (**17a** and **18a**) and γTD-OO-epoxyMeL (**17b** and **18b**) are formed as the addition products of γTD radicals (γTD˙) with the MeL-OOH-derived peroxyl radicals. Another oxidation product is tocored (**14**). Tocored has been reported as the reaction product with an alkylperoxyl radical[58] and with reactive nitrogen oxide species.[55] Thus, the unsubstituted C-5 position of γT˙ might react with MeL-OOH-derived peroxyl radicals to form tocored.

The iron-catalyzed reaction of MeL-OOH with γTH in methanol solution proceeded relatively quickly compared with that in toluene solution and yielded only γT-OO-epoxyMeL (**12b**) and γTED-OO-epoxyMeL (**17b**) as the addition products.[60] It has been reported that iron-catalyzed reaction of αTH with PLPC-OOH in methanol or in liposomes gave only the addition products of αTH with epoxylipid-peroxyl radicals originated from the PLPC-OOH-derived alkoxyl radicals.[41] In such protic solvents, the first reaction was the reduction of ferric ion by αTH to give the more reactive ferrous ion, which reduced PLPC-OOH to the corresponding alkoxyl radical.[41,51] Thus, γTH and its dimers might primarily act as reductants of ferric ion in methanol solution, and the resulting ferrous ion has decomposed MeL-OOH to the alkoxyl radicals.

The antioxidant action of γTH is usually performed in bulk oils or in emulsions.[65] Therefore, the detection of these oxidation products of γTH in bulk oils or in emulsions would prove the antioxidant function of γTH under such conditions.

13.3.3 Hemin- and Myoglobin-catalyzed Reaction of PLPC-OOH with γTH in Micelles and Liposomes

γTH is one of the principal forms of vitamin E in biological tissues and may be important to human health.[5,6] Devaraj *et al.*[66] have reported the effect of γTH supplementation alone and in combination with αTH on biomarkers of oxidative stress and inflammation in human subjects with metabolic syndrome. The supplementation of γTH alone significantly reduced biomarkers of oxidative stress, and plasma lipid peroxides as well as malondialdehyde and 4-hydroxy-2-nonenal, similar to αTH alone or the combination. Therefore, γTH, along with αTH, would also be expected to suppress the formation of secondary aldehydic products by trapping lipid-derived free radicals.

The effect of γTH on the formation of aldehydic products has been studied using the hemin- and myoglobin-catalyzed decomposition of PLPC-OOH in micellar and in POPC liposomal systems.[59] The decomposition of PLPC-OOH catalyzed by heme compounds resulted in the formation of hexanal and 4-hydroxy-2-nonenal as the major secondary oxidation products.[46,67] In micellar solutions, γTH could suppress the formation of these aldehydic products by trapping PLPC-OOH-derived free radical intermediates. The suppressive effect by γTH was almost the same as that by αTH.[59] This suggests that the reactivity of γTH with PLPC-OOH-derived free radicals in micellar solutions might be near equivalent to that of αTH. On the other hand, the suppressive effect by γTH in POPC liposomes was weak. It has been reported the reactivity of TH toward peroxyl radicals was reduced in liposomes, by one-fourth compared with those in micelles.[68] Therefore, γTH might trap a part of the free radicals produced from PLPC-OOH in POPC liposomal systems. During the heme compound-catalyzed reaction, some oxidation products of γTH were detected. They were γTQ (**13**), tocored (**14**), γTED (**15**), γTBD (**16**), the addition products with peroxyl radicals, γT-OO-epoxyPLPC (**12c**) and γTD-OO-epoxyPLPC (**17c** and **18c**), and the addition products with alkyl radicals, γT-epoxyPLPC (**19a**) and γTD-epoxyPLPC (**20a**).

Figure 13.6 shows the possible mechanisms of hemin- or myoglobin-catalyzed reaction of PLPC-OOH with γTH in micelles and liposomes. The reaction would proceed in a manner similar to that in the reaction with αTH.[45] At the first step, γTH reduces the oxidized state of heme compounds (Hem_{ox})

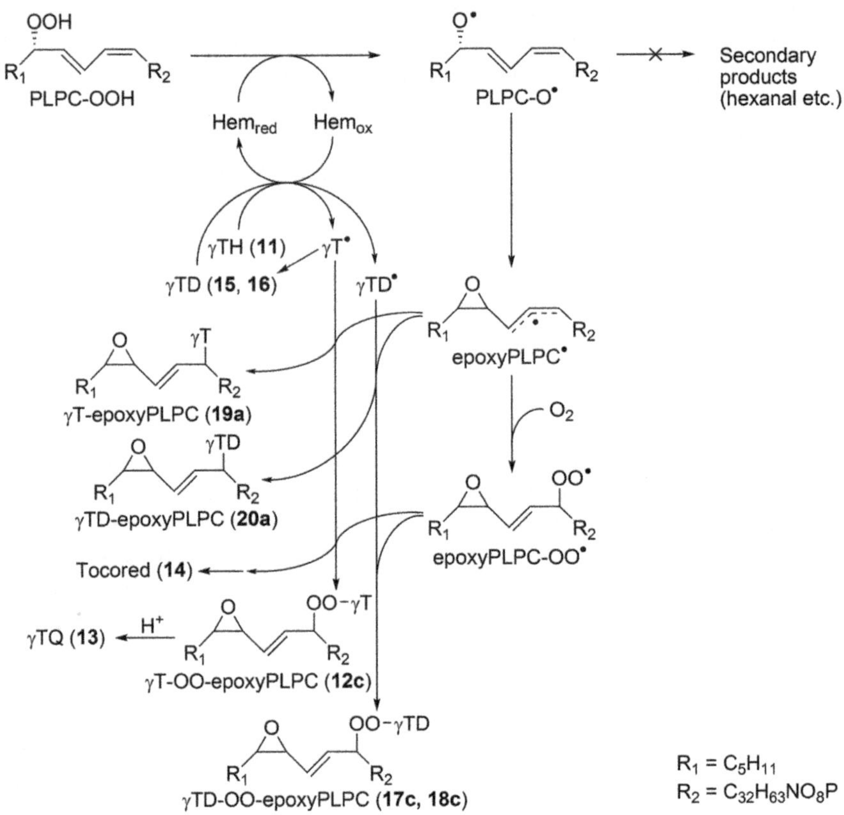

Figure 13.6 Hemin- or myoglobin-catalyzed reaction of PLPC-OOH with γTH in micelles and liposomes. Hem_{ox}, the oxidized form of hemin or myoglobin; Hem_{red}, the reduced form of hemin or myoglobin.

to their reduced state (Hem_{red}), and forms γT$^•$. The Hem_{red} then reduces PLPC-OOH to produce the alkoxyl radical (PLPC-O$^•$). Thus, the heme compound-catalyzed reaction preferentially produces PLPC-O$^•$ and γT$^•$. The PLPC-O$^•$ then rearranges intra-molecularly and forms epoxyPLPC$^•$ followed by the addition of oxygen to form epoxyPLPC-OO$^•$. The produced epoxyPLPC$^•$ and epoxyPLPC-OO$^•$ are coupled with γT$^•$ to form γT-OO-epoxyPLPC (**12c**) and γT-epoxyPLPC (**19a**), respectively. γT-OO-epoxyPLPC (**12c**) was easily hydrolyzed to form γTQ (**13**) in micelles.[59] The γT$^•$ reacts with another γT$^•$ to form γTD (**15** and **16**) if the concentration of γTH would be high.[59] These dimers are further oxidized by Hem_{ox}, and the resulting γTD$^•$ reacts with epoxyPLPC$^•$ and epoxyPLPC-OO$^•$ to form the addition products, γTD-epoxyPLPC (**20a**) and γTD-OO-epoxyPLPC (**17c**, **18c**). In addition, the C-5 position of γT$^•$ might react with PLPC-OOH-derived peroxyl radicals to form tocored (**14**) with high yields.[59]

γTH is relatively abundant in some human tissues,[5,6,69,70] and it might act as a free radical scavenger in such tissues to suppress both the primary and secondary processes of lipid peroxidation. Furthermore, γTH is known to be a better nucleophile than αTH and may scavenge electrophiles with high affinity.[5,6,56] It has been demonstrated that γTH had the capability to scavenge nitric oxide by nitration, resulting in the formation of 5-nitro-γTH.[55] 5-Nitro-γTH has been detected in human plasma[57,70] and in plant tissues,[71] and is recognized to be a marker for oxidative stress. In addition, γTH could trap hypochlorous acid, a potential biological oxidant and electrophile, to form 5-chloro-γTH and other chlorinated products.[72] 4-Oxo-2-nonenal, the bifunctional electrophile produced from the decomposition of lipid hydroperoxides, has also been trapped by γTH to form 5-substituted-γTH.[73] Therefore, γTH may trap not only free radicals but also some electrophiles to suppress oxidative stress in biological systems.

Acknowledgements

This work was supported by a Grant-in-Aid for Scientific Research (Nos. 23580162 and 26450155) from Japan Society for the Promotion of Science.

References

1. G. W. Burton and K. U. Ingold, *J. Am. Chem. Soc.*, 1981, **103**(21), 6472.
2. E. Niki, T. Saito, A. Kawakami and Y. Kamiya, *J. Biol. Chem.*, 1984, **259**(7), 4177.
3. A. Kamal-Eldin and L.-Å. Appelqvist, *Lipids*, 1996, **31**(7), 671.
4. C. M. Seppanen, Q. Song and A. S. Csallany, *J. Am. Oil Chem. Soc.*, 2010, **87**(5), 469.
5. Q. Jiang, S. Christen, M. K. Shigenaga and B. N. Ames, *Am. J. Clin. Nutr.*, 2001, **74**(6), 714.
6. K.-H. Wagner, A. Kamal-Eldin and I. Elmadfa, *Ann. Nutr. Metab.*, 2004, **48**(3), 169.

7. E. N. Frankel, *Prog. Lipid Res.*, 1980, **19**(1–2), 1.
8. N. A. Porter, S. E. Caldwell and K. A. Mills, *Lipids*, 1995, **30**(4), 277.
9. P. Franch, M. Lucarini, G. F. Pedulli, L. Valgimigli and B. Lunelli, *J. Am. Chem. Soc.*, 1999, **121**(3), 507.
10. D. C. Liebler, *Crit. Rev. Toxicol.*, 1993, **23**(2), 147.
11. R. Yamauchi, *Food Sci. Technol. Int., Tokyo*, 1997, **3**(4), 301.
12. B. Halliwell and J. M. C. Gutteridge, *Biochem. J.*, 1984, **219**(1), 1.
13. A. Sevanian and P. Hochstein, *Annu. Rev. Nutr.*, 1985, **5**, 365.
14. H. Esterbauer, R. J. Schaur and H. Zollner, *Free Radical Biol. Med.*, 1991, **11**(1), 81.
15. P. Spiteller, W. Kern, J. Reiner and G. Spiteller, *Biochim. Biophys. Acta*, 2001, **1531**(3), 188.
16. H. W. Gardner, *Free Radical Biol. Med.*, 1989, **7**(1), 65.
17. R. Yamauchi, *Vitam. Horm.*, 2007, **76**, 309.
18. G. W. Burton and M. G. Traber, *Annu. Rev. Nutr.*, 1990, **10**, 357.
19. M. G. Traber and J. Atkinson, *Free Radical Biol. Med.*, 2007, **43**(1), 4.
20. E. Niki, *Free Radical Biol. Med.*, 2014, **66**, 3.
21. C. Evans, J. C. Scaiano and K. U. Ingold, *J. Am. Chem. Soc.*, 1992, **114**(12), 4589.
22. A. S. Csallany, M. Chiu and H. H. Draper, *Lipids*, 1970, **5**(1), 63.
23. R. Yamauchi, K. Kato and Y. Ueno, *Lipids*, 1988, **23**(8), 779.
24. H. Schröder and T. Netscher, *Magn. Reson. Chem.*, 2001, **39**(11), 701.
25. R. Yamauchi, T. Matsui, K. Kato and Y. Ueno, *Lipids*, 1990, **25**(3), 152.
26. R. Yamauchi, N. Miyake, K. Kato and Y. Ueno, *Lipids*, 1993, **28**(3), 201.
27. R. Yamauchi, K. Kato and Y. Ueno, *J. Agric. Food Chem.*, 1995, **43**(6), 1455.
28. R. Yamauchi, T. Kamatani, M. Shimoyamada and K. Kato, *Biosci. Biotechnol. Biochem.*, 2002, **66**(3), 670.
29. R. Yamauchi, Y. Yagi and K. Kato, *Biochim. Biophys. Acta*, 1994, **1212**(1), 43.
30. R. Yamauchi, H. Mizuno and K. Kato, *Biosci. Biotechnol. Biochem.*, 1998, **62**(7), 1293.
31. D. C. Liebler, J. A. Burr, L. Philips and A. J. L. Ham, *Anal. Biochem.*, 1996, **236**(1), 27.
32. R. Yamauchi, H. Noro, M. Shimoyamada and K. Kato, *Lipids*, 2002, **37**(5), 515.
33. D. Pollok and H.-U. Melchert, *J. Chromatogr. A*, 2004, **1056**(1–2), 257.
34. X. Gu and R. G. Salomon, *Free Radical Biol. Med.*, 2012, **52**(3), 601.
35. I. A. Blair, *Exp. Gerontol.*, 2001, **36**(9), 1473.
36. F. Guéraud, M. Atalay, N. Bresgen, A. Cipak, P. M. Eckl, L. Huc, I. Jouanin, W. Siems and K. Uchida, *Free Radical Res.*, 2010, **44**(10), 1098.
37. K. S. Fritz and D. R. Petersen, *Free Radical Biol. Med.*, 2013, **59**, 85.
38. H. W. Gardner, K. Eskins, G. Grams and G. E. Inglett, *Lipids*, 1972, **7**(5), 324.
39. T. Kaneko and M. Matsuo, *Chem. Pharm. Bull.*, 1985, **33**(5), 1899.
40. R. Yamauchi, N. Yamamoto and K. Kato, *Lipids*, 1995, **30**(5), 395.

41. R. Yamauchi, K. Ozaki, M. Shimoyamada and K. Kato, *Chem. Phys. Lipids*, 2002, **114**(2), 193.
42. E. H. Kim and A. Sevanian, *Arch. Biochem. Biophys.*, 1991, **288**(2), 324.
43. T. Hayashi, K. Uchida, G. Takebe and K. Takahashi, *Free Radical Biol. Med.*, 2004, **36**(8), 1025.
44. C. Carlsen, J. K. S. Moller and L. H. Skibsted, *Coord. Chem. Rev.*, 2005, **249**(3–4), 485.
45. R. Yamauchi, S. Watanabe, A. S. Martin and S. Iwamoto, *Chem. Phys. Lipids*, 2014, **184**, 61.
46. P. Schieberle and W. Grosch, *J. Am. Oil Chem. Soc.*, 1981, **58**(5), 602.
47. X. Gu and R. G. Salomon, *Free Radical Biol. Med.*, 2012, **52**(3), 601.
48. H. Esterbauer, G. Jürgens, O. Quehenberger and E. Koller, *J. Lipid Res.*, 1987, **28**(4), 495.
49. R. Yamauchi, Y. Yagi and K. Kato, *Biosci. Biotechnol. Biochem.*, 1996, **60**(4), 616.
50. S. Koga, M. Nakano and K. Uehara, *Arch. Biochem. Biophys.*, 1991, **289**(2), 223.
51. K. Yamamoto and E. Niki, *Biochim. Biophys. Acta*, 1988, **958**(1), 19.
52. A. L. Wilcox and L. J. Marnett, *Chem. Res. Toxicol.*, 1993, **6**(4), 413.
53. M. Erben-Russ, W. Bors and M. Saran, *Int. J. Radiat. Biol.*, 1987, **52**(3), 393.
54. C. M. Seppanen, Q. Songand and A. S. Csallany, *J. Am. Oil Chem. Soc.*, 2010, **87**(5), 469.
55. S. Christen, A. A. Woodall, M. Shigenaga, P. T. Southwell-Keely, M. W. Duncan and B. N. Ames, *Proc. Natl. Acad. Sci. U. S. A.*, 1997, **94**(7), 3217.
56. Q. Jiang, *Free Radical Biol. Med.*, 2014, **72**, 76.
57. L. W. Morton, N. C. Ward, K. D. Croft and I. B. Puddy, *Biochem. J.*, 2002, **364**, 625.
58. R. Yamauchi, T. Matsui, K. Kato and Y. Ueno, *Agric. Biol. Chem.*, 1990, **54**(10), 2703.
59. R. Yamauchi, T. Kinoshita, Y. Hasegawa and S. Iwamoto, *Chem. Phys. Lipids*, 2017, **209**, 37.
60. R. Yamauchi, T. Kinoshita and S. Iwamoto, *J. Am. Oil Chem. Soc.*, 2018, **95**(3), 361.
61. G. W. Burton, T. Doba, E. J. Gabe, L. Hughes, F. L. Lee, L. Prasad and K. U. Ingold, *J. Am. Chem. Soc.*, 1985, **107**(24), 7053.
62. J. L. G. Nilsson, G. D. Daves, Jr and K. Folkers, *Acta Chem. Scand.*, 1968, **22**, 207.
63. J. R. Bucher, M. Tien and S. D. Aust, *Biochem. Biophys. Res. Commun.*, 1983, **111**(3), 777.
64. R. Mozuraityte, T. Rustad and I. Storrø, *J. Agric. Food Chem.*, 2008, **56**(2), 537.
65. W. Chaiyasit, R. J. Elias, D. J. McClements and E. A. Decker, *Crit. Rev. Food Sci. Nutr.*, 2007, **47**(3), 299.
66. S. Devaraj, S. Leonard, M. G. Traber and I. Jialal, *Free Radical Biol. Med.*, 2008, **44**(6), 1203.

67. K. Umegaki, M. Sano, K. Suzuki, I. Tomita and T. Esashi, *Bone Marrow Transplant.*, 1999, **23**, 173.
68. L. R. C. Barclay, K. A. Baskin, K. A. Dakin, S. J. Locke and M. R. Vinqvist, *Can. J. Chem.*, 1990, **68**(12), 2258.
69. G. W. Burton, M. G. Traber, R. V. Acuff, D. N. Walters, H. Kayden, L. Hughes and K. U. Ingold, *Am. J. Clin. Nutr.*, 1998, **67**(4), 669.
70. S. W. Leonard, R. S. Bruno, E. Paterson, B. C. Schock, J. Atkinson, T. M. Bray, C. E. Cross and M. G. Traber, *Free Radical Biol. Med.*, 2003, **35**(12), 1560.
71. C. Desel, E. M. Hubbermann, K. Schwarz and K. Krupinska, *Planta*, 2007, **226**(5), 1311.
72. Q. Nguyen and P. T. Southwell-Keely, *Lipids*, 2007, **42**(2), 171.
73. F. Saito, S. Iwamoto and R. Yamauchi, *Biosci. Biotechnol. Biochem.*, 2010, **74**(1), 168.

CHAPTER 14

Metabolism of Vitamin E

REGINA BRIGELIUS-FLOHÉ*

German Institute of Human Nutrition Potsdam-Rehbruecke, Arthur-Scheunert-Allee 114-116, 14558 Nuthetal, Germany
*E-mail: flohe@dife.de

14.1 Introduction

Vitamin E was discovered in 1922 by Evans and Bishop as a dietary factor (substance X) indispensable in the reproduction of rats.[1] It was rediscovered in the fifties as factor 2 by Klaus Schwarz together with sulfur amino acids (factor 1) and selenium (factor 3).[2] All factors were able to prevent liver necrosis mediated by a diet consisting of Torula yeast only.[3] Since then, interest in vitamin E research has increased consistently. Apart from its as yet unclear function as a vitamin, its role in human and animal health and disease, its functions beyond antioxidative ones, and also its metabolism have gained interest. This review summarizes the history of vitamin E metabolism and the findings over the past years. A number of recent reviews presenting more details are available.[4,5]

14.1.1 Basics

Vitamin E is the term for a family consisting of four tocopherols (TOH) (α, β, γ, δ) and four tocotrienols (T3) (α, β, γ, δ).[6] All forms of vitamin E consist of a chroman-6-ol ring and a hydrocarbon side chain, which is saturated in tocopherols and has three double bonds in tocotrienols. The α, β, γ, and δ

Food Chemistry, Function and Analysis No.11
Vitamin E: Chemistry and Nutritional Benefits
Edited by Etsuo Niki

Published by the Royal Society of Chemistry, www.rsc.org

forms differ in the number and position of methyl groups on the chromanol structure (Figure 14.1). α-TOH is the most abundant form in nature and, based on fetal resorption assays, has the highest biological activity. Tocopherols contain three chiral centers, two in the side chain at C4′ and C8′ and one at C2 in the chromanol ring. Natural forms of tocopherols have the *RRR* configuration at all three sites, while synthetic forms can be racemic mixtures of all eight possible *R* and *S* configurations (*all rac* or *DL*). Tocotrienols have only the chiral center at C2, which has the *R* configuration in natural forms. The racemic mixture is present in synthetic forms, reviewed in ref. 7.

14.1.2 Biosynthesis

Tocopherols and tocotrienols are synthesized exclusively in photosynthetic organisms including plants, algae and some cyanobacteria. Synthesis starts from homogentisic acid for the formation of the aromatic ring, which is condensed with phytyl diphosphate (PDP) for tocopherols or with geranylgeranyl diphosphate (GGDP) for tocotrienols (Figure 14.2). The condensation to tocopherols is catalyzed by homogentisate phytyltransferase (HPT) yielding 2-methyl-6-phytyl-1,4-benzoquinol (MPBQ). Cyclization without further methylation results in δ-TOH, which is processed to β-TOH by methylation at position 5. Methylation of MPBQ at position 3 before cyclization results in 2,3-dimethyl-6-phytyl-1,4-benzoquinol, which is cyclized to γ-TOH from which α-TOH is built by methylation at position 5.[8–10] During all cyclization reactions for tocopherols, the first three carbons of the phytyl residue are

RRR-α-Tocopherol (2*R*4′*R*8′*R*)

R-α-Tocotrienol

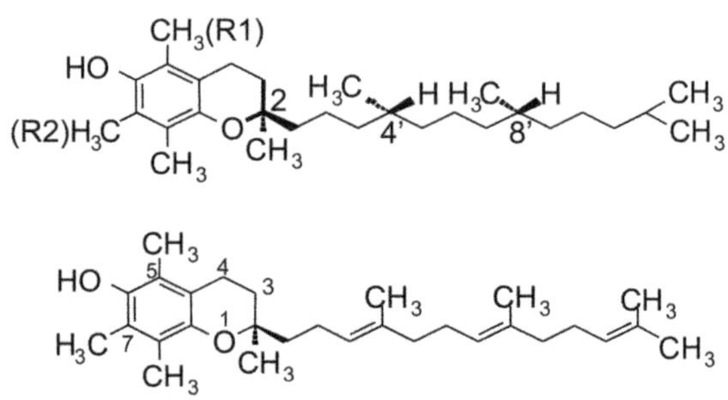

Position of methyl groups in β, γ, and δ forms of vitamin E

	R1(5)	R2(7)	
β-Toco-	CH_3	H	-pherol/-trienol
γ-Toco-	H	CH_3	-pherol/-trienol
δ-Toco-	H	H	-pherol/-trienol

Figure 14.1 Structures of tocopherols and tocotrienols with the methyl groups and position on the chroman ring shown in the table. Position numbering in the chromanol is shown in tocotrienol. Chiral centers on the side chain of tocopherol are indicated at C2, C4′ and C8′.

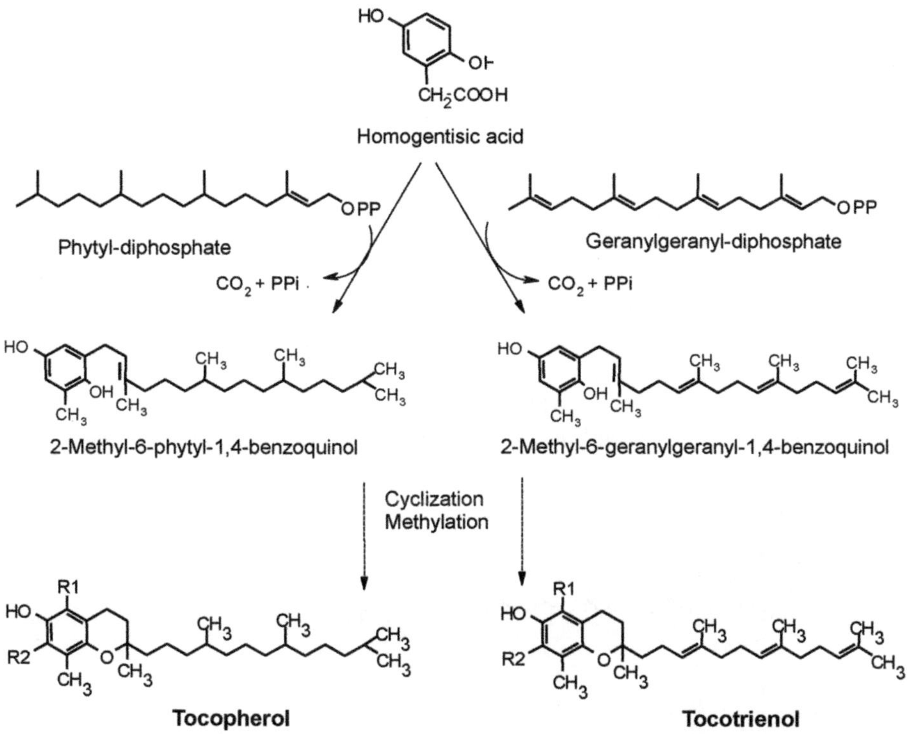

Figure 14.2 Vitamin E biosynthesis in plants. Either phytyl-diphosphate (tocopherols) or geranylgeranyl-diphosphate (tocotrienols) is attached to homogentisic acid to form 2-methyl-6-phytyl-1,4-benzoquinol or 2-methyl-6-geranylgeranyl-1,4-benzoquinol, respectively. This step is followed by cyclization and methylation reactions to end up with tocopherols or tocotrienols (for details, see text).

incorporated into the chromanol ring. Thus, taken seriously, the side chain in tocopherols is no longer a phytyl chain as often described in the literature, but rather a phytyl-derived prenyl chain. Synthesis of tocotrienols is performed respectively.

14.2 Metabolism

14.2.1 History

The first vitamin E metabolites described are the so-called Simon metabolites, α-tocopheronic acid and the lactone derived therefrom.[11] Both metabolites have an opened chromanol structure and a shortened side chain, which was taken as proof that tocopherol had reacted as an antioxidant. α-Tocopheronic acid was considered to be derived from α-tocopherylquinone formed in the reaction of the tocopheroxyl radical with a peroxyl radical.[12] The quinone could be reduced by NAD(P)H-dependent enzymes like NADPH-cytochrome

P450 reductase[13] to α-tocopherylhydroquinone, which was metabolized to α-tocopheronic acid by side-chain degradation (Figure 14.3).

The Simon metabolites were excreted as either sulfates or glucuronides *via* the urine and were found in increased amounts after increasing the α-tocopherol intake. They were, therefore, taken as indicators that vitamin E had fulfilled its destination and reacted as an antioxidant. However, this interpretation did not explain why the oxidative destruction of vitamin E increased with just an increasing intake.

Alternative metabolites were suggested already in 1965. Schmandke[14] observed that oral administration of α-tocopherol and 2,5,7,8-tetramethyl-2(β-carboxyethyl)-6-acetoxychroman to rats resulted in the excretion of α-2,5,7,8-tetramethyl-2(β-carboxyethyl)-6-hydroxychroman (α-CEHC) as well as of α-tocopheronolactone (α-TL). He postulated that the lactone was built during the extraction procedure and that α-CEHC might be the only excretion

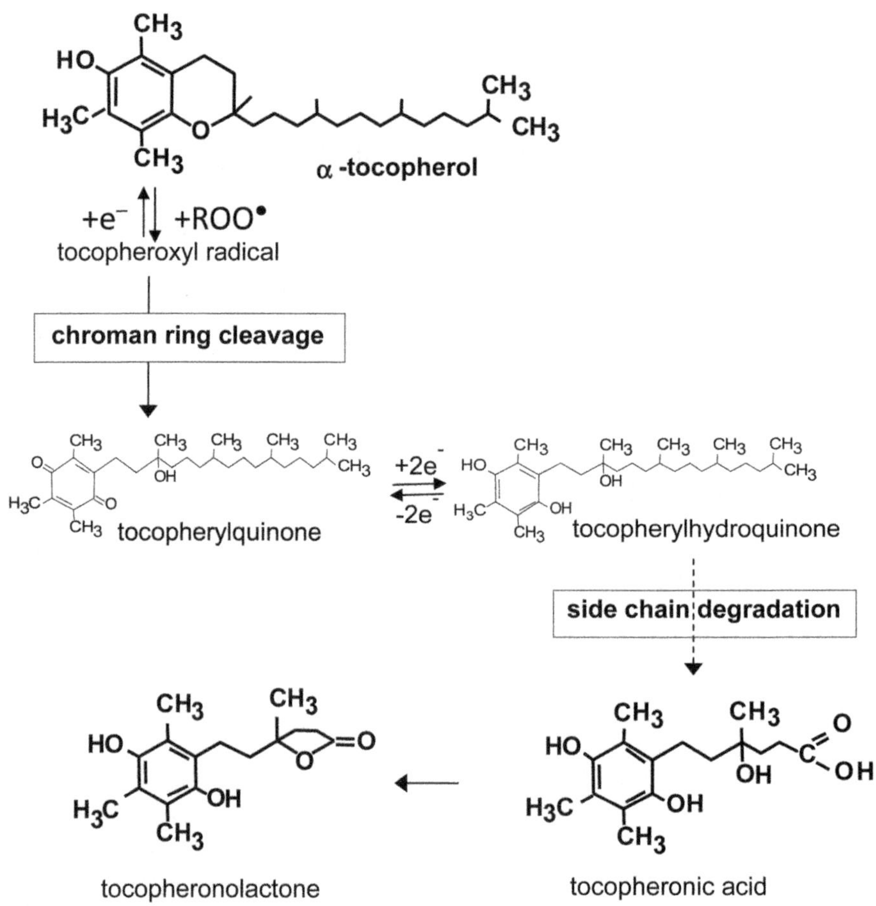

Figure 14.3 Postulated formation of Simon metabolites, tocopheronic acid and tocopheronolactone, from α-tocopherol.

product of α-tocopherol in the urine. Obviously, this work was not adequately appreciated.

About 20 years later, Chiku *et al.*[15] observed the excretion of δ-CEHC after the intravenous application of δ-tocopherol to rats. Since at that time the finding was novel and only Simon metabolites expected, the metabolic pathway of δ-tocopherol was suggested to be different from that of α-tocopherol.

About 40 years later, vitamin E metabolism in humans was reanalyzed and α-CEHC detected in the urine of male volunteers fed 650 mg *RRR*-α-tocopherol for three days.[16] In a second study undertaken to test whether Simon metabolites can serve as markers for the vitamin E supply, only α-CEHC was found in the urine of volunteers taking *RRR*-α-TOH dosages ranging from 0 to 800 mg per day for seven days.[17] Excretion started when a threshold of 7–9 μmol α-tocopherol per g total lipid was exceeded in plasma, which was reached after a daily intake of 50–150 mg α-tocopherol. Thus, α-CEHC excretion was considered to be a marker for optimal intake of vitamin E. The shortened side chain and the intact 6-hydroxychroman structure proved that vitamin E has been degraded without a previous antioxidant reaction. α-CEHC was only observed when care was taken to avoid oxygenation during sample preparation, as already suggested by Schmandke.[14] However, low amounts of α-TL postulated to not be derived from methodological artefacts were found in human urine after establishing a new analytical method.[18] In addition, glucuronidated and sulfated conjugates from α-TL were detected in children with type-1 diabetes and taken as a biomarker for oxidative stress.[19] Thus, a real excretion of α-TL under conditions of oxidative stress might contribute towards solving the question of whether vitamin E indeed has an antioxidant function *in vivo*.

γ-CEHC was first detected as an endogenous natriuretic factor, LLU-*a*,[20] and three years later was identified as a metabolite of γ-tocopherol.[21] CEHCs have also been found to be derived from β-tocopherol and all tocotrienols.

With the detection and characterization of almost all possible intermediates of side-chain degradation, the following scheme of degradation was developed.[5,22]

14.2.2 Side-chain Degradation

The metabolism of tocopherols starts with an ω-oxidation of the side chain resulting in a 13′-hydroxychromanol (13′-OH). 13′-OH is further oxidized in peroxisomes to 13′-carboxychromanol (13′-COOH) and by the first β-oxidation step to 11′-carboxydimethyldecylhydroxychroman (CDMDHC, 11′-COOH).

Subsequent β-oxidation steps lead to

- 9′-carboxydimethyloctylhydroxychroman (CDMOHC, 9′-COOH);
- 7′-carboxymethylhexylhydroxychroman (CMHHC, 7′-COOH);
- 5′-carboxymethylbutylhydroxychroman (CMBHC, 5′-COOH) and finally to
- 3′-carboxyethylhydroxychroman (CEHC) (Figure 14.4).

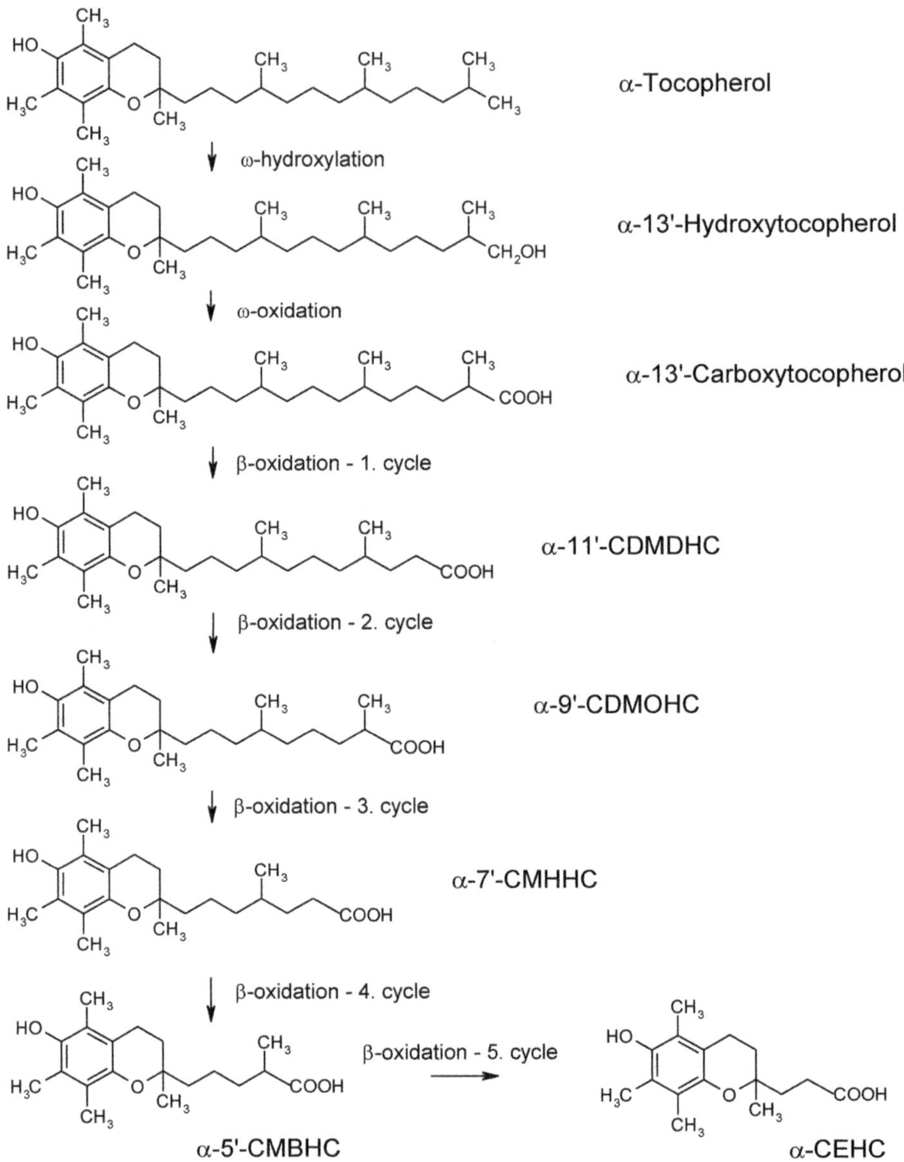

Figure 14.4 Metabolism of tocopherols with α-tocopherol as example. Degradation of tocopherols starts with an ω-oxidation of the side chain followed by five β-oxidation steps releasing two or three carbons as in the β-oxidation of fatty acids. The intermediates are: α-11′-CDMDHC, α-11′-carboxydimethyldecylhydroxychroman; α-9′-CDMOHC, α-9′-carboxydimethyloctylhydroxychroman; α-7′-CDMHHC, α-7′-carboxymethylhexylhydroxychroman; α-5′-CMBHC, α-5′-carboxymethylbutylhydroxychroman; ending in α-CEHC, α-3′-carboxyethylhydroxychroman. For further details, see text.

The metabolism of tocotrienols is comparable to that of tocopherols in principle. However, the double bonds in the side chains make it much more complex. Formation of the α,β-unsaturated fatty acid in the β-oxidation cycles leading to compounds with two conjugated double bonds have to be avoided since they are not accepted by enoyl-CoA hydratase that usually forms the β-hydroxyacyl-CoA. Also the presence of CMBHCs instead of metabolites still containing the double bond present in the precursors led to the conclusion that saturation steps have to be included in the degradation of tocotrienols.[23] Whether auxiliary enzymes such as 2,4-dienoyl-CoA reductase and 3,2-enoyl-CoA isomerase in the β-oxidation of unsaturated fatty acids are required remains to be investigated.

Degradation of tocotrienols starts with the ω-oxidation of the side chain probably resulting in 13′-hydroxytrienol, which is oxidized to 13′-carboxytrienol.

β-Oxidation cycles lead from

- 13′-carboxytrienol *via* reduction to 13′-carboxydienol to 11′-carboxy-dimethyldecadienylhydroxychroman (CDMD(en)$_2$HC);
- to 9′-carboxydimethyloctadienylhydroxychroman (CDMO(en)$_2$HC),
- *via* reduction to carboxydimethyloctenylhydroxychroman (CDMOenHC) to 7′-carboxymethylhexenylhydroxychroman (CMHenHC);
- to 5′-carboxymethylbutenylhydroxychroman (CMBenHC),
- *via* reduction to carboxymethylbutylhydroxychroman (CMBHC) finally to 3′-carboxyethylhydroxychroman (CEHC) (Figure 14.5).

Metabolites with a chain length of 13, 11 and nine carbon units are called long-chain metabolites (LCM), with seven and five carbon units intermediate-chain metabolites (ICM), and the end product CEHC is the short-chain metabolite (SCM).[5]

ω-Oxidation, the first cycle of β-oxidation and, if necessary, saturation of LCM, takes place in peroxisomes, the remaining cycles of β-oxidation and, if necessary, saturation of ICM and SCM in mitochondria.[5,24]

14.2.3 Enzymes Catalyzing the ω-Oxidation for Side-chain Degradation

ω-Hydroxylation is catalyzed by enzymes of the cytochrome P450 system (CYP). The first report linking vitamin E to this system appeared in 1972.[25] In rats fed an α-tocopherol-deficient diet for 8–12 weeks, hydroxylation capacity was reduced by 50%. Refeeding of α-tocopherol restored the CYP activity within 12 hours and could be decreased again by switching to the α-tocopherol-deficient diet. The antioxidant *N*,*N*′-diphenyl-*p*-phenylenediamine (DPPD) could not substitute for α-tocopherol. However, the restoring effect of α-tocopherol was prevented by actinomycin D, which already led the author to conclude that α-tocopherol exerts its effect at a transcriptional level.[26]

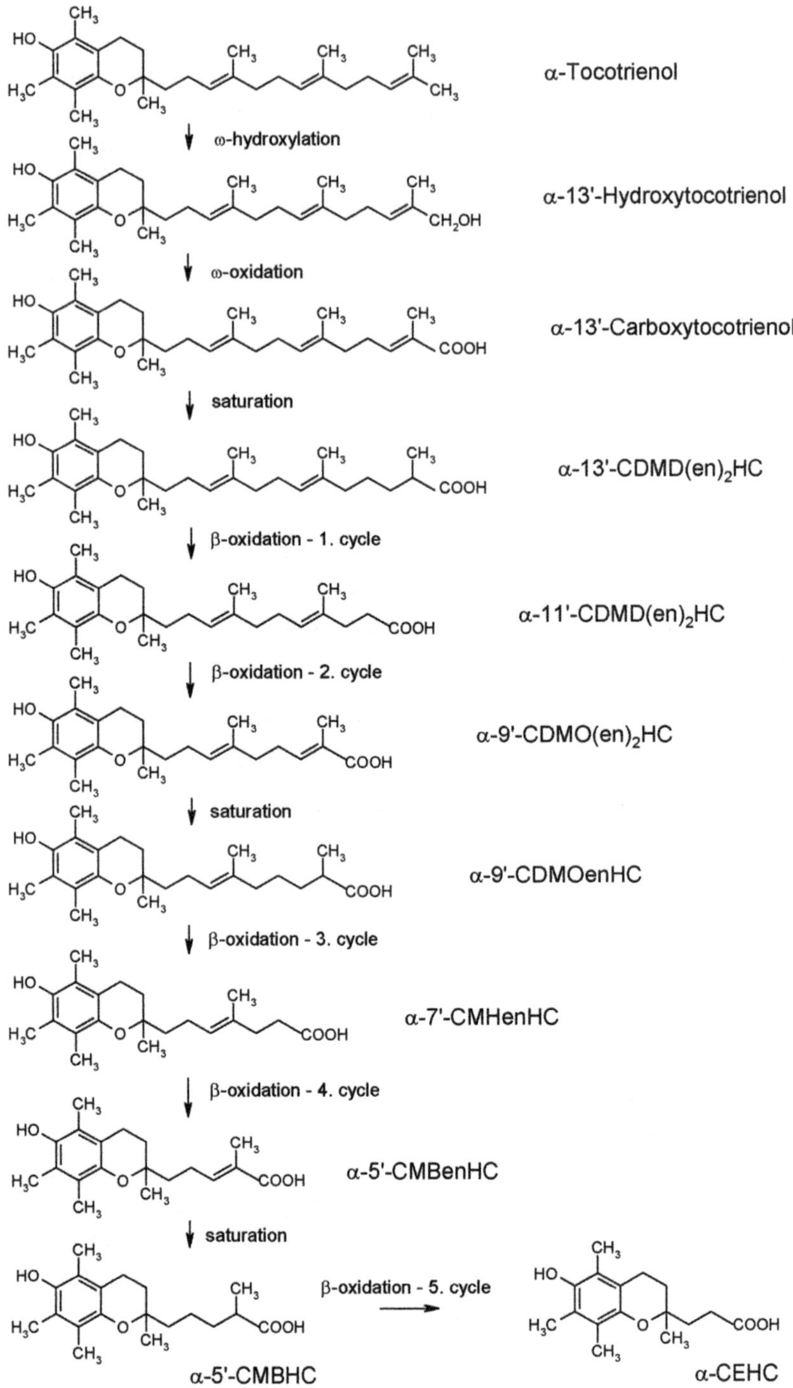

Figure 14.5 Metabolism of tocotrienols with α-tocotrienol as example. Degradation starts with an ω-oxidation of the side chain followed by five β-oxidation

Nowadays, ω-hydroxylation of vitamin E by enzymes of the CYP family is well accepted, however, which of the family members play the major role is still under discussion. Two CYPs are favored: CYP3A4 and CYP4F2.

14.2.3.1 Hints Supporting CYP3A4 as the Key Player Are:

- α- and γ-CEHC formation from α- and γ-tocopherol in primary rat hepatocytes and of γ- and δ-CEHC in HepG2/C3A cells are inhibited by ketoconazole, an inhibitor of the CYP3A family.[27]
- Non-metabolized γ-tocopherol concentrations are increased after dietary intervention with sesame oil, another inhibitor of CYP3A4, in humans[28] and in rats.[29]
- γ-CEHC excretion in rats is decreased after feeding sesame seeds.[30]
- α-TOH in rat brain is increased by feeding sesame seeds.[31]
- The urinary α-CEHC excretion after administration of α-TOH or a T3 mixture is decreased by ketoconazole in rats.[32]
- Non-metabolized γ-TOH and α- and γ-T3 are elevated by ketoconazole in rat serum while α-TOH concentrations remain unaffected.[32]
- *All rac*-α-tocopherol degradation is enhanced by the CYP3A4 inducer rifampicin in HepG2 cells but not by clofibrate, an inducer of the CYP4 type.[33] However, rifampicin did not greatly increase the metabolism of α-TOH to the level of the basic rate of γ-TOH degradation and did not influence degradation of γ-TOH at all.[34] Thus, CYP3A4 obviously does not degrade all forms of vitamin E equally well.

14.2.3.2 Regulation of CYP3A4 Expression by Vitamin E Itself

The activities of aryl hydrocarbon hydroxylase,[35] and CYP2B and CYP3A[36] were significantly decreased in vitamin E-deficient rats. CYP2C11 was increased in livers from vitamin E-supplemented rats.[37] Induction of CYP3A by rifampicin or α-tocopherol significantly increased the survival rate of paraquat-exposed rats.[38] Microarray analysis demonstrated an up-regulation of CYP4A3 transcripts in fetal brains after feeding mothers a diet enriched with α-tocopherol plus α-tocotrienol.[39] Thus, certain forms of vitamin E are obviously able to induce the CYPs responsible for their own metabolism.

---

steps. Intermediates with double bonds have to be reduced when the chain is shortened from 13 to 11, from nine to seven and from five to three carbons. The intermediates are: α-13′-CDMD(en)2HC, α-13′-carboxydimethyldecadienylhydroxychroman; α-11′-CDMD(en)2HC, α-11′-carboxydimethyldecadienylhydroxychroman; α-9′-CDMO(en)2HC, α-9′-carboxydimethyloctadienylhydroxychroman; α-9′-CDMOenHC, α-9′-carboxydimethyloctenylhydroxychroman; α-7′-CMHenHC, α-7′-carboxymethylhexenylhydroxychroman; α-5′-CMBenHC, α-5′-carboxymethylbutenylhydroxychroman; α-5′-CMBHC, α-5′-carboxymethylbutylhydroxychroman; also ending in CEHC, α-3′-carboxyethylhydroxychroman. For further details, see text.

Induction of CYP3As is mediated by the nuclear pregnane X receptor (PXR).[40] An activation of a PXR-driven chloramphenicol acetyl transferase reporter in HepG2 cells by different forms of vitamin E has been described.[41] The most active forms were α- and γ-tocotrienol, which induced the activity to an even higher extent than rifampicin, a potent PXR activator, whereas δ-, α-, and γ-tocopherol were less active. Accordingly, γ-T3 induced CYP3A4 and CYP3A5 to an extent comparable to rifampicin.[41]

In a subsequent study, mice were fed an *RRR*-α-tocopherol-deficient, -adequate or supra-nutritional diet for three months.[42] Half of each group were fed a daily dose of 250 μg γ-T3 for the last seven days. Induction of CYP3a11, the murine orthologue to human CYP3A4, correlated with α-TOH feeding. Concomitantly, α-CEHC excretion was increased. γ-T3 feeding also increased γ-CEHC, but in all three groups to the same extent, irrespective of an increase in CYP3a11 mRNA. It was concluded that CYP3a11 metabolizes α-TOH but might not be the CYP that metabolizes γ-T3 *in vivo*. In contrast to the findings in HepG2 cells, additional feeding of γ-T3 did not affect CYP3a11 expression in mice. As an explanation, a possible inability of γ-T3 to activate the murine PXR was discussed, since PXRs from different species respond differently to activators.[43,44] The confirmation of a PXR-dependent CYP3a11 induction by α-TOH in mice came about 10 years later. Pxr-null, PXR-humanized (h-PXR) and WT mice were used to test whether α-tocopherol metabolism is influenced by species-specific differences in PXR function.[45] Mice were fed 500 mg kg^{-1} *all rac*-α-tocopherol acetate per day, which corresponds to 400–600 mg vitamin E supplementation of a 70 kg human. α-Tocopherol increased CYP3a11 expression only in WT, demonstrating that α-tocopherol is indeed able to activate mouse Pxr. The failure to induce CYP3a11 in h-PXR mice may point to the probability that h-PXR activation by α-TOH is not strong enough to activate the murine CYP3a11 promoter. Up-regulation of CYP4f forms were observed only for CYP4f13, and not for CYP4f14 (Section 14.2.3.5), and only in WT mice. The authors concluded that α-tocopherol is a partial agonist of mouse Pxr and that Pxr is required for the induction of CYP3a11 by α-TOH in mice.[45]

The ability of α-tocopherol to induce CYP3a11 but not CYP4f forms was also found in the livers of mice[46] and an up-regulation of hepatic CYP3A, CYP2B and CYP2C family members but not of CYP4F was observed after subcutaneous injections of α-tocopherol to rats.[47] Studying a possible modulation of genes of hepatic xenobiotic pathways by α-TOH in mice revealed an increase of CYP3a11 whereas CYP4f isoforms were not affected.[48] In guinea pigs, expression of neither CYP3A4 nor CYP4F2 orthologues was influenced by α-tocopherol, indicating that effects of α-tocopherol on CYP expression might be species-specific.[49]

14.2.3.3 Hints Supporting CYP4F2 as the Key Player Are:

- Like the activity of CYP3A4, that of CYP4F2 is also inhibited by sesamin and ketoconazole.[50] Thus, inhibition by these compounds is not a final proof to identify the CYP involved in vitamin E degradation.

- Screening of microsomes from insects, transfected with a series of different CYP enzymes, including CYP3A4, revealed that only microsomes with CYP4F2 had tocopherol-ω-hydroxylase activity.[50]
- The specificity of tocopherol-ω-hydroxylase in microsomes from human liver and from CYP4F2-transfected insects was comparable in both sources. In human samples, the degradation rate was the lowest with α- and β-TOH followed by δ-TOH, α-T3, γ-TOH, δ- and γ-T3. Hydroxylase activity in insect samples was the lowest with α- and β-TOH, too, followed by δ- and γ-TOH, and then by α-, δ-, and γ-T3 with a clear distance.[51] The findings were explained by the methylation of α- and β-TOH at position C5 in the chromanol system. Obviously, a free position at C5 and the unsaturated side chain is required for CYP4F2 activity, which explains why γ- and δ-TOH and all T3s were more efficiently metabolized than α- and β-TOH. This points to a specificity of CYP4F2 for non-α-tocopherols.

14.2.3.4 *Regulation of CYP4F2 Expression by Vitamin E Itself*

CYP4F2 is not or only minimally induced by α-TOH. Furthermore, it is not a target for PXR, but was suggested to respond to PPAR-γ[52] and SREBP-1 and -2.[53] However, CYP4F2 could be up-regulated in HepG2 cells when *RRR*-α-tocopherol was administered by a bolus to HepG2 cells. The resulting production of α-13′-OH and α-13′-COOH identified α-13′-OH as the main metabolite, which, interestingly, was a much more effective inducer of CYP4F2 than the parent tocopherol.[54]

Ethanol has been hypothesized to inhibit the expression of certain CYP4F genes.[52] Indeed, ethanol decreased CYP4F2 mRNA and protein in HepG2 cells, which could be restored by α-TOH, whereas CYP3A4, other CYPs and α-13′-OH, α-13′-COOH and α-CEHC were increased. Thus, CYPs distinct from CYP4F2 appeared to sustain vitamin E metabolism.[55] An ethanol-mediated decrease in CYP4F2 expression was accompanied by a decrease of PPAR-γ and SREBP-1, supporting the postulated regulation of CYP4F2 by these transcription factors.[55] Collectively, these findings do not point to a preferential role of CYP4F2 in the degradation of α-tocopherol.

14.2.3.5 *Other CYPs*

To further clarify the function of CYP4F2 in the metabolism of vitamin E, the murine orthologue of human CYP4F2, CYP4f14, was deleted in mice.[56] These mice were fed a γ-TOH-rich diet as usually taken up by American people. Mice hyper-accumulated γ-TOH in plasma and tissues in concentrations about twice those of WT, whereas tissue and plasma levels of α-TOH were not changed. The metabolism of α-, γ-, and δ-TOH was decreased by 65, 90, and 70%, respectively, as measured by urinary CEHC and CMBHC excretion. The lowest decrease in α-TOH metabolism is in line with the low metabolism of α-TOH in CYP4F2-transfected insects and human liver microsomes (Section 14.2.3.3). Vitamin E-ω-hydroxylase activity in liver microsomes was not

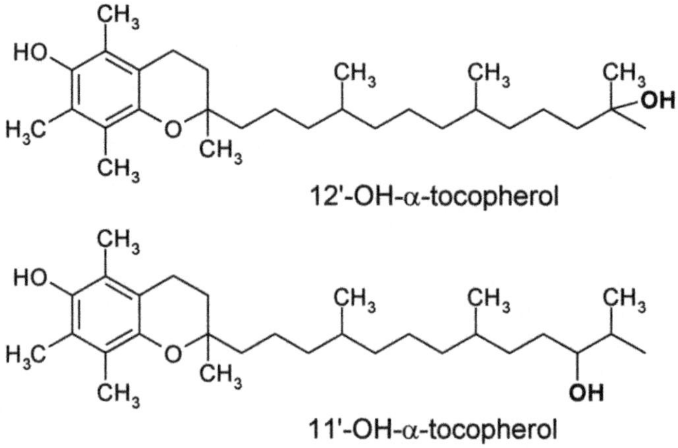

Figure 14.6 Novel ω-1- and ω-2-hydroxytocopherol metabolites.

completely decreased, indicating that one or more CYPs apart from CYP4f forms are active in mouse liver.

60% of the total metabolites were eliminated *via* fecal excretion in WT, which was increased to 85% in CYP4f14$^{-/-}$ mice together with novel hydroxyl-tocopherols, 12′-OH-γ-TOH and 11′-OH-γ-TOH, the products of ω-1 and ω-2 hydroxylation of γ- and δ-TOH (Figure 14.6). 12′-OH-TOHs cannot be oxidized to the keto form and, thus, are terminal metabolites. Thus, loss of CYP4f14 obviously resulted in a shift from SCM to LCM excreted *via* the fecal and not *via* the urinary route to prevent an accumulation of unwanted tocopherols.

Mice in which all hepatic CYPs were deleted by a disruption of CYP reductase (L-Cpr$^{-/-}$ mice) revealed the existence of extra-hepatic ω, ω-1, and ω-2 tocopherol hydroxylase activities in the intestine.[57] Extra-hepatic activity of ω-hydoxylase was already identified in A549 lung epithelial cells[58] and in prostate cancer cells.[59] Thus, ways to prevent accumulation of vitamin E appear to be many and are worthy of further investigation.

14.3 Absorption, Distribution, Excretion

14.3.1 Absorption

In the intestine, all forms of vitamin E are taken up along with dietary fats, like triglycerides, phospholipids, and cholesterol.[60] In the duodenum, vitamin E is transferred to mixed micelles, structures essential for absorption by the enterocyte. There are two pathways by which vitamin E is taken up from the micelles: (1) by passive diffusion and (2) by receptor-mediated transport. Uptake of α- and γ-TOH at the brush border membrane of enterocytes in the small intestine occurs *via* the scavenger receptor-B1 (SR-B1), of α-TOH and γ-T3 also by the Niemann-Pick C1-like protein-1 (NPC1L1).[61–66] CD36, expressed in the brush border membrane of the proximal intestine has also been discussed to play a role.[66,67] Release from enterocytes at the basolateral side either occurs with chylomicrons in the lymph or by the apolipoprotein A-1 pathway. The ATP binding cassette subfamily A member-1 (ABCA1) forms nascent HDL, which takes up vitamin E for release into the portal venous circulation.[66,68–72] An efflux *via* the SR-B1 can also take place at the basolateral membrane as well as *via* the canalicular membrane into bile.

In plasma, triglycerides in chylomicrons are cleaved by the endothelial cell-bound lipoprotein lipase resulting in the formation of chylomicron remnants. Chylomicron remnant-bound vitamin E enters liver cells *via* the LDL receptor or the LDL-R-related protein (LRP).[60] If bound to HDL, α-tocopherol is taken up by SR-B1.[73]

In the liver, α-TOH is bound to the α-tocopherol transfer protein (α-TTP) for distribution to peripheral tissues (Section 14.3.2).[74,75] There, uptake is dependent on the cell type and the transporting lipoprotein *via* LDL-R or SR-B1. For recent reviews, see refs. 66, 76 and 77 and Chapter 5.

14.3.2 Distribution and Retention of α-TOH

More than 90% of plasma and tissue vitamin E is α-tocopherol. Two processes/pathways have been discussed to be responsible for the preferential retention of α-TOH: (1) the specificity of α-TTP for α-TOH, and (2) the preferential degradation of non-α-TOH by specific CYPs.

α-TTP has been detected as an α-TOH-binding liver protein.[78,79] Relative affinities of α-TTP to the various forms of vitamin E were based on *RRR*-α-TOH taken as 100%. The ranking was: β-TOH 38%, α-T3 12%, *SRR*-α-TOH 11%, γ-TOH 9% and δ-TOH 2%.[76,80] As a consequence, the function of α-TTP was defined as specifically separating α-TOH from the dietary mixture and facilitating its release for redistribution to peripheral tissues by supporting its transfer into nascent VLDL.[75,81] For more details, see Chapter 5.

The second possibility to explain the prevalence of α-TOH is its much lower elimination compared to other forms of vitamin E by CYP-mediated pathways. If CYP pathways were inhibited by sesamin, γ-TOH levels increased and γ-CEHC decreased in mice,[82] rats,[30] and humans.[83] *Drosophila melanogaster*, which does not express an α-TTP-related protein, selectively accumulated α-TOH and eliminated δ- and γ-TOH *via* a CYP-dependent pathway.[84] In a recent study, α-TTP$^{-/-}$, α-TTP$^{+/-}$, and wildtype mice were fed α- and γ-TOH in equimolar concentrations alone or together with sesamin for six weeks.[82] α-TOH was decreased in tissues of α-TTP KO mice, confirming earlier findings.[85] α-TOH levels were not influenced by sesamin. In contrast, sesamin increased γ-TOH levels in tissues in WT mice, but not in α-TTP KO mice indicating that α-TTP is not responsible for the discrimination against γ-TOH. Furthermore, HepG2 cells, which did not express α-TTP, were transfected with human α-TTP and incubated with sesamin. γ-CEHC secretion into the culture medium was lower in α-TTP-positive cells than in controls. Sesamin decreased formation of γ-CEHC in α-TTP-lacking cells but not in α-TTP-expressing cells. When cells were cultured with increasing α-TOH concentrations, γ-CEHC formation increased in α-TTP-expressing but not in control cells, indicating that α-TOH, probably due to the much higher binding affinity, may displace γ-TOH from α-TTP and, thus, prevent retention of γ-TOH.

To sum up, most studies undertaken to figure out the regulation of vitamin E metabolism point to the more crucial role of tocopherol-ω-hydroxylase activity than of α-TTP in the preferential retention of α-TOH. Thus, the role of α-TTP might not be selection but rather distribution of α-TOH.

14.3.3 Excretion

SCMs are excreted *via* the urine as sulfates or as glucuronides.[22,86–88] Sulfation is catalyzed by sulfotransferases, from which SULT1 family members have a higher activity towards tocopherols and their metabolites and show a substrate preference for γ-TOH over α-TOH and for γ-CEHC over α-CEHC.[89] Glucuronidation is catalyzed by UDP-glucuronosyl transferase. In addition, a γ-CEHC β-D-glucoside[90] and conjugates of α-CEHC with taurine, glycine, glycine glucuronide, and glutamine exist.[91] Also ICMs are glucuronidated or sulfated. From LCMs, so far only sulfates have been found in human lung epithelial A549 cells and in the plasma and liver of rats, which obviously are formed in parallel with β-oxidation.[92,93] Non-metabolized and unconjugated metabolites are primarily found in feces excreted *via* the biliary route.[22,56,57]

In mice, hepatic forms of vitamin E are transported into the bile *via* ATP-binding cassette (ABC) transporters (MDR). A knockout of MDR2 in mice decreased biliary α-TOH levels by 75% indicating that MDR2 transports most of the α-TOH.[94] In rats, levels of MDR1, the equivalent to mouse MDR2, in canalicular membranes of hepatocytes were increased by high-dose α-TOH supplementation[47] and α-TOH injections.[95] Increasing MDR1 also decreased hepatic α-TOH.[47] Thus, α-TOH appears to regulate its excretion. Whether the elimination of other forms of vitamin E are affected by MDRs too remains to be investigated.

14.4 Possible Adverse Effects

The degradation of vitamin E by an enzyme system developed to eliminate xenobiotics was unexpected. Also, the finding that vitamin E is able to induce such enzymes (CYPs, MDRs) was unexpected, inasmuch as the induction was mediated by transcription factors regulating xenobiotic and drug metabolism, *e.g.* PXR.[41] This has raised the provocative question of whether vitamin E can interact with drug metabolism and/or whether there are conditions under which vitamin E can be considered to be foreign or even dangerous.[96–98]

Hints to raise these questions came from the regulation of CYPs and transporters by vitamin E (Sections 14.2.3 and 14.3.3). Induction of CYPs can interfere with drug metabolism and induction of ABC transporters might also lead to a limitation of absorption and distribution of drugs. Contradictory results revealed that an interaction of vitamin E with drug metabolism is much more complicated than expected. It appears to be dependent on species, tissues, transcription factors, and the forms of drugs and of vitamin E. *In vivo* trials to test a direct interaction of vitamin E and drugs have so far not been performed. Available human intervention trials and animal studies were summarized and critically reviewed.[99] The authors did not find convincing evidence for an interaction of vitamin E at regular intakes with drug metabolism. However, high doses of vitamin E, exceeding the tolerable upper intake level of 300 mg per day α-TOH, may interfere with aspirin and warfarin as anticoagulants, with tamoxifen in breast cancer treatment, and with cyclosporine A as an immunosuppressant and should not be taken without regular drug efficacy tests. The discussion about the safety of vitamin E is still going on.[100]

14.5 Concluding Remarks

Studies on the metabolism of tocopherols and tocotrienols of recent decades show that their prominent metabolites do not result from the presumed action of the vitamins as antioxidants. Typically, the chromanol ring remains intact, while the side chains are attacked by ω-oxidation followed by β-oxidation and, as a rule, the pattern of metabolites is poorly or not affected by interventions causing oxidative stress. The enzymatic systems involved appear to vary with the vitamin form, with species and even with the tissue. Obviously, CYP3A4 is more specific for α-forms, whereas CYP4F2 prefers non-α-forms. In humans, α-TOH is peculiar in that it prevails in plasma and tissues, being slowly degraded and binding avidly to α-TTP, which guides the vitamin out of the liver to the circulation for distribution to tissues of demand. Nevertheless, preferential retention of α-TOH is caused by its low degradation. The upper plasma levels of α-TOH are relatively fixed. Any excess of α-TOH is rapidly degraded, a process that involves induction of enzymes of the CYP family. As these enzymes are generally considered to be responsible for the metabolism of xenobiotics, one may speculate that excess tocopherols are considered by the organism as foreign and possibly harmful. In any case, effects of supra-nutritional dosages of vitamin E justify concern in respect to interference with drug metabolism.

Collectively, the detection of vitamin E metabolites with an intact chromanol ring, the regulation of upper plasma levels of the vitamin, and the induction processes involved point to specific actions of these vitamins, which unfortunately are yet to be identified, even almost 100 years after their detection. Detailed studies of functions of individual metabolites are needed since they are appearing to turn out to be a new class of regulatory signaling molecules.[101–107]

References

1. H. M. Evans and K. S. Bishop, *Science*, 1922, **56**, 650.
2. K. Schwarz, *Fed. Proc.*, 1965, **24**, 58.
3. K. Schwarz, *Ann. N. Y. Acad. Sci.*, 1954, **57**, 878.
4. Q. Jiang, *Free Radical Biol. Med.*, 2014, **72**, 76.
5. L. Schmölz, M. Birringer, S. Lorkowski and M. Wallert, *World J. Biol. Chem.*, 2016, **7**, 14.
6. IUPAC-IUB, *Eur. J. Biochem.*, 1982, **123**, 473.
7. R. Brigelius-Flohé and M. G. Traber, *FASEB J.*, 1999, **13**, 1145.
8. D. DellaPenna, *Trends Plant Sci.*, 2005, **10**, 574.
9. V. L. Lushchak and N. M. Semchuk, *Acta Physiol. Plant.*, 2012, **34**, 1607.
10. L. Mène-Saffrané, *Antioxidants*, 2017, **7**, 2.
11. E. J. Simon, A. Eisengart, L. Sundheim and A. T. Milhorat, *J. Biol. Chem.*, 1956, **221**, 807.
12. D. C. Liebler, *Crit. Rev. Toxicol.*, 1993, **23**, 147.
13. T. Hayashi, A. Kanetoshi, M. Nakamura, M. Tamura and H. Shirahama, *Biochem. Pharmacol.*, 1992, **44**, 489.
14. H. Schmandke, *Int. Z. Vitaminforsch.*, 1965, **35**, 346.
15. S. Chiku, K. Hamamura and T. Nakamura, *J. Lipid Res.*, 1984, **25**, 40.
16. A. Schönfeld, M. Schultz, M. Petrizka and B. Gassmann, *Nahrung*, 1993, **37**, 498.
17. M. Schultz, M. Leist, M. Petrzika, B. Gassmann and R. Brigelius-Flohé, *Am. J. Clin. Nutr.*, 1995, **62**, 1527S.
18. S. A. Pope, P. T. Clayton and D. P. Muller, *Arch. Biochem. Biophys.*, 2000, **381**, 8.
19. G. Sharma, D. P. Muller, S. M. O'Riordan, S. Bryan, M. T. Dattani, P. C. Hindmarsh and K. Mills, *Free Radical Biol. Med.*, 2013, **55**, 54.
20. J. W. Wechter, D. Kantoci, E. D. Murray, D. C. D'Amico, M. E. Jung and W.-H. Wang, *Proc. Natl. Acad. Sci. U. S. A.*, 1996, **93**, 6002.
21. D. Kantoci, W. J. Wechter, E. D. Murray, Jr, S. A. Dewind, D. Borchardt and S. I. Khan, *J. Pharmacol. Exp. Ther.*, 1997, **282**, 648.
22. Y. Zhao, M. J. Lee, C. Cheung, J. H. Ju, Y. K. Chen, B. Liu, L. Q. Hu and C. S. Yang, *J. Agric. Food Chem.*, 2010, **58**, 4844.
23. M. Birringer, P. Pfluger, D. Kluth, N. Landes and R. Brigelius-Flohé, *J. Nutr.*, 2002, **132**, 3113.
24. D. J. Mustacich, S. W. Leonard, N. K. Patel and M. G. Traber, *Free Radical Biol. Med.*, 2010, **48**, 73.
25. M. P. Carpenter, *Ann. N. Y. Acad. Sci.*, 1972, **203**, 81.
26. M. P. Carpenter and C. N. Howard, Jr, *Am. J. Clin. Nutr.*, 1974, **27**, 966.
27. R. S. Parker, T. J. Sontag and J. E. Swanson, *Biochem. Biophys. Res. Commun.*, 2000, **277**, 531.
28. M. Lemcke-Norojarvi, A. Kamal-Eldin, L. A. Appelqvist, L. H. Dimberg, M. Ohrvall and B. Vessby, *J. Nutr.*, 2001, **131**, 1195.
29. K. Yamashita, S. Ikeda, Y. Iizuka and I. Ikeda, *Lipids*, 2002, **37**, 351.
30. S. Ikeda, K. Toyoshima and K. Yamashita, *J. Nutr.*, 2001, **131**, 2892.

31. C. Abe, S. Ikeda and K. Yamashita, *J. Nutr. Sci. Vitaminol.*, 2005, **51**, 223.
32. C. Abe, T. Uchida, M. Ohta, T. Ichikawa, K. Yamashita and S. Ikeda, *Lipids*, 2007, **42**, 637.
33. M. Birringer, D. Drogan and R. Brigelius-Flohé, *Free Radical Biol. Med.*, 2001, **31**, 226.
34. R. Brigelius-Flohé, D. Kluth, N. Landes, P. Pfluger and M. Birringer, in *The Antioxidant Vitamins C and E*, ed. L. Packer, M. Traber, K. Kraemer and B. Frei, AOCS Press, Champaign, Illinois, 2002, pp. 171–179.
35. C. Gairola and L. H. Chen, *Int. J. Vitam. Nutr. Res.*, 1982, **52**, 398.
36. P. Cassand, S. Decoudu, F. Leveque, M. Daubeze and J. F. Narbonne, *Mutat. Res.*, 1993, **319**, 309.
37. M. Murray, *Biochem. Pharmacol.*, 1991, **42**, 2107.
38. H. Shimada, H. Furuno, K. Hirai, J. Koyama, J. Ariyama and E. Simamura, *Arch. Biochem. Biophys.*, 2002, **402**, 149.
39. S. Roy, B. H. Lado, S. Khanna and C. K. Sen, *FEBS Lett.*, 2002, **530**, 17.
40. J. M. Lehmann, D. D. McKee, M. A. Watson, T. M. Willson, J. T. Moore and S. A. Kliewer, *J. Clin. Invest.*, 1998, **102**, 1016.
41. N. Landes, P. Pfluger, D. Kluth, M. Birringer, R. Rühl, G. F. Böl, H. Glatt and R. Brigelius-Flohé, *Biochem. Pharmacol.*, 2003, **65**, 269.
42. D. Kluth, N. Landes, P. Pfluger, K. Müller-Schmehl, K. Weiss, C. Bumke-Vogt, M. Ristow and R. Brigelius-Flohé, *Free Radical Biol. Med.*, 2005, **38**, 507.
43. J. L. Barwick, L. C. Quattrochi, A. S. Mills, C. Potenza, R. H. Tukey and P. S. Guzelian, *Mol. Pharmacol.*, 1996, **50**, 10.
44. T. Ostberg, G. Bertilsson, L. Jendeberg, A. Berkenstam and J. Uppenberg, *Eur. J. Biochem.*, 2002, **269**, 4896.
45. C. H. Johnson, J. A. Bonzo, J. Cheng, K. W. Krausz, D. W. Kang, H. Luecke, J. R. Idle and F. J. Gonzalez, *Drug Metab. Dispos.*, 2013, **41**, 406.
46. M. G. Traber, L. K. Siddens, S. W. Leonard, B. Schock, K. Gohil, S. K. Krueger, C. E. Cross and D. E. Williams, *Free Radical Biol. Med.*, 2005, **38**, 773.
47. D. J. Mustacich, S. W. Leonard, M. W. Devereaux, R. J. Sokol and M. G. Traber, *Free Radical Biol. Med.*, 2006, **41**, 1069.
48. D. J. Mustacich, K. Gohil, R. S. Bruno, M. Yan, S. W. Leonard, E. Ho, C. E. Cross and M. G. Traber, *J. Nutr. Biochem.*, 2009, **20**, 469.
49. M. C. Podszun, N. Grebenstein, U. Hofmann and J. Frank, *Toxicol. Appl. Pharmacol.*, 2013, **266**, 452.
50. T. J. Sontag and R. S. Parker, *J. Biol. Chem.*, 2002, **277**, 25290.
51. T. J. Sontag and R. S. Parker, *J. Lipid Res.*, 2007, **48**, 1090.
52. J. P. Hardwick and J. Y. Chiang, *PPAR Res.*, 2009, **2009**, 589626.
53. M. H. Hsu, U. Savas, K. J. Griffin and E. F. Johnson, *J. Biol. Chem.*, 2007, **282**, 5225.
54. P. Torquato, D. Bartolini and D. Guisepponi, *et al.*, *Free Radical Biol. Med.*, 2016, **96**, S19.
55. A. Russo, D. Bartolini and P. Torquato, *et al.*, *Toxicol. In Vitro*, 2017, **40**, 124.

56. S. A. Bardowell, F. Duan, D. Manor, J. E. Swanson and R. S. Parker, *J. Biol. Chem.*, 2012, **287**, 26077.
57. S. A. Bardowell, X. Ding and R. S. Parker, *J. Lipid Res.*, 2012, **53**, 2667.
58. C. S. You, T. J. Sontag, J. E. Swanson and R. S. Parker, *J. Nutr.*, 2005, **135**, 227.
59. C. Conte, A. Floridi, C. Aisa, M. Piroddi, A. Floridi and F. Galli, *Ann. N. Y. Acad. Sci.*, 2004, **1031**, 391.
60. A. Rigotti, *Mol. Aspects Med.*, 2007, **28**, 423.
61. B. Abuasal, P. W. Sylvester and A. Kaddoumi, *Drug Metab. Dispos.*, 2010, **38**, 939.
62. P. Mardones and A. Rigotti, *J. Nutr. Biochem.*, 2004, **15**, 252.
63. P. Mardones, P. Strobel and S. Miranda, *et al.*, *J. Nutr.*, 2002, **132**, 443–449.
64. K. Narushima, T. Takada, Y. Yamanashi and H. Suzuki, *Mol. Pharmacol.*, 2008, **74**, 42–49.
65. E. Reboul, A. Klein and F. Bietrix, *et al.*, *J. Biol. Chem.*, 2006, **281**, 4739–4745.
66. Y. Yamanashi, T. Takada, R. Kurauchi, Y. Tanaka, T. Komine and H. Suzuki, *J. Atheroscler. Thromb.*, 2017, **24**, 347.
67. A. Goncalves, S. Roi, M. Nowicki, I. Niot and E. Reboul, *Mol. Nutr. Food Res.*, 2014, **58**, 2297.
68. K. Anwar, H. J. Kayden and M. M. Hussain, *J. Lipid Res.*, 2006, **47**, 1261.
69. K. Nagao, M. Tomioka and K. Ueda, *FEBS J.*, 2011, **278**, 3190.
70. J. F. Oram, A. M. Vaughan and R. Stocker, *J. Biol. Chem.*, 2001, **276**, 39898.
71. E. Reboul, D. Trompier, M. Moussa, A. Klein, J. F. Landrier, G. Chimini and P. Borel, *Am. J. Clin. Nutr.*, 2009, **89**, 177.
72. M. Shichiri, Y. Takanezawa, D. E. Rotzoll, Y. Yoshida, T. Kokubu, K. Ueda, H. Tamai and H. Arai, *J. Nutr. Biochem.*, 2010, **21**, 451.
73. D. Goti, H. Reicher, E. Malle, G. M. Kostner, U. Panzenboeck and W. Sattler, *Biochem. J.*, 1998, **332**, 57.
74. D. J. Mustacich, R. S. Bruno and M. G. Traber, *Vitam. Horm.*, 2007, **76**, 1.
75. M. G. Traber and H. Arai, *Annu. Rev. Nutr.*, 1999, **19**, 343.
76. P. Borel, D. Preveraud and C. Desmarchelier, *Nutr. Rev.*, 2013, **71**, 319.
77. E. Reboul, *Antioxidants*, 2017, **6**, 95.
78. G. L. Catignani, *Biochem. Biophys. Res. Commun.*, 1975, **67**, 66.
79. G. L. Catignani and J. G. Bieri, *Biochim. Biophys. Acta*, 1977, **497**, 349.
80. A. Hosomi, M. Arita and Y. Sato, *et al.*, *FEBS Lett.*, 1997, **409**, 105.
81. M. G. Traber, G. W. Burton and R. L. Hamilton, *Ann. N. Y. Acad. Sci.*, 2004, **1031**, 1.
82. N. Grebenstein, M. Schumacher, L. Graeve and J. Frank, *Mol. Nutr. Food Res.*, 2014, **58**, 1052.
83. R. V. Cooney, L. J. Custer, L. Okinaka and A. A. Franke, *Nutr. Cancer*, 2001, **39**, 66.
84. R. S. Parker and C. C. McCormick, *Biochem. Biophys. Res. Commun.*, 2005, **338**, 1537.
85. S. W. Leonard, Y. Terasawa, R. V. Farese, Jr and M. G. Traber, *Am. J. Clin. Nutr.*, 2002, **75**, 555.

86. T. Morinobu, S. Yoshikawa, K. Hamamura and H. Tamai, *Eur. J. Clin. Nutr.*, 2003, **57**, 410.
87. S. A. Pope, G. E. Burtin, P. T. Clayton, D. J. Madge and D. P. Muller, *Free Radical Biol. Med.*, 2002, **33**, 807.
88. W. Stahl, P. Graf, R. Brigelius-Flohé, W. Wechter and H. Sies, *Anal. Biochem.*, 1999, **275**, 254.
89. T. Hashiguchi, K. Kurogi and Y. Sakakibara, *et al.*, *Biosci. Biotechnol. Biochem.*, 2011, **75**, 1951.
90. J. Y. Cho, D. W. Kang and X. Ma, *et al.*, *J. Lipid Res.*, 2009, **50**, 924.
91. C. H. Johnson, O. Slanar and K. W. Krausz, *et al.*, *Am. J. Clin. Nutr.*, 2012, **96**, 818.
92. H. Freiser and Q. Jiang, *J. Nutr.*, 2009, **139**, 884.
93. Q. Jiang, H. Freiser, K. V. Wood and X. Yin, *J. Lipid Res.*, 2007, **48**, 1221.
94. D. J. Mustacich, J. Shields, R. A. Horton, M. K. Brown and D. J. Reed, *Arch. Biochem. Biophys.*, 1998, **350**, 183.
95. M. G. Traber, E. M. Labut, S. W. Leonard and K. M. Lebold, *Free Radical Biol. Med.*, 2011, **51**, 2031.
96. R. Brigelius-Flohé, *Biochem. Biophys. Res. Commun.*, 2003, **305**, 737.
97. R. Brigelius-Flohé, *Genes Nutr.*, 2007, **2**, 249.
98. M. G. Traber, *Arch. Biochem. Biophys.*, 2004, **423**, 6.
99. M. Podszun and J. Frank, *Nutr. Res. Rev.*, 2014, **27**, 215.
100. C. J. Oliver and S. P. Myers, *BMC Complementary Altern. Med.*, 2017, **17**, 408.
101. S. Ciffolilli, M. Wallert and D. Bartolini, *et al.*, *Free Radical Biol. Med.*, 2015, **89**, 952.
102. Y. Jang, N. Y. Park, A. L. Rostgaard-Hansen, J. Huang and Q. Jiang, *Free Radical Biol. Med.*, 2016, **95**, 190.
103. Q. Jiang, X. Yin, M. A. Lill, M. L. Danielson, H. Freiser and J. Huang, *Proc. Natl. Acad. Sci. U. S. A.*, 2008, **105**, 20464.
104. M. C. Podszun, M. Jakobi, M. Birringer, J. Weiss and J. Frank, *Mol. Nutr. Food Res.*, 2017, **61**, 1600605.
105. M. Schubert, S. Kluge, L. Schmölz, M. Wallert, F. Galli, M. Birringer and S. Lorkowski, *Antioxidants*, 2018, **7**, 10.
106. P. Torquato, O. Ripa and D. Giusepponi, *et al.*, *J. Pharm. Biomed. Anal.*, 2016, **124**, 399.
107. M. Wallert, S. Mosig and K. Rennert, *et al.*, *Free Radical Biol. Med.*, 2014, **68**, 43.

CHAPTER 15

Analysis of Vitamin E Metabolites

PIERANGELO TORQUATO[†a], DANILO GIUSEPPONI[b], ROBERTA GALARINI[†b], DESIRÉE BARTOLINI[a], MARTA PIRODDI[a] AND FRANCESCO GALLI[a*]

[a]Departement of Pharmaceutical Sciences, University of Perugia, Perugia, Italy; [b]Istituto Zooprofilattico Sperimentale Umbria e Marche "Togo Rosati", Perugia, Italy
*E-mail: francesco.galli@unipg.it

15.1 Vitamin E Metabolism

Emerging evidence suggests the need to expand nutritional investigation protocols on vitamin E to include other forms and derivatives of this fat-soluble vitamin, especially those derived from its metabolism in human tissues (reviewed elsewhere in ref. 1–3). In the last few years, the metabolic fate of vitamin E in animal models and humans has been characterized thoroughly, disclosing new and important aspects. This metabolism results in the formation of bioactive metabolites in the liver and other tissues (recently reviewed in ref. 2 and 3), the characteristics of which are summarized in Table 15.1. Vitamin E metabolites are commonly divided into two main groups: enzymatic (essentially, cytochrome P-450-derived forms and tocopherol

[†]Equally contributed.

Food Chemistry, Function and Analysis No.11
Vitamin E: Chemistry and Nutritional Benefits
Edited by Etsuo Niki

Published by the Royal Society of Chemistry, www.rsc.org

phosphate) and non-enzymatic compounds. The latter are generated during the free-radical-dependent oxidation of vitamin E, *i.e.* during its function as an essential anti-peroxidation factor of the cell membrane and circulating lipoproteins (reviewed in ref. 3 and 4).

Biological characteristics and analytical strategies used to identify and quantify these metabolites are presented in this chapter.

15.1.1 Non-enzymatic Metabolites

Vitamin E, in the form of α-tocopherol (α-TOH), is a well-recognized chain-breaker of lipid peroxidation reactions occurring in tissues during either physiological metabolism or under conditions of oxidative stress.[5,6] Often identified as the antioxidant function of this vitamin, this reaction proceeds *via* a H atom (or proton-coupled electron) transfer effect[7] (and further discussed in ref. 8) by the hydroxyl group in position 6 on the chroman ring toward peroxyl radicals formed by the oxidation of polyunsaturated lipids.[9] This reaction is believed to synergise with that of other lipophilic and hydrophilic antioxidants, such as ubiquinol and possibly vitamin C, and enzymes such as glutathione peroxidase-4 and 1-Cys peroxiredoxin VI, to provide a fundamental detoxification process toward toxic peroxides and a mechanism of protection against the degeneration of lipid bilayers of cellular compartments.[4]

The free-radical-mediated oxidation of α-TOH (Figure 15.1) generates two groups of unstable primary products (recently reviewed in ref. 10).

Table 15.1 Biological effects of vitamin E metabolites.

Vitamin E metabolites	Biological effect	Targets[a]	References
Short-chain metabolites (SCMs)			
γ-CEHC	Natriuretic properties	70 pS K+ channel	53
γ-CEHC	Anti-proliferative and anti-oxidant	-cyclin D1	74
γ-CEHC		NOx scavenger	22
α-CEHC; γ-CEHC	Anti-inflammatory	TNF alpha, PGE_2	24
Long-chain metabolites (LCMs)			
α-13′OH; δ-13′COOH	Anti-proliferative		75
α-13′COOH; δ-13′COOH	Pro-apoptotic	Caspase-3 and 9, PARP-1	35
α-13′COOH	Lipid metabolism and toxicity	CD36	39
α-13′COOH; α-13′OH; δ-T3-13′COOH		PPARγ, CYP4F2	50
α-13′COOH		PLIN2	2
δ-13′COOH	Anti-inflammatory	COX2, PGE_2	47
δ-T3-13′COOH; α-13′COOH		Il1β, Il6, TNF alpha	52

[a]Abbreviations: TNF alpha: tumor necrosis factor alpha; PGE_2: prostaglandin E_2; CD36: cluster of differentiation 36; PPARγ: peroxisome proliferator-activated receptor gamma; CYP4F2: cytochrome P4F2 PLIN2: perilipin 2; COX2: cyclooxygenase 2; Il1β: interleukin 1 beta; Il6: interleukin 6.

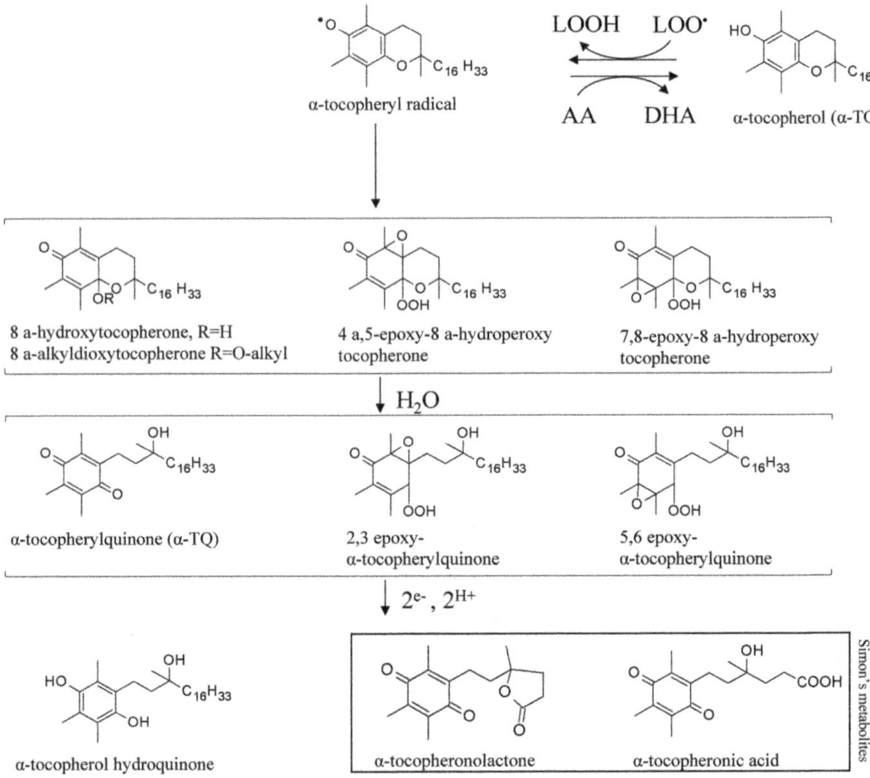

Figure 15.1 Non-enzymatic (free-radical-dependent) pathway of vitamin E (α-TOH) oxidation.

The first group consists of isomeric epoxy(hydroperoxy)-tocopherones, which are formed by an unknown mechanism; these spontaneously rearrange to epoxyquinones (α-TQE1 and α-TQE2) that are transformed to short-chain metabolites commonly identified as Simon's metabolites (described later in this section). The second group comprises 8-substituted tocopherones (hydroxy- or alkyldioxy-tocopherones), formed by one-electron oxidation of α-TOH or hydrolysis of the tocopherone cation that readily hydrolyses to α-tocopherylquinone (α-TQ). The reduction of α-TQ yields α-tocopherylhydroquinone (α-THQ).[11] The analysis of these oxidation compounds has been used to give a measure of lipid peroxidation in cellular membranes[11–13] and α-TQ appears to represent the main product of these reactions in human tissues and biological fluids.[10,14,15]

Particular attention to pre-analytical interferences must be paid in the analysis of this and the other products of the antioxidant function of vitamin E. Protocols have been developed to suppress the artefactual formation of free-radical-derived metabolites during either sample work-up or storage (see Section 15.2.1). For example, the levels of α-TQ can be affected by both the enzymatic degradation to short-chain metabolites, also identified as the Simon's metabolites discussed later in this section, and the facile redox interconversion to α-THQ.[11] Mottier and co-workers[14] elegantly demonstrated how the measure of α-TQ in biological samples can easily be overestimated if specific precautions are not adopted to avoid the spontaneous generation of this metabolite during long-term storage or work-up of an analytical sample. In human plasma, this spontaneous oxidation may lead to levels of α-TQ more than 10-fold higher compared with fresh samples.[14]

Other authors have further highlighted the importance of these aspects when vitamin E metabolites are investigated *in vivo*. In the 1950s,[16] the Simon's metabolites α-tocopheronic acid and α-tocopheronolactone (α-TL) were identified in the urine of men given large oral doses of α-tocopherol. Their formation was supposed to derive from the oxidative opening of the hydroxychroman ring with subsequent enzymatic degradation of the side-chain; but other reports claimed Simon's metabolite formation as an artefact of α-carboxyethyl-hydroxychroman (α-CEHC) auto-oxidation during sample processing for the analysis.[14,17–21] Such a disagreement in the literature as to whether the Simon's metabolites are formed *in vivo* or are produced as an artefact during analytical processing was partially sorted out by adding d_9-α-CEHC to urine samples. Though care was taken in minimising any artifactual oxidation, 5 to 10% of d_9-α-CEHC was converted to the hydroquinone form of d_9-α-tocopheronolactone. A greater percentage of α-tocopheronolactone resulted for the endogenous unlabelled metabolite, indicating possible excretion of some endogenous α-tocopheronolactone in urine not coming from the artifactual oxidation of CEHC.[20]

Early *in vitro* findings suggested biologically relevant antioxidant effects also for some metabolites of vitamin E, such as CEHC metabolites (reviewed in ref. 22 and 23). These enzymatic metabolites of vitamin E were demonstrated to act through mechanisms and reaction intermediates similar to other chromanols when investigated in transition-metal-catalysed and inflammatory cell-mediated oxidation systems.[22] However, such an antioxidant function may assume physiological relevance only if the concentrations of these metabolites reach the micromolar range, the same concentrations of other phenolic antioxidants observed, for example, in human plasma.[24] Studies by our group have demonstrated that these conditions could be reached at the intracellular level[25] and possibly in the interstitial space of tissues during high-dose vitamin E supplementation,[26] while body fluid concentrations always remain submicromolar.[27]

15.1.2 Enzymatic Metabolites

The cytochrome-P450-dependent metabolism of vitamin E was described in the early 1980s thanks to the pioneering observation made by Nakamura's group of a short-chain urinary metabolite in rat fed a diet rich in δ-TOH, the origin of which was independent from that of free-radical-derived metabolites.[28] In liver tissue, this arm of vitamin E metabolism is essentially associated with the catabolism and excretion of the dietary vitamin exceeding the physiological needs, *i.e.* the quota overcoming the capability of specialised liver proteins to bind and transfer the vitamers to nascent very-low-density lipoproteins.[29,30] α-TOH is the form of vitamin E that is selectively retained in liver cells and delivered to hepatic lipoproteins by the binding and trafficking function of α-tocopherol transfer protein and possibly of other lipid-binding proteins. Such selectivity is believed to expose desmethyl forms and tocotrienols to a higher hepatic metabolism thus lowering their tissue availability compared with α-TOH.[30,31]

In 2002, Sontag and Parker described for the first time in human hepatocytes maintained in culture, the pathway involving the cytochrome-P450-mediated omega-hydroxylation of the carbon ending of the tocopherol side chain to form the 13′-OH LCM.[32] Soon after, the molecular players of this metabolism were tentatively identified and appear to include the nuclear receptor PXR as a key regulation hub for the series of detoxification/drug metabolising genes that govern the different steps of vitamin E degradation and excretion (see ref. 33 and further described in ref. 34).

After these pioneering studies, vitamin E metabolism gained more attention and several groups demonstrated the importance, but not the exclusive role, of the liver in this metabolism. Such metabolism results in the formation of a series of SCMs and LCMs that have been identified in the liver and other solid tissue, in bile, faeces and urine, as either free or conjugated forms (reviewed in ref. 10 and 35). After the first enzymatic hydroxylation reaction at the terminal methyl group of the phytyl side-chain, the resulting LCMs undergo sequential steps of β-oxidation-like degradation to remove two or three carbon units. These enzymatic steps result in the shortening of the side-chain to form a series of medium-chain intermediates up to the formation of the final CEHC SCM.[36] CYP4F2 has been suggested to represent the main cytochrome P-450 isoenzyme with tocopherol-ω-hydroxylase activity in liver cells and rodent tissues,[37,38] but the possibility that other forms, such as CYP3A4, would play a role in this metabolism is more than probable.[33,35,39,40]

CEHCs are the main metabolites of vitamin E excreted as glucuronide and sulfate conjugates with bile or urine and are also present in the plasma of animals and human subjects as complementary indicators of vitamin E intake and metabolism at a nutritional steady-state.[17,32,41–43]

A large inter-individual variability in the rate of both the generation and clearance of CEHC has been observed, the latter process being influenced by both liver metabolism and renal excretion functions.[44–46]

δ- and γ-LCMs have also been identified in the medium used to grow human lung epithelial A549 cells. In these cells, and then *in vivo* in vitamin E supplemented rats, Jiang and Freiser identified free and sulfated forms of the intermediate forms γ- and δ-13′-(6-hydroxy-2,5,7,8,-tetramethylchroman-2-yl)-2,6,10-trimethyltridecanoic acid (γ and δ-13′-COOH), γ- and δ-11′-COOH, and γ- and δ-9′-COOH.[47–49]

The formation kinetics and activity of the α-TOH-derived LCMs α-13′-OH and α-13′-COOH have recently been characterised in human hepatocytes utilising both tumoral and non-tumoral cell lines,[50] including the hepatocarcinoma cell line used in the former studies of Sontag and Parker[32] and Landes and co-workers.[33] Treating these liver cells with α-TOH, a self-feeding mechanism of regulation was confirmed for the CYP-450 genes responsible for LCM formation as well as of membrane transport genes responsible for metabolite efflux.[51] Moreover, cell-specific differences were observed comparing tumoral with non-tumoral cells; the latter show a more efficient and complete catabolism (up to CEHC formation). At a steady state, the intracellular concentrations of LCMs were ≥1 μM, and α-13′-OH was largely the main form observed in HepG2 hepatocarcinoma cells (≥2 orders of magnitude compared with α-13′-COOH). This finding and a markedly decreased formation of α-CEHC in these tumoral cells suggest a cancer-specific defect in the second step of oxidation of vitamin E catabolism downstream of α-13′-OH metabolite formation (Figure 15.2). Furthermore, the efflux in the extracellular medium of LCMs was very efficient in these liver cells and linearly correlated with their intracellular levels.

As far as physiological levels of LCMs are concerned, the presence of α-13′-COOH in human serum was first described by Wallert *et al.* in 2014[52] and, soon after, Ciffolilli *et al.* determined that the actual concentrations of α-13′-OH in sera of healthy non-supplemented adults fed an average diet (vitamin E intake between 15 and 25 mg per day) span between 1 and 10 nM.[53] These concentrations increase several times (up to 5–7) after supplementation with supra-nutritional doses of α-TOH (1 g per day for 1 week). Recently, we described a new LC-MS/MS procedure to simultaneously measure, in human plasma, the SCM and LCM formed during the metabolism of α-TOH.[27] Metabolite levels measured with this procedure are reported in the next section.

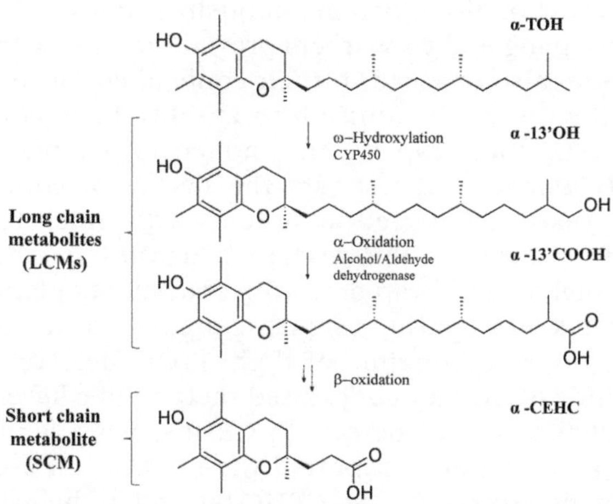

Figure 15.2 Main steps in the enzymatic oxidation of vitamin E (α-TOH). Reproduced from ref. 27 with permission from Elsevier, Copyright 2017.

15.2 Analysis of Vitamin E Metabolites

Recalling the main stages in the history of vitamin E metabolite identification and analysis, the first metabolites to be identified were the SCMs. In 1984, Chiku and co-workers reported the identification of a novel urinary metabolite, δ-CEHC, in rats based on infrared, NMR and off-line MS spectra.[28] In 1995, Schultz *et al.* identified the analogous compound of α-TOH, α-CEHC, in human urine thanks to GC-EI-MS measurements.[17] One year later, Wechter *et al.* described a new endogenous vitamin E metabolite, γ-CEHC (also identified as LLU-α), purifying about 800 liters of human uremic urine to obtain about 0.6 mg of crystalline γ-CEHC.[54] This pure material was then subjected to spectroscopic analysis for structure elucidation by ^{1}H NMR, ^{13}C NMR and MS, and the racemic form of the supposed metabolite was synthesised to confirming its molecular identity. These pioneering studies paved the way to a series of important findings obtained at the beginning of the 2000s. In the Frank Kelly laboratories in London, an isotopic dilution method was set up to measure by GC-MS analysis CEHCs in human plasma rather than in urine, and this allowed a first *in vivo* investigation of the metabolism of the two main forms of vitamin E, *i.e.* α-TOH and γ-TOH, carried out by the administration of deuterium-labelled vitamers in healthy volunteers.[43,44,55]

In the same years, based on EI-MS spectra, Pope *et al.* reported the tentative identification of the MCMs α-CMBHC and γ-CMBHC,[20] which were further characterised by Birringer *et al.*[56]

It was 2002 when Sontag and Parker identified by GC-MS in hepatocyte culture extracts the LCMs' α-13′-OH, γ-13′-OH, α-13′-COOH and γ-13′-COOH.[32] Afterwards, the conjugated forms (phase II) of these metabolites became a subject of investigation. The fact that CEHCs are mainly excreted in urine and bile as glucuronide and sulfate conjugates was well-known since the beginning of vitamin E metabolism studies,[17,28,41,57] but it was only in 2007 that Jiang and co-workers proved that the LCMs γ-13′-COOH and γ-13′-OH were also released in their conjugated forms in rat plasma and liver.[47] In their work, in addition to LC-FLD, Jiang *et al.* applied LC coupled to an ESI (−) ion trap MS and inductively coupled plasma mass spectrometry (ICP-MS) to investigate the nature of conjugated LCMs. In the last few years, MS detection systems with much higher selectivity have been introduced in this research. Identified as high-resolution systems, these include Orbitrap detectors and MS/MS platforms, such as triple quadrupoles and hybrid systems composed of quadrupoles combined with ion traps and/or time of flight (TOF) detectors. As a consequence, new kinds of urinary conjugated metabolites have recently been identified. In 2009, Cho *et al.* detected γ-CEHC-glucoside in mice urine,[58] whereas Johnson *et al.* (2012) reported the presence of α-CEHC-glycine, α-CEHC-glycine-glucuronide and α-CEHC-taurine in the urine of humans and mice.[59] Tandem MS techniques have recently been utilised in our laboratories to develop a quantitative analysis method to simultaneously determine tocopherols and a set of cytochrome P-450- and free radical-derived metabolites in human plasma and serum.[27] Furthermore, these techniques allowed us to identify new isobaric forms of α-13′-OH and α-13′-COOH in human plasma that are now under further investigation to elucidate their molecular identity.

15.2.1 Pre-analytical and Analytical Issues

Sample work-up and storage conditions are critical steps in the analysis of vitamin E-related molecules. Concerns are mainly related to auto-oxidation processes that are intimately linked with the free-radical-scavenging properties of tocols introduced in Section 15.1.1. Obviously, these aspects become more critical in the case of metabolite analysis due to the much lower concentrations incurred in biological fluids and tissues (nM instead of μM). Surprisingly, no standardised protocol has been established so far to deal with these issues. As a general precaution, ascorbic acid and butylated hydroxytoluene (BHT) are commonly used to prevent metabolite oxidation during sample processing, but different strategies are often reported without the support of validation experiments and controls. In some cases, ascorbic acid and/or BHT are added only during the extraction steps;[60,61] in other cases, ascorbic acid and/or BHT are added during both the extraction step and in the mixture used to reconstitute the final extract immediately prior to LC injection.[18,41,62–64]

A mixture of three antioxidants (BHT, pyrogallol and ascorbic acid) and the transition metal chelator deferoxamine mesylate has been utilised by Mottier *et al.* in the analysis of TQs to avoid oxidation during saponification at 60 °C prior to metabolite extraction.[14] This paper also reported an interesting and rigorous study on sample storage strategies to preserve α-TOH and α-TQ in human plasma. Based on this study, plasma samples and standard solutions should both be maintained in liquid nitrogen to minimise the artefactual oxidation of α-TOH to α-TQ. In addition, comparing GC-MS and LC-MS/MS methods, the authors also found that the derivation step prior to GC injection could produce an artefactual oxidation. Since α-TOH concentrations in human plasma are about three orders of magnitude higher than those of α-TQ, only α-TQ measurements were significantly affected by this phenomenon. This rigorous study demonstrated how the storage temperature, and possibly time, should be strictly controlled for an accurate unbiased analysis of TQs.

Two years later, in 2004, Pollok *et al.* published an analytical method to measure all four forms of TQ (α, β, γ, and δ).[18] In this study, the metabolites were measured by HPLC-FLD with post-column derivatisation; the concentrations of α-TQ were determined in fresh human serum samples of 14 healthy volunteers and values were found to range from 30 to 61 ng mL^{-1}, which were higher than the levels found by Mottier *et al.*[14] (from 4.5 to 8.4 ng mL^{-1}), although this latter study reported replicate measurements on the plasma of a single male volunteer. Ciffolilli *et al.* utilised an HPLC-ECD method to measure the serum concentrations of α-TQ in two healthy subjects and found levels of 30 and 165 ng mL^{-1}.[53] Our unpublished data are coherent with α-TQ levels determined by Pollock *et al.* and Ciffolilli *et al.*, more in detail, α-TQ concentrations measured in plasma and serum of five adults stored at −80 °C for two weeks were in the range 40–141 ng mL^{-1}. Probably only very low storage temperatures (liquid nitrogen) and short storage times can almost completely stop the radical oxidation of α-TOH, maintaining α-TQ concentrations lower than 10 ng mL^{-1}, but further studies are needed to confirm these assumptions. Finally, with regard to Simon's metabolites, analysing the urine of diabetic children, Sharma *et al.* demonstrated that α-TL can be a true metabolite and not only the artifact product of α-CEHC auto-oxidation.[19]

15.2.2 Deconjugation of Sulfated and Glucuronated Forms

Several enzymatic metabolites of vitamin E are present in biological fluids as conjugated (mainly sulfated and glucuronated) forms produced by the derivatisation activity of the hepatic transferases, a series of abundantly expressed phase II drug metabolising genes.[65]

At present, our knowledge of the relative abundances of free and conjugated forms of metabolites in human and animal biological fluids/tissues is essentially based on the same approach as the earliest studies on CEHC metabolites, that is the subtraction of the relative concentrations of the free form from the total metabolite in pairs of samples prepared without and with enzymatic hydrolysis, respectively.[41,42]

During metabolite analysis, the hydrolysis of the conjugated forms is generally carried out enzymatically during sample work-up utilising beta-glucuronidase that in some cases has been combined with sulfatase. The commercial beta-glucuronidase from *Helix pomatia* is one of the most commonly used deconjugation enzyme that owns both glucuronidase and sulfatase activity, whereas that from *Escherichia coli* shows only glucuronidase activity. Although enzymatic hydrolysis is the most used deconjugation methodology, its efficacy is rather poor especially to hydrolyse sulfates. One of the most exhaustive studies on this aspect was published by Li *et al.*[66] These authors compared various kinds of deconjugation strategies to measure CEHC metabolites in rat urine, and their findings clearly demonstrated the very low efficacy of sulfatase to hydrolyse sulfated α-CEHC. Conflicting results were obtained years before, when sulfatase was proven to efficiently hydrolyse δ-CEHC.[28] Such controversial findings led Li and coworkers to speculate that the number of methyl groups on the chromanol ring of CEHC metabolites influences the catalytic function of the sulfatase enzyme.[66] With more detail, these authors suggested that the fully methylated α-CEHC may interfere with the enzyme activity through a steric hindrance effect, while the demethylated counterpart δ-CEHC—bearing only one methyl group at the *meta*-position to the sulfate group of the chroman ring—is more easily hydrolysable by the sulfatase activity. In the same study, Li *et al.* found that the chemical hydrolysis carried out with 6 N HCl at 60 °C for one hour almost completely deconjugated α-CEHC sulfate, and since acidic conditions are known to catalyse the conversion of α-CEHC to the auto-oxidation product α-TL,[17] ascorbate was added to the reaction mixture to avoid this drawback. At present, it is not known whether a similar problem exists in the hydrolysis of α-CEHC glucuronide by the glucuronidase enzyme.

Freiser and Jiang demonstrated that an improvement of deconjugation of the sulfated forms in plasma of γ-CEHC, γ-9′-COOH, γ-11′-COOH and γ-13′-COOH can be achieved by overnight incubation with type H-1 sulfatase after metabolite extraction with MeOH–hexane (2 : 5 v/v).[49] These authors affirmed that interfering substances, such as lipids and proteins, inhibit the hydrolysis reaction and therefore their removal favours the release of the free metabolites.

To overcome the problem of an inefficient deconjugation of the different metabolites, methods have been developed to directly measure the conjugated forms of vitamin E metabolites by LC-MS or LC-FLD techniques.[19,48,49,67] These new methods that obviously provide a valid alternative to indirect protocols of enzymatic hydrolysis await validation for the quantitative analysis of conjugated metabolites of the different forms of vitamin E by the preparation of proper internal standards and evaluation of analytical performances.

15.2.3 Sample Preparation

The most used extraction/purification strategy to measure vitamin E and its metabolites in biological samples has been liquid–liquid extraction (LLE). Because vitamin E compounds are lipophilic molecules, mixtures of apolar solvents have generally been utilised. However, SCM and MCM are much more polar than their parent long-chain compounds: the log*P* of 13′-α-COOH is the same as α-TOH (*ca.* 9), whereas the log*P* of the MCM α-CMBHC and the SCM α-CEHC are 5.5 and 4.0, respectively. Accordingly, some procedures have utilised ethyl ether to extract SCMs[63,66,68,69] and others were based on a mixture of hexane and DCM (1 : 1 v/v).[41,43,69] When LCMs or TQs are the analytes of interest, hexane is generally the preferred extraction solvent.[14,18,53] Since there is an increasing trend toward the development of multi-parameter analysis methods for the simultaneous determination of vitamin E and its LCMs and SCMs, some studies have described separate extraction procedures, carried out in parallel, to optimise the recovery of the different compounds. For example, Lee *et al.* extracted vitamin E from rat serum, urine and tissues with hexane and CECH/CMBHC metabolites using ethyl acetate.[61] Differently, Zhao *et al.* reported a protocol of extraction with hexane of tissue homogenates or biological fluids, prepared in ethanol and water, that was suitable for recovering less polar analytes, *i.e.* vitamin E and LCMs.[60] In a second step, they re-extracted the remaining ethanolic phase with ethyl acetate to recover the more hydrophilic metabolites, *i.e.* MCMs and SCMs.

However, these studies in which parallel extraction protocols have been investigated were not designed to obtain quantitative results for each analyte (vitamers and metabolites). Such a quantitative goal was recently achieved in our laboratories simultaneously investigating vitamin E and its SCMs and LCMs in human plasma and serum.[64] Taking into account that tocopherols occurred at much higher concentrations than metabolites in these types of samples (Table 15.5), a mixture of hexane–TBME (4 : 1, v/v) was used to extract tocopherols, while hexane–TBME (2 : 1 v/v) was used to extract SCMs/LCMs in a separate vial. Thus, the two extracts were processed in separate chromatographic runs. Such an analytical strategy also allowed the addition, prior to sample work-up, of more suitable concentrations of each deuterium-labelled internal standard in the two separate extraction samples, *i.e.* in the order of ppm for tocopherols and ppb for their metabolites, resulting in a rigorous quantification approach for all the main vitamin E-related compounds so far identified in human blood.

15.2.4 Chromatographic Separation and Detection

Since the eight forms of vitamin E (α-, γ-, β-, δ-tocopherols and α-, γ-, β-, δ-tocotrienols) and all their metabolites cannot be directly separated by GC, liquid chromatography (LC) carried out utilising reversed-phase (RP) columns has been the most employed separation technique so far. Detection techniques employed for metabolite analysis include electrochemistry with coulometric detection (ECD), fluorescence (FLD) and, more recently, mass spectrometry (MS).

The most significant analytical methods used so far to measure metabolites (also including vitamin E precursors) are summarised in Tables 15.2–15.4; the most applied no-MS detector has been ECD (Table 15.2). The most sensitive detector used so far to measure vitamin E metabolites by LC methods is surely ECD, which is also intrinsically selective because it exploits the specific redox properties of the different chemical configurations of the chromanol ring in the various metabolites. However, ECD cannot provide selective identification and structural information on individual molecules, and this technique is intrinsically applicable only to the analysis of oxidisable forms. Therefore, ECD cannot detect metabolites in which the hydroxyl group on the chroman ring has been derivatised, such as the conjugated forms, or modified by auto-oxidation processes (such as in the case of TQs).

Only Pollok and coworkers have applied an HPLC-FLD method, but with a particular configuration (post-column photoreaction).[18] Fluorimetric detection is less sensitive than ECD and the response factor for some of the physiological metabolites is largely unfavourable to expect applications in animal and human studies.[70] Furthermore, TQs do not have native fluorescence.

GC-MS has played an important historical role in metabolite analysis (Table 15.3). Although GC involves a derivatisation step that might affect, to some extent, metabolite stability,[14] GC-MS systems were widely applied in the years that preceded the advent of modern LC-MS technology in the

Table 15.2 LC-FLD and LC-ECD methods for vitamin E metabolite analysis.

Reference	Analytes	Matrices[a]	Sample preparation	Detector	Mobile phases	Validation scheme (metabolites)	LOD
Stahl *et al.* 1999[41]	α-CEHC, γ-CEHC	Serum	EH (β-glucuronidase) LLE: hex–DCM (5:2, v/v): discarded; acidification of aqueous residue; LLE: hex–DCM (1:1, v/v)	ECD	ACN–water 35:65 v/v containing EDTA and tetraethylammonium hydroxide (pH = 4.5)	Linearity/precision/recovery	—
Lodge *et al.* 2000[b 36]	α-CEHC, γ-CEHCs	Urine	EH (β-glucuronidase/sulfatase); LLE: diethyl ether	ECD	(a) ACN (1% Li perchlorate) (b) McIlvaine buffer pH 4.15 (1% Li perchlorate)	Linearity/recovery/precision	—
Morinobu *et al.* 2003[69]	α-TOH, γ-TOH; α-CEHC, γ-CEHCs	Serum (?)/urine	EH (β-glucuronidase); urine: LLE: diethyl ether serum: LLE: hex–DCM (1:1 v/v)	ECD	ACN–water 40:60 v/v (50 mM Na perchlorate, pH 3.6)	Recovery/precision	—
Pollok *et al.* 2004[18]	α-TQ, γ-TQ, β-TQ, δ-TQ	Serum	LLE: hex; purification on a silica gel minicolumn	FLD[b]	MeOH	Linearity/recovery/precision (intra- and inter-day)	*ca.* 5 mol
Li *et al.* 2008[66]	α-CEHC, α-TL	Rat urine	CH (HCl); LLE: diethyl ether	ECD[c]	MeOH–H_2O (43:57, v/v) with 50 mM Na acetate (pH 4.5)	Linearity/recovery/precision	0.45 pmol
Lee *et al.* 2018[61]	Vitamin E (tocopherols/tocotrienols); α-, γ-, δ-CEHC; α-, γ-, δ-CMBHC	Rat serum, urine, tissues	EH (β-glucuronidase/sulfatase); LLE: hex (vitamin E), LLE: ethyl acetate (CEHCs, CMBHCs)	ECD	ACN–MeOH–water (15 mM of Li acetate) (a) 28:4:68 v/v/v (b) 84:13:3 v/v/v	Recovery/precision (intra- and inter-day)/matrix effect	5 nM[d]

[a]Unless specified differently, biological matrices belong to humans.
[b]On-line post column photochemical derivatization prior to fluorescence detection. Identity confirmation by GC-MS.
[c]α-TL was measured *via* HPLC-UV.
[d]LOD (metabolites) has been roughly estimated from linearity data.

Table 15.3 GC-MS methods for vitamin E metabolite analysis.

Reference	Metabolites	Matrices[a]	Sample preparation	Validation scheme	LOD
Swanson *et al.* 1999[42]	γ-CEHC	Urine	EH (β-glucuronidase); LLE: hex–MTBE 99:1 v/v	—	—
Pope *et al.* 2000[20]	α-CEHC, γ-CEHC,	Urine	SPE: C4 cartridge of acidified urine (pH = 2.5);	Recovery/precision (intra- and inter-day)	—
	α-CMBHC		EH (β-glucuronidase/sulfatase)		
	α-TL		SPE: C4 cartridge		
Galli *et al.* 2002[b][43]	α-TOH, γ-TOH; α-CEHC, γ-CEHC	Plasma/Urine	Plasma: EH (β-glucuronidase) (a) LLE: hex–dichloromethane (5:2, v/v): for α-T and γ-T; (b) acidification of the aqueous residue (pH = 3.5–4.0); LLE: hex–dichloromethane (1:1, v/v) for CEHCs; urine: EH (β-glucuronidase); acidification (pH = 1.5); LLE: hex–MTBE (99:1, v/v) for CEHCs	Linearity/recovery evaluation of labeled analytes	2–5 nM
Ciffolilli *et al.* 2015[c][53]	α-13′-OH	Serum	MeOH protein precipitation; LLE: hex	Linearity/recovery	3 nM

[a]Unless specified differently, biological matrices belong to humans.
[b]The analysis of α-T and γ-T was carried out with LC-UV.
[c]Samples were also analysed with HPLC-ECD.

identification and structural characterisation of several vitamin E-related compounds. Derivatisation agents used in metabolite and vitamin E analysis are *N,O*-bis(trimethylsilyl) trifluoroacetamide (BSTFA) with or without trimethylchlorosilane (TMCS) or *N,O*-bis(trimethylsilyl) acetamide (BSA). A double derivatisation procedure was recently set up in our laboratories in an attempt to improve the MS response of the carboxylated (COOH) forms of LCMs and SCMs[10,40] In the first step, the metabolite's carboxylic group is methylated using MeOH–HCL 37% (95 : 5 v/v). The second step consists of a silylation procedure.

In the last few years, LC-MS platforms became the gold standard in metabolite analysis. Different kinds of MS analysers can be coupled to LC: single quadrupole (MS), triple quadrupoles (MS-MS), ion trap (LTQ) and, recently, high-resolution MS analysers (HR-MS) such as time of flight (TOF) and Orbitrap (Table 15.4). HR-MS allows higher selectivity than traditional low-resolution MS thanks to acquisition of accurate mass/charge (*m/z*). Moreover, recent hybrid MS analysers (Q-TOF and Q-Orbitrap) combine accurate *m/z* and structural information derived from fragmentation experiments. In the next few years, the application of these powerful hybrid platforms are awaited to add new and relevant insights to the metabolome of vitamin E.

15.2.5 Levels of Vitamin E Metabolites in Human Blood

In the last year, we have focused our efforts on developing a quantitative LC-MS/MS procedure to simultaneously measure the main forms of vitamin E and their SCMs and LCMs in human plasma and serum.[10,15,64] The levels of these compounds in two groups of healthy controls are reported in Table 15.5A and B, also including the levels of three isobaric forms of α-13′-OH and α-13′-COOH detected for the first time with this analytical procedure, and identified as M1-3; the actual molecular identity of these metabolites remains unknown. In these subjects, metabolite levels were characterised by a large inter-individual variability that could be explained by the interference of both nutritional and metabolic factors. Moreover, these levels were compared with those observed in two human pathologies associated with disorders of blood lipid and vitamin E metabolism, and with oxidative stress,[3,71] such as chronic kidney disease (CKD) and non-alcoholic fatty liver disease (NAFLD). CKD patients showed decreased levels of α-13′-OH, but not of its isobaric form M3. Conversely, mean levels of α-13′-COOH were not modified in CKD patients, while the isobaric metabolites M1 and M2 decreased markedly compared with healthy controls (approx. by 20 and 2.5 times, respectively). These changes in LCM levels were associated with a marked accumulation of CEHCs, by the impaired renal excretion of these hydrosoluble SCMs;[46,62] slightly decreased levels of α-TOH were also observed in these patients (Table 15.5A), which further decreased if corrected for total lipids (not shown). NAFLD patients were characterised by significantly higher levels of α-TQ (Table 15.5B), while this free radical-derived metabolite was only slightly increased in CKD patients (Table 15.5A). In NAFLD, the increased levels of α-TQ were associated with increased levels of the lipid peroxidation

Table 15.4 LC-MS(MS) methods for vitamin E metabolite analysis.

Reference	Analytes	Matrices[a]	Sample preparation	Detector[b]	Mobile phases	Validation parameters (metabolites)	LOD of metabolites[c]
Mottier *et al.* 2002[d][14]	α-TQ	Plasma	Saponification with KOH LLE: hex	MS/MS	(a) Water (0.1% FA) (b) ACN (0.1% FA)	Recovery/ precision	1 nM
Himmelfarb *et al.* 2003[e][62]	α-, γ-CEHCs	Serum	EH (β-glucuronidase); MeOH protein precipitation; membrane filtration	MS	(a) Water (0.05% AcOH) (b) MeOH (0.05% AcOH)	Linearity	3.3–1.6 nM
Leonard *et al.* 2005 (JLR)[63]	α-, γ-CEHCs	Rat liver/ serum	EH (β-glucuronidase/sulfatase); LLE: diethyl ether	MS	(a) Water (0.05% AcOH) (b) MeOH (0.05% AcOH)	Linearity; recovery/ precision (liver)	?
Zhao *et al.* 2010[60]	α-, γ-, δ-TOHs; α-, γ-, δ-CEHCs; α-, γ-, δ-CMBHCs; other LCMs	Faeces, urine, serum, tissues	EtOH-water mixture or homogenate; (a) LLE: hex (vitamin E, LCMs) (b) aqueous residue: EH (β-glucuronidase/sulfatase); LLE: ethyl acetate (MCMs, SCMs)	LTQ	ACN–EtOH–H_2O (20 mM NH_4Ac, pH 4.4) (a) 165 : 135 : 700 v/v/v (b) 539 : 441 : 20 v/v/v	Linearity/ recoveries/ precision[f]	100–200 nM[g]
Sharma *et al.* 2013[19]	α-CEHC, α-TL (direct measure of conjugated forms)	Urine	Centrifugation	MS/MS	(a) MeOH–ACN (2 : 1, v/v) (b) 4 mM NH_4Ac (0.01% FA)	Full validation	60–1300 nM
Giusepponi *et al.* 2017[27]	α-TOH, γ-TOH, α-CEHC, γ-CEHC, α-13′-OH, α-13′-COOH	Serum/ plasma	LLE: hex–MTBE 4 : 1, v/v (for TOHs); (for metabolite analysis) EH (β-glucuronidase/sulfatase); LLE: hex–MTBE 2 : 1, v/v	MS/MS	(a) Water (0.1% FA) (b) MeOH (0.1% FA)	Full validation	0.4–0.1 nM

[a]Unless specified differently, biological matrices belong to humans.
[b]All the methods are developed using ESI(−) except Giusepponi *et al.* (2017) who applied ESI(+).
[c]If only the limit of quantification (LOQ) is furnished, it is reported as dividing by 3.
[d]Comparison between LC-MS/MS and GC-MS methods.
[e]In the same work, α-T and γ-T were determined with a different extraction protocol and technique (HPLC-UV).
[f]Recovery and precision are not available for each analyte included in the method scope.
[g]LOD has been roughly estimated from linearity data.[27]

Table 15.5 Plasma concentrations of vitamin E metabolites in healthy controls and CKD and NAFLD patients.

	A[a]		B	
Vitamin E molecule	Healthy controls[b]	CDK[b]	Healthy controls[b]	NAFLD[b]
γ-CEHC (nM)	135 ± 40.4	399 ± 187[e]	88.6 ± 38.3	112 ± 126
α-CEHC (nM)	23 ± 6.9	104 ± 42.6[f]	24 ± 14	29 ± 17
α-13′COOH (nM)	1.8 ± 0.9	1.0 ± 0.2	3.0 ± 0.6	3.3 ± 0.5
M1[c] (nM)	68 ± 40	3.8 ± 3.0[f]	25 ± 14	28 ± 9.5
M2[c] (nM)	5.3 ± 2.9	1.9 ± 2.4[e]	6.1 ± 3.1	6.5 ± 1.7
α-13′COOH + M1 + M2 (nM)	75 ± 41	6.8 ± 3.6[f]	34 ± 17	38 ± 11
α-13′OH (nM)	3.4 ± 1.4	1.7 ± 0.8[e]	2.2 ± 1.3	2.3 ± 2.5
M3[d] (nM)	0.6 ± 1.0	0.8 ± 0.3	1.8 ± 0.6	2.3 ± 0.9
α-13′OH + M3 (nM)	4.0 ± 1.7	2.5 ± 0.8[e]	4.0 ± 1.7	4.7 ± 2.6
α-TOH (μM)	29 ± 5.9	21 ± 6.4[e]	27 ± 5.2	34 ± 8.8
γ-TOH (μM)	0.6 ± 0.1	0.6 ± 0.2	0.5 ± 0.3	0.5 ± 0.3
α-TQ (nM)	143 ± 38.9	176 ± 62.2	143 ± 38.9	237 ± 61.7[g]

[a]Table A: data from ref. 27.
[b]Results expressed as mean ± SD.
[c]As α-13′-COOH equivalents.
[d]As α-13′-OH equivalents.
[e]Significant differences between healthy control ($n = 6$) and CKD patient ($n = 6$) groups (Mann–Witney test, $p < 0.05$).
[f]Significant differences between healthy control ($n = 6$) and CDK patient ($n = 6$) groups (Mann–Witney test, $p < 0.01$).
[g]Significant differences between healthy control ($n = 6$) and NAFLD patient ($n = 6$) groups (Mann–Witney test, $p < 0.01$).

marker protein-bound 4-hydroxynonenal.[15] In these patients, enzymatic metabolites and α-TOH showed a trend toward increased levels. These findings suggest disease-specific changes in the human metabolome of vitamin E that are worth investigating further to explore underlying mechanisms and possible clinical consequences.

15.3 Discussion

Reviewing analytical methods that have been published so far to quantify vitamin E metabolites, there is an evident lack of validated procedures, *i.e.* protocols with well-characterised performance characteristics that include: linearity, selectivity, trueness, repeatability, intra-lab reproducibility, LOD and LOQ. This probably occurred because the majority of these methods were developed in the 1990s and early 2000s in research environments without specialisms in analytics. This represents a potential source of bias that is nowadays avoided by the application of standardised pre-analytical and analytical conditions.[72] As a consequence, the published levels of metabolites in various biological matrices are often affected by high levels of uncertainty; in fact, the applied procedures in several studies are almost qualitative, lacking suitable calibration approaches, assessment of extraction recovery of individual analytes and matrix effects when LC-MS techniques are applied.

There are many examples of these limitations and of the quantitation gap between different studies on metabolite analysis have been published in recent years.[48,49,60,67,73] Undoubtedly, these papers add excellent and original information on the metabolism of vitamin E, but they claim to furnish quantitative data without adequate methodological bases. Last but not least, pre-analytical conditions (storage/deconjugation) have not been sufficiently explored in the majority of papers that have appeared in the literature so far.

Notwithstanding, a series of analytical advancements have been introduced since the first identification of the different vitamin E metabolites in 1984.[28] Nowadays, the "endometabolome" of vitamin E can be explored in detail and individual metabolites can be identified and quantitated in a variety of biological samples that include human serum and plasma, tissues and cell culture specimens.[10,40,64,74] High-performance chromatography methods coupled with MS and the advancement provided by the last generation of high-resolution MS, are now available in several laboratories to accurately measure the entire series of vitamin E metabolites and also to identify new molecules generated during the metabolism of this vitamin.

Acknowledgements

Part of this work was supported by grants from the Italian Ministry of Education, Universities and Research (MIUR), and National Technology Agrifood Cluster, Health and Nutrition program, PROS.IT project (CTN01_00230_413096).

We are indebted to Prof. Marc Birringer and Dr. Francesco Mazzini for providing some analytical standards used in the LC-MS/MS analysis of vitamin E metabolites.[75]

The authors acknowledge Mr. Sebastiani Bartolomeo for technical support.

References

1. F. Galli, M. C. Polidori, W. Stahl, P. Mecocci and F. J. Kelly, *Vitam. Horm.*, 2007, **76**, 263–280.
2. L. Schmolz, M. Schubert, J. Kirschner, S. Kluge, F. Galli, M. Birringer, M. Wallert and S. Lorkowski, *Biochim. Biophys. Acta*, 2018, **1863**, 919–927.
3. F. Galli, A. Azzi, M. Birringer, J. M. Cook-Mills, M. Eggersdorfer, J. Frank, G. Cruciani, S. Lorkowski and N. K. Ozer, *Free Radical Biol. Med.*, 2017, **102**, 16–36.
4. M. G. Traber and J. Atkinson, *Free Radical Biol. Med.*, 2007, **43**, 4–15.
5. K. U. Ingold, G. W. Burton, D. O. Foster, M. Zuker, L. Hughes, S. Lacelle, E. Lusztyk and M. Slaby, *FEBS Lett.*, 1986, **205**, 117–120.
6. J. M. Upston, A. C. Terentis and R. Stocker, *FASEB J.: Off. Publ. Fed. Am. Soc. Exp. Biol.*, 1999, **13**, 977–994.
7. T. Inagaki and T. Yamamoto, *J. Phys. Chem. B*, 2014, **118**, 937–950.
8. A. Galano and J. R. Alvarez-Idaboy, *Int. J. Quantum Chem.*, 2018, **118**, e25665.

9. D. Marquardt, J. A. Williams, N. Kucerka, J. Atkinson, S. R. Wassall, J. Katsaras and T. A. Harroun, *J. Am. Chem. Soc.*, 2013, **135**, 7523–7533.
10. P. Torquato, O. Ripa, D. Giusepponi, R. Galarini, D. Bartolini, M. Wallert, R. Pellegrino, G. Cruciani, S. Lorkowski, M. Birringer, F. Mazzini and F. Galli, *J. Pharm. Biomed. Anal.*, 2016, **124**, 399–412.
11. D. C. Liebler, J. A. Burr, L. Philips and A. J. Ham, *Anal. Biochem.*, 1996, **236**, 27–34.
12. S. K. Jain, R. Wise and J. J. Bocchini, Jr, *J. Am. Coll. Nutr.*, 1996, **15**, 44–48.
13. A. C. Terentis, S. R. Thomas, J. A. Burr, D. C. Liebler and R. Stocker, *Circ. Res.*, 2002, **90**, 333–339.
14. P. Mottier, E. Gremaud, P. A. Guy and R. J. Turesky, *Anal. Biochem.*, 2002, **301**, 128–135.
15. P. Torquato, D. Bartolini, D. Giusepponi, M. Piroddi, R. Galarini and F. Galli, *Free Radical Biol. Med.*, 2018, **120**, S122.
16. A. Eisengart, A. T. Milhorat, E. J. Simon and L. Sundheim, *J. Biol. Chem.*, 1956, **221**, 807–817.
17. M. Schultz, M. Leist, M. Petrzika, B. Gassmann and R. Brigelius-Flohe, *Am. J. Clin. Nutr.*, 1995, **62**, 1527S–1534S.
18. D. Pollok and H. U. Melchert, *J. Chromatogr. A*, 2004, **1056**, 257–262.
19. G. Sharma, D. P. Muller, S. M. O'Riordan, S. Bryan, M. T. Dattani, P. C. Hindmarsh and K. Mills, *Free Radical Biol. Med.*, 2013, **55**, 54–62.
20. S. A. Pope, P. T. Clayton and D. P. Muller, *Arch. Biochem. Biophys.*, 2000, **381**, 8–15.
21. R. Brigelius-Flohe and M. G. Traber, *FASEB J.: Off. Publ. Fed. Am. Soc. Exp. Biol.*, 1999, **13**, 1145–1155.
22. F. Galli, M. Piroddi, A. Lannone, S. Pagliarani, A. Tomasi and A. Floridi, *Int. J. Vitam. Nutr. Res. Int. Z. Vitam. Ernahrungsforsch. J. Int. Vitaminol. Nutr.*, 2004, **74**, 362–373.
23. F. Galli and A. Azzi, *Biofactors*, 2010, **36**, 33–42.
24. K. Hensley, E. J. Benaksas, R. Bolli, P. Comp, P. Grammas, L. Hamdheydari, S. Mou, Q. N. Pye, M. F. Stoddard, G. Wallis, K. S. Williamson, M. West, W. J. Wechter and R. A. Floyd, *Free Radical Biol. Med.*, 2004, **36**, 1–15.
25. C. Conte, A. Floridi, C. Aisa, M. Piroddi and F. Galli, *Ann. N. Y. Acad. Sci.*, 2004, **1031**, 391–394.
26. P. Torquato, D. Bartolini, D. Giusepponi, G. Saluti, A. Russo, C. Barola, M. Birringer, R. Galarini and F. Galli, *Free Radical Biol. Med.*, 2016, **96**, S19–S20.
27. D. Giusepponi, P. Torquato, D. Bartolini, M. Piroddi, M. Birringer, S. Lorkowski, C. Libetta, G. Cruciani, S. Moretti, G. Saluti, F. Galli and R. Galarini, *Talanta*, 2017, **170**, 552–561.
28. S. Chiku, K. Hamamura and T. Nakamura, *J. Lipid Res.*, 1984, **25**, 40–48.
29. M. Hacquebard and Y. A. Carpentier, *Curr. Opin. Clin. Nutr. Metab. Care*, 2005, **8**, 133–138.
30. S. Chung, M. Ghelfi, J. Atkinson, R. Parker, J. Qian, C. Carlin and D. Manor, *J. Biol. Chem.*, 2016, **291**, 17028–17039.

31. W. Aeschimann, S. Staats, S. Kammer, N. Olieric, J. M. Jeckelmann, D. Fotiadis, T. Netscher, G. Rimbach, M. Cascella and A. Stocker, *Sci. Rep.*, 2017, **7**, 4970.
32. T. J. Sontag and R. S. Parker, *J. Biol. Chem.*, 2002, **277**, 25290–25296.
33. N. Landes, P. Pfluger, D. Kluth, M. Birringer, R. Ruhl, G. F. Bol, H. Glatt and R. Brigelius-Flohe, *Biochem. Pharmacol.*, 2003, **65**, 269–273.
34. M. C. Podszun, M. Jakobi, M. Birringer, J. Weiss and J. Frank, *Mol. Nutr. Food Res.*, 2017, **61**, 1600605.
35. M. Birringer, *Mol. Nutr. Food Res.*, 2010, **54**, 588–598.
36. J. K. Lodge, M. G. Traber, A. Elsner and R. Brigelius-Flohe, *J. Lipid Res.*, 2000, **41**, 148–154.
37. S. A. Bardowell, F. Duan, D. Manor, J. E. Swanson and R. S. Parker, *J. Biol. Chem.*, 2012, **287**, 26077–26086.
38. S. A. Bardowell, D. E. Stec and R. S. Parker, *J. Nutr.*, 2010, **140**, 1901–1906.
39. M. Wallert, L. Schmolz, F. Galli, M. Birringer and S. Lorkowski, *Redox Biol.*, 2014, **2**, 495–503.
40. A. Russo, D. Bartolini, P. Torquato, D. Giusepponi, C. Barola, R. Galarini, M. Birringer, S. Lorkowski and F. Galli, *Toxicol. In Vitro*, 2017, **40**, 124–133.
41. W. Stahl, P. Graf, R. Brigelius-Flohe, W. Wechter and H. Sies, *Anal. Biochem.*, 1999, **275**, 254–259.
42. J. E. Swanson, R. N. Ben, G. W. Burton and R. S. Parker, *J. Lipid Res.*, 1999, **40**, 665–671.
43. F. Galli, R. Lee, C. Dunster and F. J. Kelly, *Free Radical Biol. Med.*, 2002, **32**, 333–340.
44. F. Galli, R. Lee, J. Atkinson, A. Floridi and F. J. Kelly, *Free Radical Res.*, 2003, **37**, 1225–1233.
45. S. W. Leonard, E. Paterson, J. K. Atkinson, R. Ramakrishnan, C. E. Cross and M. G. Traber, *Free Radical Biol. Med.*, 2005, **38**, 857–866.
46. F. Galli, A. G. Floridi, A. Floridi and U. Buoncristiani, *Clin. Nutr.*, 2004, **23**, 205–212.
47. Q. Jiang, H. Freiser, K. V. Wood and X. Yin, *J. Lipid Res.*, 2007, **48**, 1221–1230.
48. H. Freiser and Q. Jiang, *J. Nutr.*, 2009, **139**, 884–889.
49. H. Freiser and Q. Jiang, *Anal. Biochem.*, 2009, **388**, 260–265.
50. P. Torquato, D. Bartolini, D. Giusepponi, G. Saluti, A. Russo, C. Barola, M. Birringer, R. Galarini and F. Galli, *Free Radical Biol. Med.*, 2016, **96**, S15–S20.
51. R. Brigelius-Flohe, *Genes Nutr.*, 2007, **2**, 249–256.
52. L. Schmolz, M. Wallert, J. Heise, F. Galli, O. Werz, M. Birringer and S. Lorkowski, *Free Radical Biol. Med.*, 2014, **75**(Suppl 1), S48.
53. S. Ciffolilli, M. Wallert, D. Bartolini, V. Krauth, O. Werz, M. Piroddi, B. Sebastiani, P. Torquato, S. Lorkowski, M. Birringer and F. Galli, *Free Radical Biol. Med.*, 2015, **89**, 952–962.
54. W. J. Wechter, D. Kantoci, E. D. Murray, Jr, D. C. D'Amico, M. E. Jung and W. H. Wang, *Proc. Natl. Acad. Sci. U. S. A.*, 1996, **93**, 6002–6007.

55. F. Galli, R. Lee, C. Dunster, J. Atkinson, A. Floridi and F. J. Kelly, *Biofactors*, 2001, **15**, 65–69.
56. M. Birringer, D. Drogan and R. Brigelius-Flohe, *Free Radical Biol. Med.*, 2001, **31**, 226–232.
57. M. G. Traber, A. Elsner and R. Brigelius-Flohe, *FEBS Lett.*, 1998, **437**, 145–148.
58. J. Y. Cho, D. W. Kang, X. Ma, S. H. Ahn, K. W. Krausz, H. Luecke, J. R. Idle and F. J. Gonzalez, *J. Lipid Res.*, 2009, **50**, 924–937.
59. C. H. Johnson, O. Slanar, K. W. Krausz, D. W. Kang, A. D. Patterson, J. H. Kim, H. Luecke, F. J. Gonzalez and J. R. Idle, *Am. J. Clin. Nutr.*, 2012, **96**, 818–830.
60. Y. Zhao, M. J. Lee, C. Cheung, J. H. Ju, Y. K. Chen, B. Liu, L. Q. Hu and C. S. Yang, *J. Agric. Food Chem.*, 2010, **58**, 4844–4852.
61. M. J. Lee, W. Feng, L. Yang, Y. K. Chen, E. Chi, A. Liu and C. S. Yang, *J. Food Drug Anal.*, 2018, **26**, 318–329.
62. J. Himmelfarb, J. Kane, E. McMonagle, E. Zaltas, S. Bobzin, S. Boddupalli, S. Phinney and G. Miller, *Kidney Int.*, 2003, **64**, 978–991.
63. S. W. Leonard, E. Gumpricht, M. W. Devereaux, R. J. Sokol and M. G. Traber, *J. Lipid Res.*, 2005, **46**, 1068–1075.
64. D. Giusepponi, P. Torquato, D. Bartolini, M. Piroddi, S. Moretti, G. Saluti, M. Birringer, S. Lorkowski, F. Galli and R. Galarini, *Free Radical Biol. Med.*, 2017, **108**, S105.
65. F. Galli, *Curr. Drug Metab.*, 2007, **8**, 830–838.
66. Y. J. Li, S. C. Luo, Y. J. Lee, F. J. Lin, C. C. Cheng, Y. S. Wein, Y. H. Kuo and C. J. Huang, *J. Agric. Food Chem.*, 2008, **56**, 11105–11113.
67. Q. Jiang, T. Xu, J. Huang, A. S. Jannasch, B. Cooper and C. Yang, *J. Lipid Res.*, 2015, **56**, 2217–2225.
68. J. K. Lodge, *J. Plant Physiol.*, 2005, **162**, 790–796.
69. T. Morinobu, S. Yoshikawa, K. Hamamura and H. Tamai, *Eur. J. Clin. Nutr.*, 2003, **57**, 410–414.
70. P. Torquato, D. Bartolini, D. Giusepponi, G. Saluti, A. Russo, C. Barola, M. Birringer, R. Galarini and F. Galli, *Free Radical Biol. Med.*, 2016, **96**, S19–S20.
71. F. Galli, M. Piroddi, C. Annetti, C. Aisa, E. Floridi and A. Floridi, *Contrib. Nephrol.*, 2005, **149**, 240–260.
72. A. A. Albahrani and R. F. Greaves, *Clin. Biochem. Rev.*, 2016, **37**, 27–47.
73. M. G. Traber, E. Mah, S. W. Leonard, G. Bobe and R. S. Bruno, *Am. J. Clin. Nutr.*, 2017, **105**, 571–579.
74. D. Bartolini, P. Torquato, C. Barola, A. Russo, C. Rychlicki, D. Giusepponi, G. Bellezza, A. Sidoni, R. Galarini, G. Svegliati-Baroni and F. Galli, *J. Nutr. Biochem.*, 2017, **47**, 120–131.
75. F. Mazzini, M. Betti, T. Netscher, F. Galli and P. Salvadori, *Chirality*, 2009, **21**, 519–524.

CHAPTER 16

Essentiality, Bioavailability, and Health Benefits of α-Tocopherol Stereoisomers

RICHARD S. BRUNO*

The Ohio State University, Human Nutrition Program, 1787 Neil Ave, 325 Campbell Hall, Columbus, OH 43210, USA
*E-mail: Bruno.27@osu.edu

16.1 Introduction

Of the eight forms of vitamin E found in nature (Figure 16.1), only α-tocopherol is essential for human health. However, in the United States, only 2–8% of women and men meet the recommended intake of α-tocopherol from food alone.[1] This is attributed to the relatively few food sources that are rich in α-tocopherol (*e.g.* certain nuts and oils) in combination with most adults obtaining dietary α-tocopherol from energy-dense foods that contain relatively little α-tocopherol (*e.g.* desserts, beef, breads). These trends are further complicated by the use of dietary supplements. Unlike natural sources of α-tocopherol, most supplements contain synthetic or *all racemic*-α-tocopherol. Although supplements are a potentially important tool to bridge the gap between inadequate and adequate intakes, *all racemic*-α-tocopherol contains an equimolar mixture of α-tocopherol stereoisomers; only 50% of these stereoisomers have appreciable bioavailability and contribute to

Food Chemistry, Function and Analysis No.11
Vitamin E: Chemistry and Nutritional Benefits
Edited by Etsuo Niki

Published by the Royal Society of Chemistry, www.rsc.org

vitamin E requirements in humans. This chapter will therefore provide an overview of vitamin E structure, function, bioavailability and metabolism relative to dietary requirements as well as the trafficking of α-tocopherol stereoisomers.

16.2 Structure and Function

Plants have the ability synthesize each of the eight vitamin E forms, resulting in four tocopherols and four tocotrienols that are found in nature (Figure 16.1).[2] Tocotrienols are unique from tocopherols in that they have an unsaturated side chain whereas tocopherols have a saturated phytyl tail. Within each class, there are α, β, γ, and δ forms of tocopherols and tocotrienols. Their designation is based on the number and position of methyl groups on the chromanol head. In this regard, α-tocopherol and α-tocotrienol have a fully methylated chromanol head (Figure 16.1). γ-Tocopherol and γ-tocotrienol have two methyl group substitutions on the chromanol head. β-Tocopherol and β-tocotrienol also have two methyl group substitutions but differ in their position on the chromanol head. Finally, δ-tocopherol and δ-tocotrienol have a single methyl group substitution.

The best-known function of vitamin E is that of a lipophilic antioxidant that terminates the cyclic progression of lipid peroxidation (Figure 16.2). Indeed, vitamin E is largely embedded in phospholipid membranes to provide antioxidant function and limit the oxidation of polyunsaturated fatty acids.[3] Any number of exogenous factors (*e.g.* smoking, radiation) or endogenous factors (*e.g.* poor diet) can provoke oxidative stress and lead to the

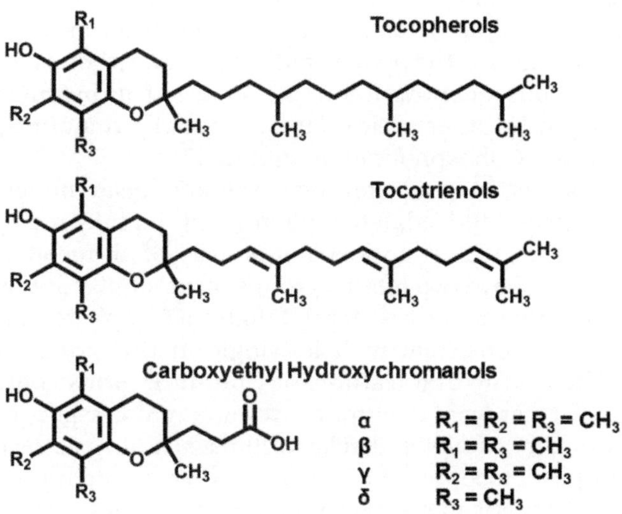

Figure 16.1 Chemical structures of tocopherols, tocotrienols, and carboxyethyl-hydroxychromans. α-, β-, γ, and δ- forms are based on the number and position of methyl groups at R_1, R_2, and R_3.

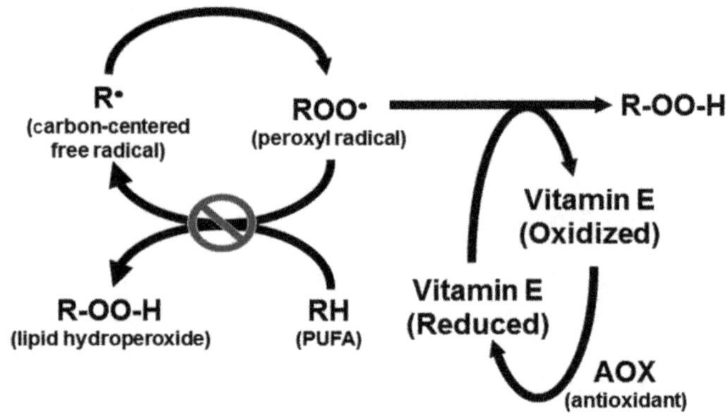

Figure 16.2 Vitamin E antioxidant function terminates the cyclic progression of lipid peroxidation. In the absence of adequate vitamin E, peroxyl radicals oxidize polyunsaturated fatty acids. This results in the generation of a lipid hydroperoxide and the formation of a new carbon-centered radical that propagates the cycle. In the presence of vitamin E, peroxyl radicals are scavenged to terminate the cyclic progression of lipid peroxidation since a new carbon-centered radical is not generated.

generation of a carbon-centered radical. Following their reaction with oxygen, peroxyl radicals are formed. In the absence of adequate vitamin E, peroxyl radicals lead to the generation of lipid hydroperoxides by inducing the peroxidation of polyunsaturated fatty acids. Importantly, this process also results in the generation of a new carbon-centered radical that can continue the cyclic propagation of lipid peroxidation. However, in the presence of vitamin E, peroxyl radicals are preferentially scavenged by vitamin E to terminate the chain reaction of lipid peroxidation.[4] While this forms a hydroperoxide following the reduction of the peroxyl radical, a new carbon-centered radical is not generated. Consequently, vitamin E is oxidized during this process, but can be regenerated by other antioxidants, especially vitamin C, that are in close proximity to the phospholipid membrane.[3,5]

The number of methyl groups on the chromanol head influences vitamin E antioxidant activity.[6] Indeed, α-tocopherol with its fully methylated chromanol head has the greatest antioxidant activity of all tocopherols. γ- and β-Tocopherol, which have two methyl groups, have similar antioxidant activity but lower than that of α-tocopherol. Finally, δ-tocopherol has the lowest antioxidant activity consistent with its single methyl group on the chromanol head. Thus, structural features of vitamin E forms contribute to its antioxidant activity, at least *in vitro*. However, a major aspect that must be considered is dietary intake and relative bioavailability of each vitamin E form. For example, γ-tocopherol is the major dietary vitamin E form in the American diet.[7] While it might be reasonable to suggest that γ-tocopherol contributes substantially to the total antioxidant activity of vitamin E *in vivo*, this form actually accumulates considerably less than that of α-tocopherol. Thus, despite dietary intakes of α-tocopherol being relatively low compared with γ-tocopherol, α-tocopherol has substantially greater bioavailability that results in it being the major circulating vitamin E form in humans. It therefore contributes most greatly to overall vitamin E antioxidant activity *in vivo*, and thus an appreciation of intestinal and hepatic trafficking is needed to understand the physiological benefits and requirements of vitamin E relative to human health.

16.3 Intestinal and Hepatic Trafficking

Dietary-rich sources of vitamin E include nuts, seeds, and vegetable oils. Although these differ in the relative abundance of vitamin E forms, all tocopherols and tocotrienols are absorbed equally at the small intestines (Figure 16.3). Their absorption is dependent upon the hydrolysis of vitamin

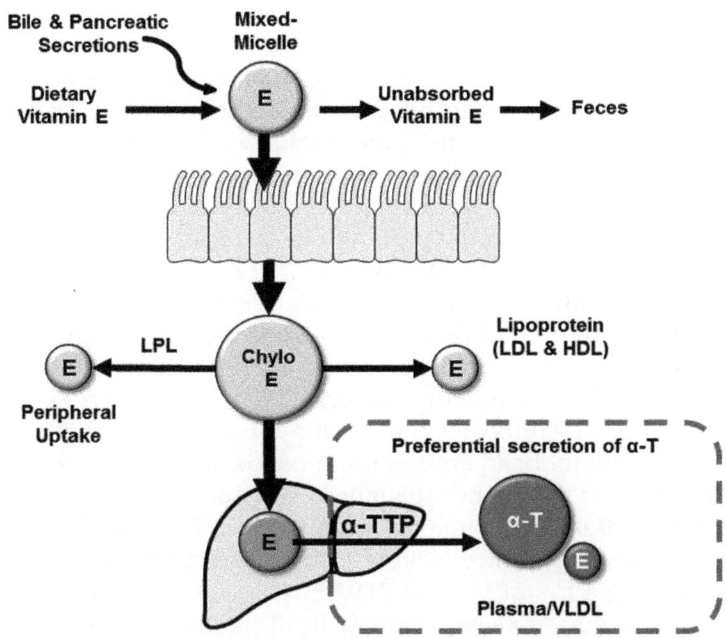

Figure 16.3 Vitamin E trafficking along the gut–liver axis. All forms of dietary vitamin E (tocopherols, tocotrienols) are equally absorbed at the small intestine through the actions of bile acids and pancreatic secretions that promote micellularization of vitamin E. In the enterocyte, all forms of vitamin E are packaged without discrimination into chylomicrons and secreted basolaterally to the lymphatic system. Chylomicron remnants containing vitamin E are taken up at the liver in a receptor-mediated process. The liver, through the actions of the α-tocopherol transfer protein (α-TTP), discriminates among the various vitamin E forms. α-Tocopherol is preferentially enriched into VLDL and secreted into the general circulation to deliver vitamin E to peripheral tissues.

E from the food matrix and the concerted actions of bile and pancreatic secretions that facilitate micelle formation within the lumen of the intestine. In this regard, the co-ingestion of dietary fat has been established to promote α-tocopherol absorption.[8] However, recent studies show that α-tocopherol bioavailability is relatively high when co-ingested with dairy milk (~30% of the dose administered), and that no further benefit in bioavailability occurred by the incremental addition of dairy milk fat.[9] These contrasting findings may be attributed to the differing forms of α-tocopherol that were administered (*i.e.* α-tocopheryl acetate *versus* unesterified α-tocopherol) or physiochemical properties of dairy milk independent of its fat content that promote micelle formation. Regardless, similar to dietary lipids, the vitamin E content of mixed micelles is transferred to the enterocyte at the brush border of the small intestines without any discrimination between dietary vitamin E forms. Likewise, all vitamin E forms are packaged into chylomicrons and secreted into the lymphatic system without any discrimination.

Once absorbed at the intestines, relatively small proportions of vitamin E are transferred to other lipoproteins (*e.g.* HDL and LDL) or are taken up at peripheral tissues following lipoprotein lipase-mediated hydrolysis of chylomicron lipids. Consequently, the majority of vitamin E is delivered to the liver as part of a chylomicron remnant, which is taken up in a receptor-mediated manner. At the liver, significant discrimination occurs among vitamin E forms resulting in the preferential secretion of α-tocopherol into the circulation as part of VLDL (Figure 16.3). This is mediated through the actions of the hepatic α-tocopherol transfer protein (α-TTP).[10,11] Indeed, relative to 100% binding affinity for α-tocopherol, the binding affinity of α-TTP towards β-tocopherol (38%), γ-tocopherol (9%), δ-tocopherol (2%), and α-tocotrienol is considerably lower.[10] That α-tocopherol's preferential affinity with α-TTP indicates that both a saturated phytyl tail and fully methylated chromanol head are required; one or more of these structural aspects are not met with other tocopherols or tocotrienols (Figure 16.1). The importance of α-TTP for maintaining vitamin E status is emphasized by studies in transgenic mice lacking α-TTP expression and observations in persons having a mutation in α-TTP.[12,13] In mice, increased atherosclerosis is observed due to elevated aortic lipid peroxidation that occurs in association with nearly absent circulating α-tocopherol.[12] In humans with a mutation in α-TTP, poor α-tocopherol status leads to ataxia with vitamin E deficiency (*i.e.* AVED).[13]

Synthetic α-tocopherol is widely found in dietary supplements, in some fortified foods (*e.g.* ready-to-eat breakfast cereals), and as a food additive for preservation purposes. Structural differences between natural and synthetic α-tocopherol are important to consider because, in nature, α-tocopherol exists solely as 2*R*,4′*R*,8′*R* or *RRR* (Figure 16.4). If added as a food ingredient, this is commonly labeled as "natural" or "d" α-tocopherol. However, because the α-tocopherol molecule has three chiral centers, its synthetic manufacture results in the formation of an *all racemic* mixture of α-tocopherol

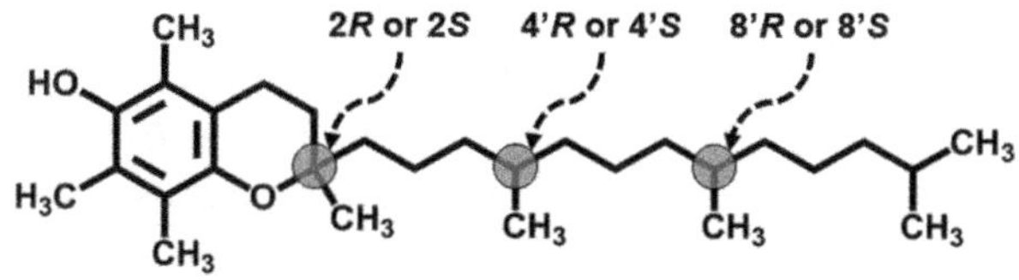

Figure 16.4 Stereochemistry of α-tocopherol. The α-tocopherol molecule has a chiral center at each C-2, C-4′, and C-8′. Because each chiral center has *R* and *S* configurations, eight stereoisomers of α-tocopherol can be formed. An equimolar mixture of stereoisomers (*RRR, RRS, RSR, RSS, SSS, SRR, SRS, SSR*) occurs in the synthetic preparation of α-tocopherol (*i.e. all racemic* or dl-α-tocopherol) whereas *RRR*-α-tocopherol (*i.e.* natural or D-α-tocopherol) is found in nature.

stereoisomers. These are: *RRR, RRS, RSR, RSS, SSS, SRR, SRS, SSR*. On food labels, this is commonly reported as "dl" α-tocopherol. Importantly, of these eight stereoisomers, only the four having the 2*R* stereoisomer configuration (*i.e. RRR, RRS, RSR, RSS*) are appreciably recognized by α-TTP.[2] For example, *SRR*-α-tocopherol only has 11% binding affinity towards α-TTP relative to *RRR*-α-tocopherol.[10] Thus, in addition to α-TTP binding affinity being driven by the specificity of a saturated phytyl tail (*i.e.* tocopherols *vs.* tocotrienols) and a fully methylated chromanol head (*i.e.* α-tocopherol), the C-2 position is required to be in the 2*R* configuration.[10] These factors help to explain how *RRR*- and *SRR*-α-tocopherol are equally absorbed and packaged into chylomicrons at the intestines, but that *RRR*-α-tocopherol is preferentially secreted from the liver as part of VLDL.[14] Further, determinants of α-TTP binding affinity are important in relation to human dietary requirements of α-tocopherol.

Consistent with the actions of α-TTP, non-α-tocopherol forms of vitamin E are secreted limitedly from the liver. However, they do not accumulate excessively at the liver because they actively undergo xenobiotic metabolism to facilitate their elimination. In addition to phase III protein transporters that promote biliary elimination of intact vitamin E,[15] phase I metabolism that is mediated through the cytochrome P450 (CYP) system facilitates the catabolism of vitamin E.[16,17] In this regard, the terminal methyl group on the vitamin E side chain is hydroxylated, likely by CYP4F2. The side chain is then truncated step-wise, likely at the mitochondria,[18] until a terminal water-soluble metabolite is formed as carboxyethyl-hydroxychroman (CEHC; Figure 16.1). This process results in a series of intermediates of varying chain length until α-CEHC is formed from either α-tocopherol or α-tocotrienol. These catabolites can also be subjected to phase II metabolism that results in their glucuronidation or sulfation to enhance their polarity and elimination. This catabolic process also results in the generation of β-, γ-, and δ-CEHCs generated from each respective pair of tocopherols or tocotrienols. Because α-tocopherol has greater α-TTP binding affinity, and hence, greater bioavailability, this explains why its xenobiotic metabolism occurs to a lesser extent than other vitamin E forms. For example, the co-ingestion of equal amounts of α-tocopherol and γ-tocopherol results in lower bioavailability and a shorter half-life of γ-tocopherol in association with greater γ-CEHC formation.[5] It has therefore been suggested that α-CEHC is generated when adequate or excess hepatic α-tocopherol occurs.[19] This implies that α-CEHC is not only a detoxification product of excess hepatic α-tocopherol, but also a possible biomarker of adequate α-tocopherol status that can be used in conjunction with circulating α-tocopherol to assess vitamin E status in humans. This premise is under active investigation,[19] and has been conceptually applied in the study of smokers and persons with metabolic syndrome.[9,20]

16.4 Vitamin E Requirements in Humans

In the United States, dietary requirements of vitamin E are limited solely to α-tocopherol because this is the only form of vitamin E that reverses clinical deficiency.[2] Further, none of the other vitamin E forms are interconverted to α-tocopherol by humans. Requirements are also limited to α-tocopherol stereoisomers occurring in the 2*R* form because those in the 2*S* configuration have poor binding affinity to the hepatic α-TTP. Consistent with this concept, clinical studies using isotopically labeled α-tocopherols show that co-ingestion of 75 mg each of *RRR*- and *all racemic*-α-tocopherol for six consecutive days results in approximately two-times greater circulating levels of α-tocopherol not only during the supplementation period but also during the post-supplementation period (Figure 16.5).[21] This difference in bioavailability is consistent with *all racemic*-α-tocopherol having 50% of its α-tocopherol stereoisomers in the 2*R* configuration whereas the other 50% are in the 2*S* configuration. Further, the lower bioavailability of *all racemic*-α-tocopherol occurred in association with increased α-CEHC generation compared with relatively lesser α-CEHC generated from *RRR*-α-tocopherol.[20] Thus, the limited bioavailability of *all racemic*-α-tocopherol is attributed to its greater propensity to undergo xenobiotic metabolism. This is even apparent in smokers who were provided isotopically labeled *RRR*- and *all racemic*-α-tocopherol simultaneously.[20] Smokers generated less α-CEHC compared with non-smokers, likely because α-tocopherol was utilized consistent with its antioxidant function. However, regardless of smoking status, α-CEHC generation occurred to a greater extent in response to *all racemic*-α-tocopherol compared with α-CEHC derived specifically from *RRR*-α-tocopherol.

The current recommended dietary allowance (RDA) of α-tocopherol is based on the susceptibility of hydrogen-peroxide-induced hemolysis of red blood cells *ex vivo*.[2] In studies conducted in the mid-20th century by Horwitt's group, vitamin E depletion was induced in patients by restricting dietary α-tocopherol in combination with feeding rancid fat.[2] After several years on this regimen, circulating α-tocopherol levels decreased considerably, but overt clinical signs of deficiency (*i.e.* ataxia) were not observed. However, it was

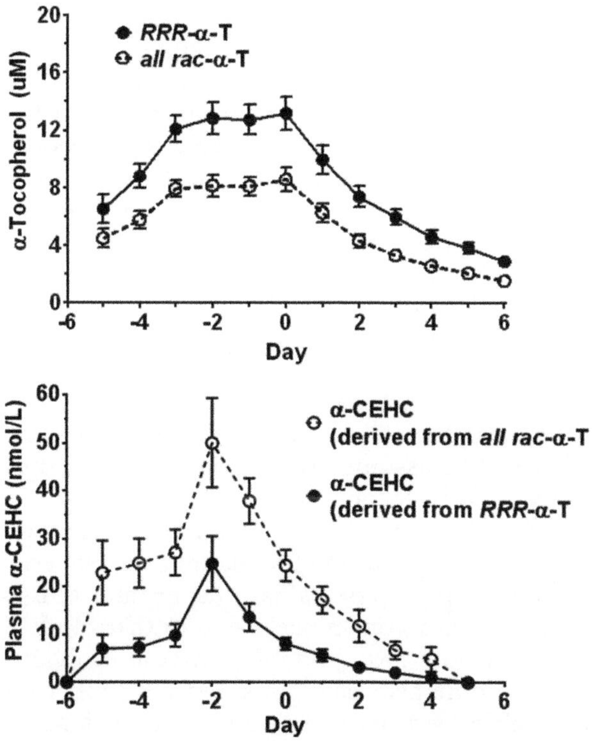

Figure 16.5 Bioavailability and metabolism of *RRR*- and *all racemic*-α-tocopherol. Healthy participants were provided 75 mg each of *RRR*- and *all racemic*-α-tocopherol from day-6 to day-1. Circulating concentrations were measured in the fasting state on each day during the supplementation period and post-supplementation period (days 0–6). Despite being provided equal amounts of *RRR*- and *all racemic*-α-tocopherol, circulating levels of *RRR*-α-tocopherol were approximately double those from *all racemic*-α-tocopherol. In these same persons, circulating α-carboxyethyl-hydroxychroman (α-CEHC) was greater in response to *all racemic*-α-tocopherol compared with *RRR*-α-tocopherol. This indicates that cytochrome P450-mediated catabolism of *all racemic*-α-tocopherol occurs to a greater extent than that of *RRR*-α-tocopherol.

observed that red blood cells were more susceptible to hydrogen-peroxide-induced hemolysis. Thus, the repletion phase of the investigation was aimed at defining the dietary α-tocopherol intake level needed to restore circulating levels and minimize red blood cell hemolysis *ex vivo*. From this work, it was established that 15 mg d^{-1} of α-tocopherol was needed to achieve adequate vitamin E status.

Because of the difference in binding affinity to α-TTP, dietary recommendations are specific to those forms of α-tocopherol having the 2*R* stereoisomer configuration (*i.e. RRR-, RSR-, RRS-, RSS-*). This emphasizes the need to be cognizant of the form of α-tocopherol ingested to accurately quantify dietary intakes in relation to meeting dietary recommendations. For example, 1 mg of *all racemic* α-tocopherol (as α-tocopherol acetate) corresponds to 1 IU, but this equates to 0.45 mg of 2*R* α-tocopherol.[2] Similarly, 0.91 mg of *all racemic* α-tocopherol (as unesterified tocopherol) corresponds to 1 IU or 0.45 mg 2*R* α-tocopherol. Other important conversions between α-tocopherol forms are published.[2] Nonetheless, the RDA (15 mg) can be achieved from synthetic α-tocopherol by ingesting approximately double the amount of *all racemic* α-tocopherol compared with that from *RRR*-α-tocopherol alone. This is consistent with guidance from the US Food and Drug Administration indicating that the biopotency ratio of *RRR*-α-tocopherol relative to *all racemic*-α-tocopherol is 2 : 1.[22] However, there remains scientific debate in this area because the ratio is based on bioavailability rather than a biologic endpoint that specifically assesses biopotency. Accordingly, when new food labeling laws are implemented, the use of IU will be eliminated *in lieu* of the use of milligrams of α-tocopherol. It will be assumed that 2 mg of *all racemic*-α-tocopherol is equivalent to 1 mg of *RRR*-α-tocopherol.

At present, dietary α-tocopherol recommendations do not differ by gender, nor do they differ among specialized populations despite evidence supporting otherwise. For example, smokers have faster rates of α-tocopherol disappearance compared with non-smokers due to their underlying oxidative stress; this is true for both *RRR*- and *all racemic*-α-tocopherols.[5,21] Persons with metabolic syndrome also have impaired α-tocopherol bioavailability compared with healthy persons due to altered trafficking of α-tocopherol along the gut–liver axis.[9,23] Thus, oxidative stress and metabolic abnormalities likely justify higher dietary requirements, but these aspects have not yet been formally evaluated nor have considerations for α-tocopherol stereoisomers been addressed adequately.

16.5 Cancer and Cardiovascular Health

Numerous prospective intervention trials have examined α-tocopherol to reduce the risk of certain cancers and cardiovascular-related morbidity and mortality. The challenge innate to both investigative areas is that the dose and frequency differ among trials, and no large-scale interventions have simultaneously compared *RRR*-α-tocopherol *versus all racemic*-α-tocopherol relative to placebo. Many studies also do not report which stereoisomer of α-tocopherol was administered, thereby limiting interpretation of the outcomes regarding the relative biopotencies of α-tocopherol stereoisomers.

The preponderance of literature does not support α-tocopherol, regardless of stereoisomer form, to prevent or treat cancer. For example, supplementation of *RRR*-α-tocopherol failed to influence the incidence of breast, lung, or colon cancer in the Women's Health Study.[24] *RRR*-α-Tocopherol also did not reduce the incidence of total cancer or cancer-related mortality.[25] Similarly, *all racemic*-α-tocopherol supplementation did not affect overall mortality due to cancer or the incidence of total cancer in randomized placebo-controlled trials.[26,27] Thus, the potential differential biopotency of α-tocopherol stereoisomers is of limited concern in the absence of any clear benefit of α-tocopherol in cancer risk.

Perhaps the most promising evidence for α-tocopherol in health relates to its possible role in managing cardiovascular disease risk. The premise of this scientific area is that α-tocopherol, consistent with its antioxidant function and high accumulation in circulating LDL, may potentially mitigate the progression of atherosclerosis. The concept of atherosclerosis prevention by α-tocopherol has been evidenced in mice that were bred to lack α-TTP expression in combination with high circulating cholesterol.[12] α-Tocopherol deficiency and hypercholesterolemia in these mice increased atherosclerosis in association with greater lipid peroxidation in the vasculature. In humans, however, the benefits of α-tocopherol in cardiovascular disease risk have been equivocal with some studies showing positive effects and others showing null effects. Whether these disparities are attributed to the use of *RRR- versus all racemic*-α-tocopherol also remains unclear because no studies have directly investigated these differing α-tocopherol stereoisomers. Thus, comparisons must be made between studies with different stereoisomer forms administered. However, fundamental differences between studies make comparisons challenging. Some studies focused on primary prevention whereas others investigated secondary prevention. Further, doses and duration of supplementation differed, and participant compliance based on changes in circulating α-tocopherol was not evaluated in most large-scale studies. Meta-analyses showing favorable[28] or neutral effects[29] of α-tocopherol on cardiovascular-related outcomes also have not segregated the studies based on stereoisomer form. In one controlled study in which 200–800 IU of either *RRR*- or *all racemic*-α-tocopherol was provided for two months, circulating α-tocopherol increased to a similar extent regardless of supplement form and dose.[30] The resistance of LDL to oxidation was also attenuated at doses above 400 IU regardless of α-tocopherol supplement form. The lack of difference between *RRR*- and *all racemic*-α-tocopherol in these outcomes are likely attributed to the relatively high dose administered compared with usual dietary intakes and the likely need to develop biomarkers that are responsive to α-tocopherol therapy at intakes lower than those only obtainable from supplements. This will help to address the ongoing controversy regarding biopotency of *RRR*- compared with *all racemic*-α-tocopherol.

16.6 Infant and Maternal Health

There has been considerable interest in studying the trafficking of maternal α-tocopherol to the fetus in an effort to improve health outcomes of infants. Studies in pregnant women have been conducted to examine the relationship between α-tocopherol in maternal blood and cord blood. Pregnant women received 15–300 mg d^{-1} of a 1 : 1 mixture of isotopically-labeled *RRR*-α-tocopherol and *all racemic*-α-tocopherol for five days prior to scheduled infant delivery.[31] Analysis of the maternal circulation showed that the ratio of α-tocopherol derived from natural relative to *all racemic* forms was approximately 2 : 1. This was consistent with other studies showing that the hepatic α-TTP likely discriminates against 2*S* forms of α-tocopherol. However, an interesting observation was that the ratio of natural relative to *all racemic* α-tocopherol forms in cord blood was closer to 3.5 : 1. This suggests that an extrahepatic mechanism, possibly at the placenta where α-TTP is also expressed, provides additional discrimination against α-tocopherol stereoisomers. Thus, *RRR*-α-tocopherol may contribute more substantially to infant α-tocopherol status than *all racemic*-α-tocopherol.

In a longer-term intervention, infants were randomized to receive infant formula that was prepared with either high natural α-tocopherol (High-NAT), low natural α-tocopherol (Low-NAT), or *all racemic*-α-tocopherol (SYN). Not surprising, the High-NAT treatment that contained 20 IU α-tocopherol (provided as 14.5 mg *RRR*-α-tocopherol acetate; equivalent to 13.4 mg 2*R*-α-tocopherol) resulted in greater circulating α-tocopherol concentrations compared with infants randomized to the Low-NAT (10 IU) or SYN (13.5 IU) formulas. The Low-NAT formula contained 7.3 mg *RRR*-α-tocopherol acetate (equivalent to 6.7 mg L^{-1} 2*R*-α-tocopherol) whereas the SYN formula contained 13.5 IU (13.5 mg *all rac*-α-tocopherol acetate or 6.1 mg L^{-1} 2*R*-α-tocopherol). Despite the SYN formula containing approximately double the amount of *total* α-tocopherol, circulating α-tocopherol did not differ among infants randomized to the Low-NAT group compared with the SYN group. This supports the fact that infants, like adults, are capable of discriminating against certain stereoisomers derived from *all racemic*-α-tocopherol.

The influence of α-tocopherol stereoisomers has also been examined in breast milk from lactating women in response to dietary supplementation of *RRR*- or *all racemic*-α-tocopherol.[32] In a randomized study, lactating women were assigned to a daily supplement containing *all racemic*-α-tocopherol (45.5 mg; ARAC-SUPP), 22.8 mg *all racemic*-α-tocopherol plus 20.1 mg *RRR*-α-tocopherol (MIX-SUPP), or 40.2 mg *RRR*-α-tocopherol (RRR-SUPP). In both breast milk and the circulation, *RRR*-α-tocopherol was the predominant stereoisomer regardless of treatment group. Dietary supplementation altered *RRR*-α-tocopherol in milk and circulation such that the greatest proportion of *RRR*-α-tocopherol was observed in response to RRR-SUPP followed by MIX-SUPP and then ARC-SUPP. However, the proportion of *RRR*-α-tocopherol only increased in the milk and plasma among lactating women provided RRR-SUPP; it decreased from pre-supplementation levels among those in the MIX-SUPP group and was most substantially lowered in the ARAC-SUPP group. These changes were accompanied by an increase in the proportion of non-*RRR*-α-tocopherol stereoisomers in both breast milk and plasma among those receiving ARAC-SUPP. These findings indicate that the approach undertaken with dietary supplementation can substantially alter α-tocopherol stereoisomer profiles, thereby affecting the form of α-tocopherol delivered to a breastfed infant.

Whether shifts in α-tocopherol stereoisomer profiles impact short- or long-term infant health remains an open question. However, detailed analytical studies in the human infant brain (from infants who died of sudden infant death syndrome) show that *RRR*-α-tocopherol is the predominant stereoisomer that accumulates in the brain.[33] This was true for all brain regions analyzed, including the frontal cortex, hippocampus, and visual cortex. For example, *RRR*-α-tocopherol in these regions were 5.5–10.5 μg g^{-1}, whereas concentrations of *RRS*-, *RSR*-, and *RSS*-α-tocopherol were 1–1.5, 0.8–1.0, and 0.7–0.9 μg g^{-1}, respectively. By contrast, the sum of all 2*S* α-tocopherol stereoisomers was substantially lower at 0.2–0.3 μg g^{-1}. Thus, not only does the brain discriminate against 2*S* stereoisomers of α-tocopherol, consistent with brain tissue expressing α-TTP,[34] it appears to provide some discrimination against 2*R* forms that are not in the *RRR*- stereoisomer configuration.

16.7 Conclusions

α-Tocopherol is clearly essential for human health. Although bioavailability studies have provided clear evidence that *all racemic*-α-tocopherol has approximately half the bioavailability of *RRR*-α-tocopherol, a critical gap remains to define whether both forms can be effective to alleviate the risk of chronic disorders. On the one hand, the greater bioavailability of *RRR*-α-tocopherol compared with *all racemic*-α-tocopherol suggests that it would have greater health benefits. On the other hand, there is limited information regarding the relative biopotency of these α-tocopherol forms. Meeting dietary α-tocopherol recommendations from foods is feasible but challenging,[35] so dietary supplements may be needed by some populations to achieve such intakes. Thus, carefully controlled studies are needed to test the independent and/or differential health benefits and natural and synthetic α-tocopherol forms. This will also likely require new technologies and/or biomarkers that can rigorously assess α-tocopherol bioactivity or biopotency across a range of intakes from usual dietary levels through the use of high-dose supplements. Importantly, it is critical for the scientific field that investigators detail the composition of α-tocopherol supplements used in studies to help reconcile this controversial area of vitamin E research.

References

1. J. E. Maras, O. I. Bermudez, N. Qiao, P. J. Bakun, E. L. Boody-Alter and K. L. Tucker, *J. Am. Diet. Assoc.*, 2004, **104**, 567.
2. Food and Nutrition Board, Institute of Medicine, *Dietary Reference Intakes for Vitamin C, Vitamin E, Selenium, and Carotenoids*, National Academy Press, Washington DC, 2000.
3. M. Afri, B. Ehrenberg, Y. Talmon, J. Schmidt, Y. Cohen and A. A. Frimer, *Chem. Phys. Lipids*, 2004, **131**, 107.

4. G. R. Buettner, *Arch. Biochem. Biophys.*, 1993, **300**, 535.
5. R. S. Bruno, S. W. Leonard, J. Atkinson, T. J. Montine, R. Ramakrishnan, T. M. Bray and M. G. Traber, *Free Radical Biol. Med.*, 2006, **40**, 689.
6. G. W. Burton, K. H. Cheeseman, T. Doba, K. U. Ingold and T. F. Slater, *Ciba Found. Symp.*, 1983, **101**, 4.
7. Q. Jiang, S. Christen, M. K. Shigenaga and B. N. Ames, *Am. J. Clin. Nutr.*, 2001, **74**, 714.
8. R. S. Bruno, S. W. Leonard, S. Park, Y. Y. Zhao and M. G. Traber, *Am. J. Clin. Nutr.*, 2006, **83**, 299.
9. E. Mah, T. N. Sapper and C. Chitchumroonchokchai, *et al.*, *Am. J. Clin. Nutr.*, 2015, **102**, 1070.
10. A. Hosomi, M. Arita and Y. Sato, *et al.*, *FEBS Lett.*, 1997, **409**, 105.
11. K. C. Min, R. A. Kovall and W. A. Hendrickson, *Proc. Natl. Acad. Sci. U. S. A.*, 2003, **100**, 14713.
12. Y. Terasawa, Z. Ladha and S. W. Leonard, *et al.*, *Proc. Natl. Acad. Sci. U. S. A.*, 2000, **97**, 13830.
13. T. Gotoda, M. Arita and H. Arai, *et al.*, *N. Engl. J. Med.*, 1995, **333**, 1313.
14. M. G. Traber, G. W. Burton, K. U. Ingold and H. J. Kayden, *J. Lipid Res.*, 1990, **31**, 675.
15. D. J. Mustacich, J. Shields, R. A. Horton, M. K. Brown and D. J. Reed, *Arch. Biochem. Biophys.*, 1998, **350**, 183.
16. T. J. Sontag and R. S. Parker, *J. Biol. Chem.*, 2002, **277**, 25290.
17. T. J. Sontag and R. S. Parker, *J. Lipid Res.*, 2007, **48**, 1090.
18. D. J. Mustacich, S. W. Leonard, N. K. Patel and M. G. Traber, *Free Radical Biol. Med.*, 2010, **48**, 73.
19. K. M. Lebold, A. Ang, M. G. Traber and L. Arab, *Am. J. Clin. Nutr.*, 2012, **96**, 801.
20. R. S. Bruno, S. W. Leonard, J. Li, T. M. Bray and M. G. Traber, *Am. J. Clin. Nutr.*, 2005, **81**, 1052.
21. R. S. Bruno, R. Ramakrishnan, T. J. Montine, T. M. Bray and M. G. Traber, *Am. J. Clin. Nutr.*, 2005, **81**, 95.
22. K. M. Ranard and J. W. Erdman, Jr, *Nutr. Rev.*, 2018, **76**, 141.
23. M. G. Traber, E. Mah, S. W. Leonard, G. Bobe and R. S. Bruno, *Am. J. Clin. Nutr.*, 2017, **105**, 571.
24. I. M. Lee, N. R. Cook and J. M. Gaziano, *et al.*, *JAMA*, 2005, **294**, 56.
25. E. Lonn, J. Bosch and S. Yusuf, *et al.*, *JAMA*, 2005, **293**, 1338.
26. S. M. Lippman, E. A. Klein and P. J. Goodman, *et al.*, *JAMA*, 2009, **301**, 39.
27. J. M. Gaziano, R. J. Glynn and W. G. Christen, *et al.*, *JAMA*, 2009, **301**, 52.
28. P. J. Joris and R. P. Mensink, *Nutrients*, 2015, **7**, 1728.
29. Q. Bin, X. Hu, Y. Cao and F. Gao, *Thromb. Haemostasis*, 2011, **105**, 579.
30. S. Devaraj, B. Adams-Huet, C. J. Fuller and I. Jialal, *Arterioscler. Thromb. Vasc. Biol.*, 1997, **17**, 2273.
31. R. V. Acuff, R. G. Dunworth, L. W. Webb and J. R. Lane, *Am. J. Clin. Nutr.*, 1998, **67**, 459.

32. S. Gaur, M. J. Kuchan, C. S. Lai, S. K. Jensen and C. L. Sherry, *J. Nutr.*, 2017, **147**, 1301.
33. M. J. Kuchan, S. K. Jensen, E. J. Johnson and J. C. Lieblein-Boff, *Br. J. Nutr.*, 2016, **116**, 126.
34. R. P. Copp, T. Wisniewski, F. Hentati, A. Larnaout, M. Ben Hamida and H. J. Kayden, *Brain Res.*, 1999, **822**, 80.
35. X. Gao, P. E. Wilde, A. H. Lichtenstein, O. I. Bermudez and K. L. Tucker, *J. Nutr.*, 2006, **136**, 1021.

CHAPTER 17

Vitamin E Deficiency and Inadequacy; Insights Using Zebrafish, Lipidomics and Metabolomics

MARET G. TRABER*

Linus Pauling Institute, Oregon State University, Corvallis, OR, 97331, USA
*E-mail: maret.traber@oregonstate.edu

17.1 Introduction

This chapter is dedicated to Lester Packer, who had an unwavering interest and enthusiasm for studies of aging, free radicals and antioxidants.

Vitamin E (α-tocopherol, VitE) was discovered in 1922 because it prevented embryonic mortality in rats.[1] Importantly, VitE deficiency has been recognized since the 1960s to cause cerebellar encephalomalacia in humans.[2] In the 1980s, VitE was shown to be necessary for normal neurologic function in humans[3] and in mice.[4–7] Genetic deficits in the α-tocopherol transfer protein (α-TTP) in humans[8] result in ataxia with VitE deficiency, which is caused by degeneration of peripheral nerves and results in a spinocerebellar ataxia and Purkinje cell death.[9,10] α-TTP is highly expressed in the liver, where it plays a critical role in regulating plasma VitE (*e.g.* α-tocopherol) concentrations and is present in several brain regions, where it is important in VitE trafficking.[11,12] Thus, VitE is needed for membrane stability[13] and is known to

Food Chemistry, Function and Analysis No.11
Vitamin E: Chemistry and Nutritional Benefits
Edited by Etsuo Niki

Published by the Royal Society of Chemistry, www.rsc.org

protect sensory neurons,[14] but the necessity for VitE and its role in embryogenesis remains unknown. This is an important problem because estimated rates of miscarriage in the US are ~20% of pregnancies.[15] We have proposed that inadequate VitE intakes in women may be among the leading contributing factors for miscarriage both because of its known role in maintaining pregnancy in rodents and because there is increased risk of miscarriage associated with low plasma VitE during the first trimester of pregnancy.[16] These observations pose public health concerns since estimates of *inadequate* dietary VitE intake[17] (*e.g.* estimated average requirements, EAR, 12 mg d^{-1}) exceed 80% of the global ≥14 year-old population,[18] as well as 96% of American women.[19]

VitE is a lipophilic antioxidant,[20] protecting long-chain polyunsaturated fatty acids (PUFA), especially docosahexaenoic acid (DHA)—a high oxidizable PUFA—from lipid peroxidation.[21,22] VitE is a "chain-breaking antioxidant";[20,23] it reacts with peroxyl radicals 1000 times faster than peroxyl radicals (the propagating species) react with PUFA.[24] Is lipid peroxidation the cause of VitE-deficiency-dependent fetal death? Generally, the accumulation of oxidized lipids or biomarkers of oxidized lipids *in vivo* is relatively low.[25,26] Thus, there have remained questions as to whether there might be more specific VitE molecular or signaling functions.[27] My laboratory group has approached the question of VitE function in embryogenesis by studying zebrafish embryos. In brief, we found that zebrafish embryonic VitE deficiency results in a progression of secondary metabolic impairments.[28] These metabolic impairments will be discussed in further detail and related to the role of VitE in embryogenesis in this chapter.

17.2 Is the Zebrafish Embryo an Appropriate Model for Human Embryogenesis?

The zebrafish is an unique vertebrate model for evaluation of nutrient-dependent pathways in embryogenesis. The embryo progresses from a fertilized egg to a swimming fish in five days, thus, zebrafish embryos are good models for high-throughput screening and rapid hypothesis testing. Zebrafish molecular responses have a high degree of relevance to human biology because the signal transduction mechanisms, anatomy, and physiology of zebrafish are homologous to those of humans.[29] Their lipid and lipoprotein metabolism also mirrors that of humans. Specifically, zebrafish have the capability to synthesize triglycerides,[30,31] elongate PUFA,[32–36] and have all the major pathways for lipid trafficking and lipoprotein metabolism.[37–40] With regards to neurologic development, zebrafish require folate, express orthologs of 12 human folate metabolic genes and disruption of the folate pathway causes developmental defects, including neural tube defects (NTDs).[41]

Zebrafish, like humans, require both vitamins E and C,[42] so these animals are highly relevant for antioxidant research. We found that the embryo (separately from the mother) requires VitE and insufficient VitE causes developmental malformations or even death.[43] Zebrafish embryos express α-TTP and its mRNA is detectable in fertilized eggs, and increases dramatically by 48 hpf.[43] This VitE transfer protein is expressed in the human yolk sac[44] and in the zebrafish yolk syncytial layer (YSL),[43] suggesting a common function of α-TTP in embryogenesis to its function in the liver—facilitating VitE trafficking. Moreover, α-TTP knockdown in zebrafish caused impaired head and eye formation within the first 15 h of egg fertilization and death by 24 h.[43] Recent studies using single-cell RNAseq in mice embryos at day 8.25 have shown that the *Ttpa* gene is expressed in the developing neural tissues, especially in the neural placodes region (https://marionilab.cruk.cam.ac.uk/organogenesis/).[45] Again, these data emphasize the necessity for VitE and α-TTP in embryonic development.

We have pioneered the use of VitE-deficient (E−) or -sufficient (E+) diets that contain a limited fatty acid composition[28,46] fed to adult zebrafish to obtain E− and E+ embryos.[43,47] These fish, irrespective of diet group, lay and fertilize eggs in similar numbers and nutrient composition (other than VitE).[47] We have used targeted mass spectrometry[47,48] and other analytical techniques,[43] as well as lipidomics[49] and metabolomics,[28,46,50] to assess lipid peroxidation and metabolic changes during embryogenesis. We have found that embryonic VitE deficiency results in increased lipid peroxidation,[46,47,50] depletion of phospholipids with increased phospholipid remodeling,[49] dysregulated energy and methylation status,[28,46] organelle damage, ferroptosis (indicated by glutathione (GSH) depletion)[46] and altered DNA methylation.[50] Importantly, choline[28] is one of the earliest dysregulated metabolites and is observed prior to gross morphologic changes.[28] Thus, the model allows us to evaluate in a very consistent manner the consequences of lipid peroxidation and metabolic dysfunction due to VitE deficiency.

17.3 Pregnancy, Embryogenesis and Neurodevelopment

The maternal nutritional status directly affects prenatal development and an inadequate nutritional status can adversely affect pregnancy outcomes. Of potential importance, we have shown in zebrafish that a severe VitE deficiency causes embryonic defects during neural tube formation, resulting in impaired brain formation[43] and mortality.[47] Our discoveries in zebrafish embryos suggest that VitE is critical for normal embryonic neurodevelopment.

VitE deficiency in experimental animals was found to be associated with NTDs as early as the 1950s and 60s.[51,52] NTDs are severe congenital malformations of the central nervous system, which can be lethal (*e.g.* anencephaly and craniorachischisis).[53] NTDs are largely prevented in humans in the US by food fortification with folic acid,[54] which has halved NTD incidence, but NTDs still occur (5–6 per 10 000 births).[55,56] Potentially, NTDs in humans can also be caused by inadequate VitE. For example, increased incidence of early miscarriage is associated with low circulating VitE concentrations,[16] while higher VitE intakes are associated with lower NTD incidence.[57]

VitE is necessary to prevent NTDs in zebrafish and in rodent embryos at the analogous times that NTDs occur in human embryos with inadequate folate status (hours post fertilization (hpf) for zebrafish,[58] days (d) for rats[59] and humans,[60–62] Table 17.1). Thus, α-TTP and VitE are critical molecules during embryonic development, especially during neural plate and tube formation.[43] Food fortification with folic acid has decreased the rate of NTDs, but has not abolished the problem,[63] suggesting that additional causes need to be investigated. No NTD-responsible genes have been found,[55] but numerous candidate genes, epigenetic, environmental and dietary factors are proposed as causative.[54,64–66]

In addition to folate, another nutrient important for one-carbon metabolism that is critical for brain development is choline,[67] which is also dysregulated in zebrafish E− embryos.[50] Depletion of choline may be a key factor in neurogenesis in E− embryos. Zeisel[67,68] has argued that choline is required for optimal brain function. We have found behavioral defects in E− zebrafish embryos.[28,50] Critically, choline depletion occurs by 24 hpf and prior to morphologic abnormalities (Table 17.2).[28,50] Folate-insensitive NTD-related genes include those involved in mitochondrial choline metabolism, *de novo* purine synthesis and GSH generation.[69] Our data in E− zebrafish embryos suggest that all three of these pathways are dysregulated as a result of changes in embryonic metabolism to repair the damage caused by increased lipid peroxidation during VitE deficiency. Further, mouse embryos exhibit NTDs when VitE delivery to target tissues is impaired due to abnormal lipoprotein production[70–72] or deletion of lipoprotein transporters, such as SR-B1.[73] Additionally, SR-B1 is expressed in

Table 17.1 Comparisons between zebrafish, rats and humans of the timing of key developmental stages. VitE is necessary for human embryonic development at a time prior to when a woman knows she is pregnant, shown in bold. This timing is analogous to the situation in humans with inadequate folic acid status (hours post fertilization (hpf) for zebrafish,[58] days (d) for rats,[59] and humans.[60–62]).

Developmental stage	Zebrafish	Rat	Human
Blastula/blastocyst	2–5 h	3–5 d	4–6 d
Implantation	n/a	6 d	8–10 d
Neural plate formation	**10 h**	**9.5 d**	**17–19 d**
First somite	10–11 h	9–10 d	19–21 d
10 Somite stage	14 h	10–11 d	22–23 d
Neural tube formation	**18–19 h**	**9–12 d**	**22–30 d**
First pharyngeal arch	24 h	10 d	22–23 d
Organogenesis	48 h	5–6 d	21–56 d
First heartbeat	24 h	10.2 d	22 d
Birth/hatching	48–72 h	21 d	253 d

Table 17.2 Metabolite depletion in E− embryos begins at 24 hpf with choline depletion. Metabolomics analyses show relative-fold differences at 24 and 48 hpf between E− and E+ embryos;[28,46] *Q*-values <0.05 shown as up or down arrows. Most metabolites are elevated in E− at 24 hpf, suggesting eggs laid by E− female fish contain extra nutrients. A major exception is choline, which is significantly depleted at 24 hpf.

	E− *vs.* E+	
Metabolite	24 hpf	48 hpf
CDP-choline pathway		
Choline	↓	↓
CDP-choline	↑	↓
Glycerophosphocholine	↑	—
Ethanolamine	↑	↑
Betaine	↑	—
Energy-related		
ATP	↓	↓
ADP	↓	—
GDP	↓	↑
Riboflavin	↓	—
FAD+	↓	↓
NAD+	↓	↓
NADPH	↑	↓
NADP+	↑	—
Coenzyme-A	—	↓
D-Pantothenic acid	—	↓
Creatine	—	↓
Antioxidant/redox sensitive		
Ascorbic acid	↑	↓
Glutathione	—	↓
Cystathionine	↑	—
Amino acids		
Glutamine	—	↓
Asparagine	—	↓
Methionine	—	↓
Homoserine	—	↓
S-Adenosyl-methionine	↑	—
SAM/SAH ratio	↑	—
Ornithine	↑	—
Citrulline	↑	—
Glycolysis		
Glucose	↑	—
Fructose-6-phosphate	—	↓
Fructose-1-6-bisphosphate	↑	—
3-Phosphoglycerate	↑	—
Phosphoenolpyruvate	↑	—
Pyruvate	↑	↓
Lactate	↑	↑
Pentose phosphate pathway		
6-Phosphogluconate	↑	↑
Ribose-5-phosphate	↑	↑
Tricarboxylic acid cycle		
Aconitate	—	↓
Alpha-Ketoglutarate	↑	—
Malate	↑	↓

Fumarate	↑	↓
Oxaloacetate	↑	↓
Free fatty acids		
DHA (22 : 6 ω-3)	↑	↓
EPA (22 : 5 ω-3)	↑	—
Palmitic acid (16 : 0)	↑	↑
Stearic acid (18 : 0)	↑	↑

human first trimester trophoblast cells,[74] suggesting a critical role during pregnancy for this transporter to deliver key lipids and nutrients. As might be predicted, low SR-B1 expression levels in humans are associated with infertility.[75,76] Thus, SR-B1 may not only be important for cholesterol and essential fatty acid delivery, but also VitE delivery, to protect the fetus and the developing nervous system.

17.4 VitE, Polyunsaturated Fatty Acids (PUFAs) and Neurologic Function

VitE is necessary to prevent peroxidative damage to critical long-chain PUFAs,[24] and this may be its role in neurodevelopment during embryogenesis. Both DHA—ω-3 PUFA—and arachidonic acid (ARA, ω-6) are needed early in embryogenesis. The brain acquires DHA as lysoPL-DHA,[77] primarily in the form of lysoPC-DHA. A transporter from the major facilitator superfamily, MFSD2A, facilitates brain lysoPL-DHA uptake[78] and has high specificity towards lysoPC-DHA. Knockdown of MFSD2A causes microcephaly.[79,80] Lyso-PL-DHA is also required for blood–brain barrier formation.[78,81] Thus, DHA is a key molecule for neurogenesis and its protection from peroxidative damage is critical.

Is VitE essential because of its antioxidant function? A leading theory argues VitE is an integral part of an "antioxidant network",[82] where the one-electron VitE oxidation product (generated by VitE scavenging peroxyl radicals) is reduced by ascorbic acid. Reduction of the ascorbate radical utilizes other antioxidants, especially GSH (Figure 17.1). The subsequent reduction of GSSG (the GSH oxidization product) uses endogenous reducing equivalents, primarily NADPH (Figure 17.1).[82] This close association between GSH and VitE is further emphasized by studies of GPx4, a selenium-dependent peroxidase that reduces phospholipid LOOH.[83] Maternal VitE supplementation in mice prevents GPx4 knockdown-induced lethality in offspring.[84]

To evaluate the hypothesis that VitE is necessary to protect these long-chain PUFAs in the brain, we developed some new methodologies using lipidomics approaches.[28,46,49,50,85,86] Because so little material was available from VitE-deficient embryos, initially we evaluated the damage to lipids in brains from adult VitE-deficient zebrafish. We found that VitE deficiency

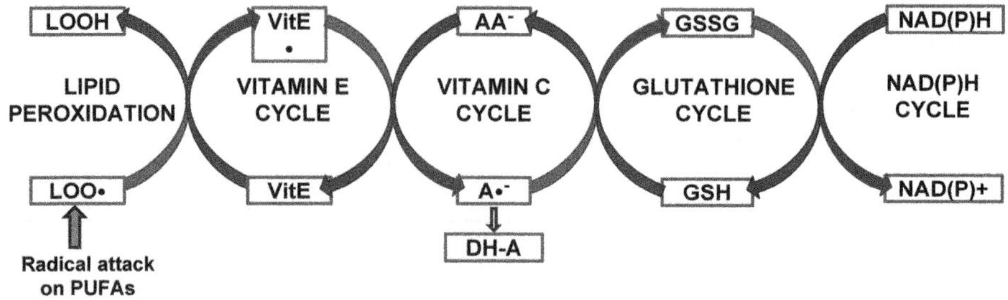

Figure 17.1 Linkage of antioxidant defense mechanisms. A leading theory argues that VitE is an integral part of an "antioxidant network",[82] where the one-electron VitE oxidation product [VitE$^{\bullet}$ (tocopheroxyl radical) generated by VitE scavenging the peroxyl radical, LOO$^{\bullet}$] is reduced by ascorbic acid (AA$^-$). Reduction of the ascorbyl radical (A$^{-\bullet}$) utilizes other antioxidants, such as glutathione (GSH). The subsequent reduction of GSSG (the GSH oxidization product) uses endogenous reducing equivalents, primarily NADPH. Additionally, GSH is necessary as a substrate for GSH peroxidase 4 (GPx4) to detoxify the lipid hydroperoxide (LOOH).[90]

led to depletion of four specific PLs and 19 different lysoPLs.[85] The major depleted PL was 1-palmitoyl-2-docosahexaenoyl-*sn*-glycero-3-phosphocholine (DHA-PC).[85] The wide variety of lysoPLs that are depleted suggests that the entire lysoPL substrate population is affected. It has been recognized since the 1950s that lysoPL are used for PL remodeling during membrane synthesis, repair, and replacement.[87] Follow-up studies in E− embryos also showed similar lipid changes, as well as increased PL-remodeling, as were found in the adult E− brains.[49]

We also found that inadequate VitE allows increased lipid peroxidation and leads to metabolic dysregulation with glucose depletion (Table 17.2). These metabolic changes are likely caused by the depletion of NADPH,[28] which is used for detoxification of LOOH during the regeneration of GSH from GSSG by glutathione reductase. NADPH can be produced by the oxidative phase of the pentose phosphate pathway, by malic enzyme, and from the serine-driven one-carbon metabolism where methylene tetrahydrofolate is oxidized to 10-formyl tetrahydrofolate and produces NADPH. Fan *et al.*[88] demonstrated that 40% of NADPH production is *via* folate metabolism, while the pentose phosphate pathway and malic enzyme each contribute 30% of NADPH production. They reported that knockdown of the enzyme, methylene-tetrahydrofolate dehydrogenase (MTHFD), decreased both NADPH–NADP$^+$ and GSH–GSSG ratios and increased cell sensitivity to oxidative stress.[88] Furthermore, choline restriction both compromised energy status and caused a rapid onset of oxidative stress in liver.[89] Thus, alterations in NADPH production and consumption are complex and dependent on a variety of pathways and substrate sources. Importantly, ferroptosis is prevented by adequate GSH, GPX4 and sufficient NADPH.[90]

17.5 VitE, Anti-ferroptotic Agent?

VitE can prevent ferroptosis and its sequelae in numerous systems and, as a corollary, it can be surmised that VitE deficiency should promote ferroptosis. Ferroptosis also provides a framework for the integration between cellular metabolism, modulation of mitochondrial function, and cell death. Thus, the question arises, does the body require VitE to regulate ferroptosis? The sequence of events in E− zebrafish embryos suggests the following:

- The primary trigger is oxidation of membrane PL, especially those PL containing DHA.
- A consequence of lipid peroxidation is the modification of the metabolic pathways engaged in energy- and redox-homeostasis.
- The coordinated "dysfunctions" lead to ferroptosis and/or other forms of cell death.

Ferroptosis is an iron-dependent, regulated cell death associated with increased lipid peroxidation, decreased antioxidant defenses and accumulation of oxidized lipids, but the molecular mechanisms causing ferroptosis remain unclear.[91,92] The process is distinct from apoptosis, necrosis and autophagy, but mitochondrial failure due to lipid peroxidation is a key feature.[93] GSH is necessary as a substrate for GPx4 to detoxify lipid hydroperoxides (LOOH).[94] Regeneration of GSH from its oxidation product, GSSG, by glutathione reductase (GR) requires NADPH (Figure 17.1).

Ferroptosis has been defined largely using drugs to inhibit GSH/GPX4 and/or in tissue culture studies where antioxidants in the medium are virtually non-existent.[95] By contrast, our model of E− embryos with high levels of DHA that are placed in the eggs by the mothers, and are subject to oxidative damage because VitE is limited, are a more physiologically relevant model system. At 24–48 hpf, E+ embryos have ARA–DHA ratios of ~2.5,[46] which is similar to erythrocyte ARA–DHA ratios in non-supplemented pregnant women of ~3 : 1.[96–98] Over a few days, E− embryos become GSH- and NADPH-depleted and become metabolically dysregulated.[28,46,49] DHA-PC depletion and up-regulation of PC synthesis or remodeling further suggests derangement in choline recycling.[28,46,50] Choline is clearly a critical molecule in the developing embryo and is one of the first hallmarks of VitE deficiency in the zebrafish embryos (Table 17.2). During folate restriction, cancer growth and cellular metabolism were impaired, but GSH was unchanged.[99] Choline restriction, however, both compromised energy status and caused a rapid onset of oxidative stress in the liver.[89] The E− embryos have no access to additional nutrients, so they are a sensitive model to assess repair mechanisms in response to increased lipid peroxidation.

Accumulation of LOOH is a hallmark of ferroptosis, but the lipid sources and the means by which they are generated are not fully known.[100] In addition to spontaneous formation, LOOH can be generated by lipoxygenases (*e.g.* 5-, 12-, or 15-LOXs).[90,94] Some specific PLs (oxidized arachidonyl-phosphatidyl ethanolamine) have been proposed as ferroptosis signaling molecules,[101] but arachidonic acid (ARA) was added to the test systems; other oxidized lipids are also present. For example, DHA is a more oxidizable PUFA than ARA,[102] and has not been evaluated. Thus, the specific oxidized lipids involved, as well as the precise mechanisms of ferroptosis-regulated cell death, are not well-defined.

To repair the damage to PL caused by lipid peroxidation, the DHA-PC must be replaced (Figure 17.2). The PC generating pathways differ with regards to mechanisms by which the fatty acids present in the PC can be replaced. The CDP-choline pathway uses choline and generates PC

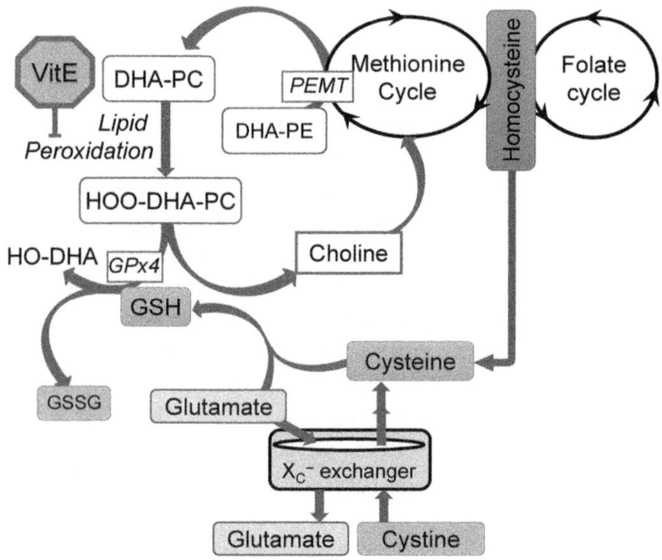

Figure 17.2 Inter-relationships affected by choline depletion. VitE deficiency leads to increased peroxidation of docosahexaenoic acid (DHA) and depletion of phospholipids (PLs) containing DHA (*e.g.* DHA-PC) with disruption of PL remodeling, decreased choline and glucose concentrations, dysregulated oxygen consumption and compromised bioenergetic capacity. Choline can be used to form phosphatidylcholine (PC) from phosphatidylethanolamine (PE) by *phosphatidyl ethanolamine methyltransferase (PEMT)*, using three *S*-adenosyl methionine (SAM) from the methionine cycle. GSH is a necessary substrate for GPx4 in the detoxification of lipid hydroperoxides (such as HOO-DHA-PC), but GSH synthesis is dependent on cysteine availability. As shown, cysteine can be obtained from either the methionine or folate cycles by conversion of homocysteine or *via* the Xc− exchanger and uptake and reduction of cystine. Note that conversion of homocysteine to cysteine depletes methionine and depresses availability of one-carbon groups for the methionine and folate cycles. Adapted from ref. 105 [https://doi.org/10.3390/antiox6030062] under the terms of a CC BY 4.0 license [https://creativecommons.org/licenses/by/4.0/].

with mono- and di-unsaturated fatty acids. The phosphatidyl ethanolamine methyl transferase (PEMT) pathway generates PC containing PUFA,[103] especially DHA.[104] Thus, when VitE is inadequate and lipid peroxidation increases causing PC-DHA depletion,[49] the mechanism to replace the peroxidized DHA-PC is the PEMT pathway, which is *S*-adenosyl methionine (SAM)-dependent. Unfortunately, the PEMT pathway further depletes methyl donors (*e.g.* SAM), which are in increasingly limited supply in the E− embryos, as we showed using a longitudinal, metabolomics approach.[28] Further protection is supplied by GSH and GPx4, which also deplete both GSH and NADPH. Moreover, GSH depletion dysregulates methionine and folate cycles by depleting cysteine, which is required for GSH synthesis (Figure 17.2). Our findings show E− embryos have increases in lipid peroxidation and NTD prevalence with both depleted choline levels and impaired methylation status, as well as impaired energy metabolism.[43] These changes can be observed by measuring metabolite patterns between 24 and 48 hours post-fertilization (hpf, Table 17.2). Further studies are needed to evaluate the metabolite flux and identify the key molecular pathways affected during VitE deficiency in embryogenesis.

17.6 Conclusion

Our data in zebrafish embryos show that insufficient VitE leads to increased DHA peroxidation and PC depletion,[36,46,49,85] which subsequently disrupts PL remodeling and depletes choline, glucose and tricarboxylic acid (TCA) cycle intermediates.[28,46,50] Thus, defects in the VitE-deficient, developing embryo are not solely the result of increased lipid peroxidation, but importantly, are also highly dependent upon secondary consequences, including increased inflammation resulting from DHA depletion, depletion of choline-dependent methyl donors, as well as dysregulation of energy metabolism, all with a common link, namely: inadequate antioxidant protection by VitE. Our findings show E− embryos have increases in NTD prevalence and mortality with both depleted choline levels and impaired methylation status, as well as impaired energy metabolism.[43] Thus, there is a critical need to identify genes involved and molecular mechanism(s) causing NTDs with regard to nutritional inadequacies, especially with regards to folate-insensitive mechanisms. Without knowledge of genetic and regulatory mechanisms involved in NTD during embryogenesis with identification of appropriate biomarkers, progress in the development of therapeutic and nutritional interventions during human pregnancy is limited.

References

1. H. M. Evans and K. S. Bishop, *Science*, 1922, **56**, 650.
2. P. Bailey, *Am. J. Clin. Nutr.*, 1963, **12**, 275.
3. D. P. Muller, *Mol. Nutr. Food Res.*, 2010, **54**, 710.

4. L. Ulatowski, C. Dreussi, N. Noy, J. Barnholtz-Sloan, E. Klein and D. Manor, *Free Radical Biol. Med.*, 2012, **53**, 2318.
5. K. Gohil, B. C. Schock and A. A. Chakraborty, *et al.*, *Free Radical Biol. Med.*, 2003, **35**, 1343.
6. T. Yokota, K. Igarashi and T. Uchihara, *et al.*, *Proc. Natl. Acad. Sci. U. S. A.*, 2001, **98**, 15185.
7. Y. Yoshida, N. Itoh and M. Hayakawa, *et al.*, *J. Nutr. Biochem.*, 2010, **21**, 66.
8. K. Ouahchi, M. Arita and H. Kayden, *et al.*, *Nat. Genet.*, 1995, **9**, 141.
9. I. Di Donato, S. Bianchi and A. Federico, *Neurol. Sci.*, 2010, **31**, 511.
10. M. Schuelke, Ataxia with Vitamin E Deficiency. GeneReviews(R) [serial on the Internet]. 1993 Initial Posting: May 20, 2005; Last Update: October 13, 2016: Available from: https://www.ncbi.nlm.nih.gov/pubmed/20301419.
11. L. Ulatowski, R. Parker, G. Warrier, R. Sultana, D. A. Butterfield and D. Manor, *Neuroscience*, 2014, **260**, 120.
12. P. Marcos, J. Gonzalez-Fuentes and L. Castro-Vazquez, *et al.*, *J. Anat.*, 2018, **232**, 699.
13. P. M. Farrell, J. G. Bieri, J. F. Fratantoni, R. E. Wood and P. A. di Sant'Agnese, *J. Clin. Invest.*, 1977, **60**, 233.
14. M. G. Traber, *Adv. Nutr.*, 2014, **5**, 503.
15. L. Ammon Avalos, C. Galindo and D. K. Li, *Birth Defects Res., Part A*, 2012, **94**, 417.
16. A. A. Shamim, K. Schulze and R. D. Merrill, *et al.*, *Am. J. Clin. Nutr.*, 2015, **101**, 294.
17. Food and Nutrition Board and Institute of Medicine, *Dietary Reference Intakes for Vitamin C, Vitamin E, Selenium, and Carotenoids*, National Academy Press, Washington, 2000.
18. S. Peter, A. Friedel, F. F. Roos, A. Wyss, M. Eggersdorfer, K. Hoffmann and P. Weber, *Int. J. Vitam. Nutr. Res.*, 2015, **85**, 261–281.
19. J. E. Maras, O. I. Bermudez, N. Qiao, P. J. Bakun, E. L. Boody-Alter and K. L. Tucker, *J. Am. Diet. Assoc.*, 2004, **104**, 567.
20. G. W. Burton, A. Joyce and K. U. Ingold, *Lancet*, 1982, **2**, 327.
21. R. Zanetti and A. Catala, *Prostaglandins, Leukotrienes Essent. Fatty Acids*, 2000, **62**, 379.
22. B. A. Wagner, G. R. Buettner and C. P. Burns, *Biochemistry*, 1994, **33**, 4449.
23. G. W. Burton and K. U. Ingold, *J. Am. Chem. Soc.*, 1981, **103**, 6472.
24. G. R. Buettner, *Arch. Biochem. Biophys.*, 1993, **300**, 535.
25. J. M. Galano, Y. Y. Lee, T. Durand and J. C. Lee, *Free Radical Res.*, 2015, **49**, 583.
26. N. Noguchi, H. Yamashita, J. Hamahara, A. Nakamura, H. Kuhn and E. Niki, *Biol. Chem.*, 2002, **383**, 619.
27. F. Galli, A. Azzi and M. Birringer, *et al.*, *Free Radical Biol. Med.*, 2017, **102**, 16.

28. M. McDougall, J. Choi and H. K. Kim, *et al.*, *Free Radical Biol. Med.*, 2017, **104**, 324.
29. V. Tropepe and H. L. Sive, *Genes, Brain Behav.*, 2003, **2**, 268.
30. D. A. Gross, E. L. Snapp and D. L. Silver, *PLoS One*, 2010, **5**, e10796.
31. B. Kadereit, P. Kumar and W. J. Wang, *et al.*, *Proc. Natl. Acad. Sci. U. S. A.*, 2008, **105**, 94.
32. O. Monroig, J. Rotllant, E. Sanchez, J. M. Cerda-Reverter and D. R. Tocher, *Biochim. Biophys. Acta*, 2009, **1791**, 1093.
33. M. Buzzi, R. J. Henderson and J. R. Sargent, *Comp. Biochem. Physiol., Part B: Biochem. Mol. Biol.*, 1997, **116**, 263.
34. N. Hastings, M. Agaba, D. R. Tocher, M. J. Leaver, J. R. Dick, J. R. Sargent and A. J. Teale, *Proc. Natl. Acad. Sci. U. S. A.*, 2001, **98**, 14304.
35. M. Agaba, D. R. Tocher, C. A. Dickson, J. R. Dick and A. J. Teale, *Mar. Biotechnol.*, 2004, **6**, 251.
36. K. M. Lebold, D. B. Jump and G. W. Miller, *et al.*, *J. Nutr.*, 2011, **141**, 2113.
37. M. Holtta-Vuori, V. T. Salo, L. Nyberg, C. Brackmann, A. Enejder, P. Panula and E. Ikonen, *Biochem. J.*, 2010, **429**, 235.
38. P. J. Babin and J. M. Vernier, *J. Lipid Res.*, 1989, **30**, 467.
39. J. P. Otis, E. M. Zeituni and J. H. Thierer, *et al.*, *Dis. Models Mech.*, 2015, **8**, 295.
40. I. Avraham-Davidi, Y. Ely and V. N. Pham, *et al.*, *Nat. Med.*, 2012, **18**, 967.
41. M. S. Lee, J. R. Bonner and D. J. Bernard, *et al.*, *BMC Dev. Biol.*, 2012, **12**, 12.
42. Committee on Animal Nutrition Board on Agriculture National Research Council, *Nutrient Requirements of Fish*, National Academies Press, Washington, D.C., 1993.
43. G. W. Miller, E. M. Labut, K. M. Lebold, A. Floeter, R. L. Tanguay and M. G. Traber, *J. Nutr. Biochem.*, 2012, **23**, 478.
44. E. Jauniaux, T. Cindrova-Davies, J. Johns, C. Dunster, J. Hempstock, F. J. Kelly and G. J. Burton, *J. Clin. Endocrinol. Metab.*, 2004, **89**, 1452.
45. X. Ibarra-Soria, W. Jawaid and B. Pijuan-Sala, *et al.*, *Nat. Cell Biol.*, 2018, **20**, 127.
46. M. McDougall, J. Choi and H. K. Kim, *et al.*, *Data Brief*, 2017, **11**, 432.
47. K. M. Lebold, J. S. Kirkwood and A. W. Taylor, *et al.*, *Redox Biol.*, 2013, **2**, 105.
48. I. Motorykin, M. G. Traber, R. L. Tanguay and C. S. Maier, *J. Proteome Res.*, 2014, **13**, 1647.
49. M. Q. McDougall, J. Choi, J. F. Stevens, L. Truong, R. L. Tanguay and M. G. Traber, *Redox Biol.*, 2016, **8**, 165.
50. M. McDougall, J. Choi, L. Truong, R. Tanguay and M. G. Traber, *Free Radical Biol. Med.*, 2017, **110**, 250.
51. D. W. Cheng, L. F. Chang and T. A. Bairnson, *Anat. Rec.*, 1957, **129**, 167.
52. K. Verma and D. Wei King, *Acta Anat.*, 1967, **67**, 623.
53. A. J. Copp and N. D. Greene, *J. Pathol.*, 2010, **220**, 217.

54. A. M. Molloy, F. Pangilinan and L. C. Brody, *Annu. Rev. Nutr.*, 2017, **37**, 269.
55. I. Zaganjor, A. Sekkarie and B. L. Tsang, *et al.*, *PLoS One*, 2016, **11**, e0151586.
56. A. Ornoy, E. A. Reece, G. Pavlinkova, C. Kappen and R. K. Miller, *Birth Defects Res., Part C*, 2015, **105**, 53.
57. A. L. Chandler, C. A. Hobbs and B. S. Mosley, *et al.*, *Birth Defects Res., Part A*, 2012, **94**, 864.
58. C. B. Kimmel, W. W. Ballard, S. R. Kimmel, B. Ullmann and T. F. Schilling, *Dev. Dyn.*, 1995, **203**, 253.
59. Growth including reproduction and morphological development, in *Biological Handbooks*, ed. P. L. Altman and D. S. Dittmer, Committee on Biological Handbooks, Federation of American Societies for Experimental Biology, Washington, D.C., 1962, vol. xiv, p. 608.
60. R. O'Rahilly, *Eur. J. Obstet. Gynecol. Reprod. Biol.*, 1979, **9**, 273.
61. A. J. Wilcox, D. D. Baird and C. R. Weinberg, *N. Engl. J. Med.*, 1999, **340**, 1796.
62. S. F. Gilbert, *Developmental Biology*, Sinauer Associates, Inc., Sunderland, MA, 9th edn, 2010.
63. K. S. Au, A. Ashley-Koch and H. Northrup, *Dev. Disabil. Res. Rev.*, 2010, **16**, 6.
64. H. B. Heseker, J. B. Mason, J. Selhub, I. H. Rosenberg and P. F. Jacques, *Br. J. Nutr.*, 2009, **102**, 173.
65. A. J. Agopian, S. C. Tinker, P. J. Lupo, M. A. Canfield, L. E. Mitchell and National Birth Defects Prevention Study, *Birth Defects Res., Part A*, 2013, **97**, 42.
66. F. Pangilinan, A. M. Molloy and J. L. Mills, *et al.*, *BMC Med. Genet.*, 2012, **13**, 62.
67. S. H. Zeisel, *Int. J. Women's Health*, 2013, **5**, 193.
68. S. H. Zeisel, *J. Pediatr.*, 2006, **149**, S131.
69. X. Gao, R. H. Finnell, H. Wang and Y. Zheng, *Birth Defects Res.*, 2018, **110**, 982–993.
70. E. Kim, C. M. Cham, M. M. Veniant, P. Ambroziak and S. G. Young, *J. Clin. Invest.*, 1998, **101**, 1468.
71. G. E. Homanics, N. Maeda, M. G. Traber, H. J. Kayden, D. B. Dehart and K. K. Sulik, *Teratology*, 1995, **51**, 1.
72. M. Raabe, L. M. Flynn, C. H. Zlot, J. S. Wong, M. M. Veniant, R. L. Hamilton and S. G. Young, *Proc. Natl. Acad. Sci. U. S. A.*, 1998, **95**, 8686.
73. N. Santander, C. Lizama and M. J. Parga, *et al.*, *Sci. Rep.*, 2017, **7**, 5182.
74. C. Wadsack, A. Hammer and S. Levak-Frank, *et al.*, *Placenta*, 2003, **24**, 131.
75. M. Velasco, C. Alexander, J. King, Y. Zhao, J. Garcia and A. Rodriguez, *Fertil. Steril.*, 2006, **85**, 1391.
76. A. Kolmakova, J. Wang, R. Brogan, C. Chaffin and A. Rodriguez, *Endocrinology*, 2010, **151**, 5519.

77. M. Lagarde, N. Bernoud, N. Brossard, D. Lemaitre-Delaunay, F. Thies, M. Croset and J. Lecerf, *J. Mol. Neurosci.*, 2001, **16**, 201, discussion 215.
78. L. N. Nguyen, D. Ma and G. Shui, *et al.*, *Nature*, 2014, **509**, 503.
79. V. Alakbarzade, A. Hameed and D. Q. Quek, *et al.*, *Nat. Genet.*, 2015, **47**, 814.
80. A. Guemez-Gamboa, L. N. Nguyen and H. Yang, *et al.*, *Nat. Genet.*, 2015, **47**, 809.
81. A. Ben-Zvi, B. Lacoste, E. Kur, B. J. Andreone, Y. Mayshar, H. Yan and C. Gu, *Nature*, 2014, **509**, 507.
82. E. Cadenas, L. Packer and M. G. Traber, *Arch. Biochem. Biophys.*, 2016, **595**, 94.
83. F. Ursini, M. Maiorino, M. Valente, L. Ferri and C. Gregolin, *Biochim. Biophys. Acta*, 1982, **710**, 197.
84. B. A. Carlson, R. Tobe and E. Yefremova, *et al.*, *Redox Biol.*, 2016, **9**, 22.
85. J. Choi, S. W. Leonard, K. Kasper, M. McDougall, J. F. Stevens, R. L. Tanguay and M. G. Traber, *J. Lipid Res.*, 2015, **56**, 1182.
86. M. McDougall, J. Choi, K. Magnusson, L. Truong, R. Tanguay and M. G. Traber, *Free Radical Biol. Med.*, 2017, **112**, 308.
87. H. Shindou, D. Hishikawa, T. Harayama, M. Eto and T. Shimizu, *J. Biochem.*, 2013, **154**, 21.
88. J. Fan, J. Ye, J. J. Kamphorst, T. Shlomi, C. B. Thompson and J. D. Rabinowitz, *Nature*, 2014, **510**, 298.
89. K. Hensley, Y. Kotake and H. Sang, *et al.*, *Carcinogenesis*, 2000, **21**, 983.
90. M. Maiorino, M. Conrad and F. Ursini, *Antioxid. Redox Signaling*, 2017, **29**, 61.
91. S. J. Dixon, K. M. Lemberg and M. R. Lamprecht, *et al.*, *Cell*, 2012, **149**, 1060.
92. H. Feng and B. R. Stockwell, *PLoS Biol.*, 2018, **16**, e2006203.
93. A. Jelinek, L. Heyder and M. Daude, *et al.*, *Free Radical Biol. Med.*, 2018, **117**, 45.
94. S. Doll and M. Conrad, *IUBMB Life*, 2017, **69**, 423.
95. M. Leist, B. Raab, S. Maurer, U. Rosick and R. Brigelius-Flohe, *Free Radical Biol. Med.*, 1996, **21**, 297.
96. D. Much, S. Brunner and C. Vollhardt, *et al.*, *Eur. J. Clin. Nutr.*, 2013, **67**, 282.
97. W. Schlormann, R. Kramer, A. Lochner, C. Rohrer, E. Schleussner, G. Jahreis and K. Kuhnt, *Food Nutr. Res.*, 2015, **59**, 29348.
98. U. Enke, A. Jaudszus, E. Schleussner, L. Seyfarth, G. Jahreis and K. Kuhnt, *Lipids Health Dis.*, 2011, **10**, 247.
99. Z. Ashkavand, C. O'Flanagan, M. Hennig, X. Du, S. D. Hursting and S. A. Krupenko, *Mol. Cancer Res.*, 2017, **15**, 189.
100. O. Zilka, R. Shah, B. Li, J. P. Friedmann Angeli, M. Griesser, M. Conrad and D. A. Pratt, *ACS Cent. Sci.*, 2017, **3**, 232.
101. V. E. Kagan, G. Mao and F. Qu, *et al.*, *Nat. Chem. Biol.*, 2017, **13**, 81.
102. L. Xu, T. A. Davis and N. A. Porter, *J. Am. Chem. Soc.*, 2009, **131**, 13037.

103. C. J. DeLong, Y. J. Shen, M. J. Thomas and Z. Cui, *J. Biol. Chem.*, 1999, **274**, 29683.
104. C. J. Pynn, N. G. Henderson, H. Clark, G. Koster, W. Bernhard and A. D. Postle, *J. Lipid Res.*, 2011, **52**, 399.
105. G. J. McBean, *Antioxidants*, 2017, **6**, 62.

CHAPTER 18

Interference Effect of Vitamin E on Vitamin K Metabolism

SAIKO IKEDA*

Department of Nutritional Sciences, Nagoya University of Arts and Sciences, 57 Takenoyama, Iwasaki, Nisshin, Aichi 470-0196, Japan
*E-mail: saiko@nuas.ac.jp

18.1 Introduction

Vitamin E and K have many similarities. First, vitamin E and K are both fat soluble and have a phytyl chain (tocopherol and PK) or a prenyl chain (tocotrienol and menaquinone MK) in the structure (Figure 18.1). Second, transport of vitamin E and K in the blood is dependent on the lipoprotein metabolism since vitamin E and K do not have specific transport proteins in the blood. Third, vitamin E and K are both degraded by shortening side chains by ω-hydroxylation followed by β-oxidation and excreted in urine as a conjugate. These similarities make it possible to imagine interactions of vitamin E and K metabolism.[1] Actually, there are some reports that vitamin K deficiency bleeding and a decrease in blood clotting activity due to overdose of vitamin E occurs.[2–4] However, there are still many uncertainties about the conditions under which they occur and the details of the interactions between vitamin E and K. Therefore, to clarify the interaction between vitamin E and K, rats were fed diets with different contents of vitamin E and K, and the concentrations in various tissues were measured. We also examined

Food Chemistry, Function and Analysis No.11
Vitamin E: Chemistry and Nutritional Benefits
Edited by Etsuo Niki

Published by the Royal Society of Chemistry, www.rsc.org

blood coagulation and bone metabolism, markers of vitamin K status, in rats with reduced extrahepatic PK concentrations due to excess intake of α-tocopherol.

18.2 Effect of α-Tocopherol Intake on Vitamin K Concentration

18.2.1 Effect of α-Tocopherol in Rats Fed a Diet Containing Phylloquinone with α-Tocopherol

Among vitamin K isoforms, the one we get most of from dietary sources is PK, which is found in green vegetables. Therefore, the effect of α-tocopherol intake on vitamin K concentration in rats fed PK was examined. Rats were fed a diet containing 0.75 mg PK per kg diet with 0, 10, 50, or 500 mg *RRR*-α-tocopherol per kg diet for six weeks. The contents of 0.75 mg PK per kg diet and 50 mg α-tocopherol per kg diet are equivalent to the vitamin K and E content, respectively, of the AIN93 rodent diet.[5] α-Tocopherol intake lowered PK concentration in extrahepatic tissues such as the kidney, lung, heart, muscle, testis, bone, and brain but not in the serum and liver (Figure 18.2).[6,7]

Interestingly, the PK concentration in the extrahepatic tissues was lower in rats fed a diet containing 50 mg α-tocopherol per kg, a basic content for rodent diets, than in rats fed a vitamin E-free diet. This suggests that

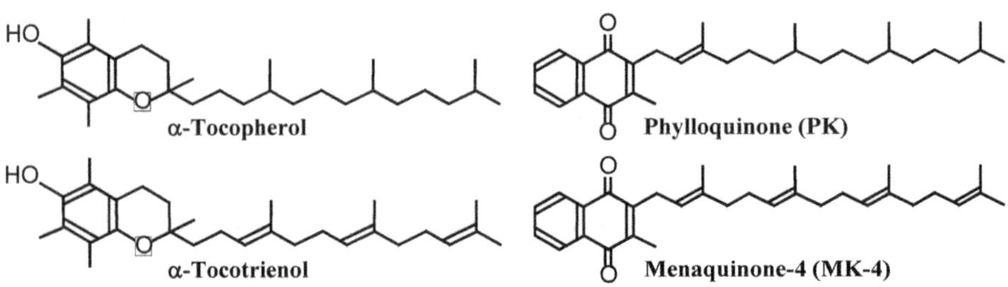

Figure 18.1 Chemical structures of vitamin E and K.

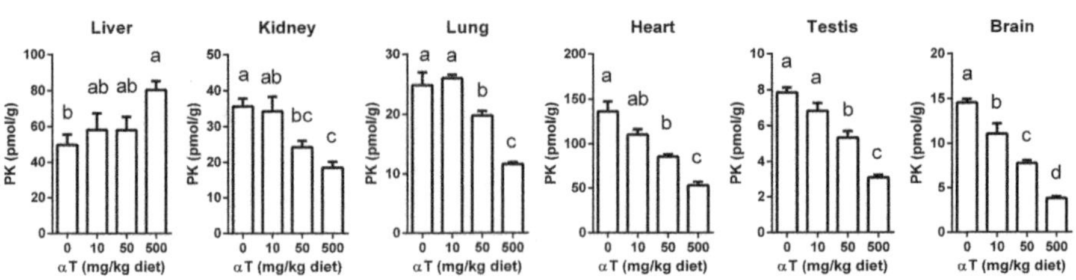

Figure 18.2 Phylloquinone (PK) concentration in tissues of rats fed a diet containing PK with α-tocopherol (αT) for six weeks. Values are mean + SEM, $n = 6$. Means not sharing a letter differ at $P < 0.05$.

vitamin E deficiency elevated extrahepatic PK concentration. Because PK is converted to MK-4, another vitamin K isoform, and used in the body,[8,9] the MK-4 concentration of the rats was also measured. α-Tocopherol intake had little effect on MK-4 concentration in the extrahepatic tissues of rats fed a diet containing PK as vitamin K. These data suggest that α-tocopherol intake lowered PK but not MK-4 concentration in the tissues except for the liver of rats fed PK.[6,7]

18.2.2 Effect of α-Tocopherol in Rats Administered Phylloquinone with α-Tocopherol

To clarify the effect of α-tocopherol on transport of ingested PK to extrahepatic tissues, PK concentration in the tissues of rats administered PK with α-tocopherol was examined. Vitamin E- and K-deficient rats were orally administered 0.2 mg PK with 0, 1, or 10 mg *RRR*-α-tocopherol. α-Tocopherol administration lowered PK concentration in the kidney, adrenal gland, lung, testis, and brain but not in the serum and liver at six hours after oral administration of PK with or without α-tocopherol (Figure 18.3).[6]

However, α-tocopherol administration had no effect on MK-4 concentration in the serum and tissues of the rats. Therefore, it was revealed that α-tocopherol suppressed an increase in PK concentration due to ingestion of PK, indicating the possibility of interference with the transfer of PK to the tissue.[6]

18.2.3 Effect of α-Tocopherol in Rats Fed a Diet Containing Menaquinone-4 with α-Tocopherol

MK-4 is one of the major vitamin Ks responsible for the physiological action of vitamin K in the body. Although dietary intake of MK-4 is lower than that of other vitamin Ks,[10] MK-4 is also used as a medicine for osteoporosis in Japan. Therefore, the effect of α-tocopherol intake on MK-4 concentration in rats fed MK-4 as vitamin K was examined. Rats were fed a diet containing 0.75 mg MK-4 per kg diet with 0, 10, 50, or 500 mg *RRR*-α-tocopherol per kg diet for six weeks. α-Tocopherol intake slightly increased the MK-4 concentration

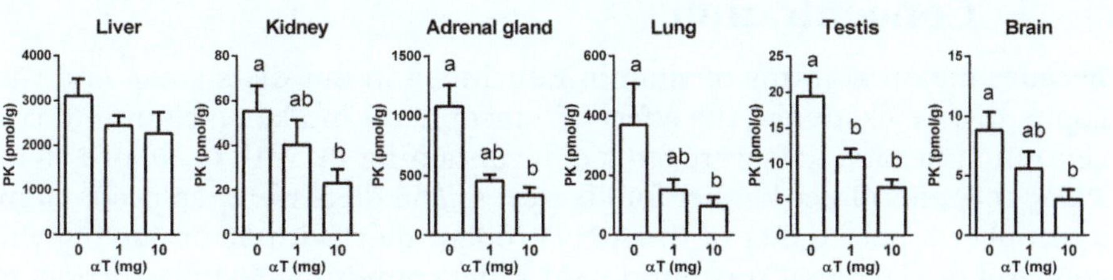

Figure 18.3 Phylloquinone (PK) concentration in tissues of rats at 6 h after PK oral administration with or without α-tocopherol (αT). Values are mean + SEM, $n = 6$. Means not sharing a letter differ at $P < 0.05$.

in the liver but did not affect MK-4 concentration in other tissues and serum. Vitamin E- and K-deficient rats were administered 0.2 mg MK-4 with 0, 1 or 10 mg α-tocopherol. Oral administration of MK-4 elevated MK-4 concentration in the serum, liver, adrenal gland, heart, and lung but not in the kidney, testis, and brain at 6 hours after the administration. α-Tocopherol did not affect the MK-4 concentration in the serum, liver, adrenal gland, heart, and lung. These data suggest that α-tocopherol did not interfere with the metabolism of ingested MK-4, unlike PK.[6]

18.2.4 Effect of α-Tocopherol in Rats Administered Menaquinone-7 with α-Tocopherol

Menaquinone-7 (MK-7) is another vitamin K isoform and is often contained in fermented foods. Of the vitamin K ingested as part of a typical Japanese diet, the most common is PK and the second most common is MK-7.[10] Therefore, the effect of α-tocopherol on MK-7 concentration in tissues of rats administered MK-7 with α-tocopherol was examined. Vitamin E- and K-deficient rats were administered 0.288 mg MK-7 with 0 or 10 mg *RRR*-α-tocopherol. A lot of MK-7 accumulated in the liver at 24 hours after oral administration, and MK-7 concentration in the kidney increased at 6 and 24 hours. α-Tocopherol did not affect the MK-7 concentration in the serum and tissues. Because of conversion of MK-7 to MK-4 in the body, the MK-4 concentration in the tissues of the rats was also measured. MK-7 administration elevated MK-4 concentration in the serum, kidney, heart, lung, testis, and brain at 24 hours after oral administration; however, α-tocopherol did not affect the MK-4 concentration. Thus, α-tocopherol did not interfere with the metabolism of MK-7, unlike PK.[23]

These results reveal that α-tocopherol lowered the concentration of PK among vitamin K, and did not affect the concentrations of MK-4 and MK-7. α-Tocopherol might interfere somewhere in PK metabolism at ingestion of PK.

18.3 Effect of γ-Tocopherol Intake on Vitamin K Concentration

Because major isoforms of vitamin E included in our diets are α- and γ-tocopherols, we examined the effect of γ-tocopherol intake on vitamin K concentration in rats. Rats were fed a diet containing PK with 0, 50, or 500 mg *RRR*-γ-tocopherol per kg diet for six weeks. The diets were replaced: 50 mg α-tocopherol per kg diet of the AIN93 rodent diet[5] with 50 or 500 mg γ-tocopherol per kg diet. Consumption of a diet containing 50 mg γ-tocopherol per kg diet did not lower PK concentration in the liver, kidney, lung, heart, testis, and brain as compared to that of the rats fed a vitamin E-free diet (Figure 18.4).[11]

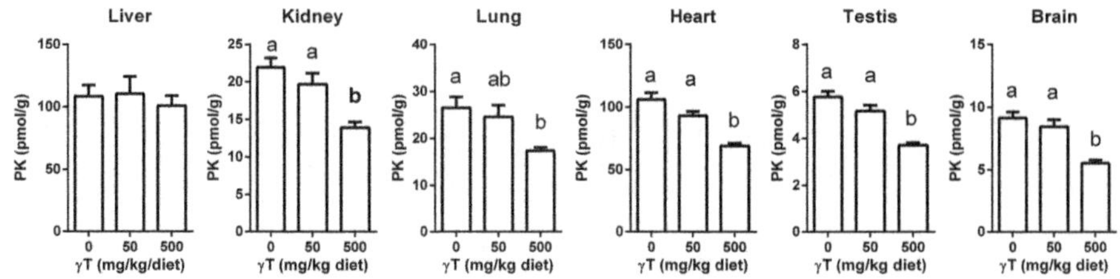

Figure 18.4 Phylloquinone (PK) concentration in tissues of rats fed a diet containing PK with γ-tocopherol (γT) for six weeks. Values are mean + SEM, $n = 6$. Means not sharing a letter differ at $P < 0.05$.

Consumption of a γ-tocopherol-rich diet containing 500 mg γ-tocopherol per kg diet significantly lowered PK concentration in the kidney, lung, heart, testis, and brain as compared with that in rats fed a vitamin E-free diet. However, consumption of the γ-tocopherol-rich diet did not lower MK-4 concentration in all tissues we measured in those rats.[11] In another study, consumption of a diet containing 100 mg α-tocopherol per kg diet but not 100 mg γ-tocopherol per kg diet lowered PK concentration in the kidney, heart, muscle, and brain.[6] Furthermore, in our preliminary study, consumption of a diet containing 140 mg of tocotrienol per kg diet for eight weeks resulted in a slight decrease in kidney PK concentration, but no decrease in MK-4 concentration. Thus, the PK-lowering effects of γ-tocopherol and tocotrienol were weaker than that of α-tocopherol.

18.4 Effect of Vitamin K Intake on Vitamin E Concentration

Because the most common dietary source of vitamin K is PK, the effect of PK intake on tocopherol and tocotrienol concentrations was examined. To examine the effect of PK intake on α-tocopherol concentration, rats were fed a diet containing 50 mg *RRR*-α-tocopherol per kg diet with or without 0.75 mg PK per kg diet for four months. PK intake had no effect on α-tocopherol concentration in the liver, kidney, lung, heart, and brain (Figure 18.5A).

In another study, PK intake did not affect α-tocopherol concentration in the liver and bone of rats fed a diet containing α-tocopherol with or without PK.[7] To clarify the effect of PK intake on γ-tocopherol concentration, rats were fed a diet containing 70 mg *RRR*-γ-tocopherol per kg diet with or without 1 mg PK per kg diet for two months. PK intake did not affect γ-tocopherol concentration in the liver and kidney (Figure 18.5B). Moreover, the effect of PK intake on tocotrienol concentration was examined. Rats were fed a diet containing 140 mg tocotrienol mixture per kg diet with or without 1 mg PK per kg diet for two months. The tocotrienol mixture contained 39% α-tocotrienol. PK intake did not affect α-tocotrienol concentration in the liver and

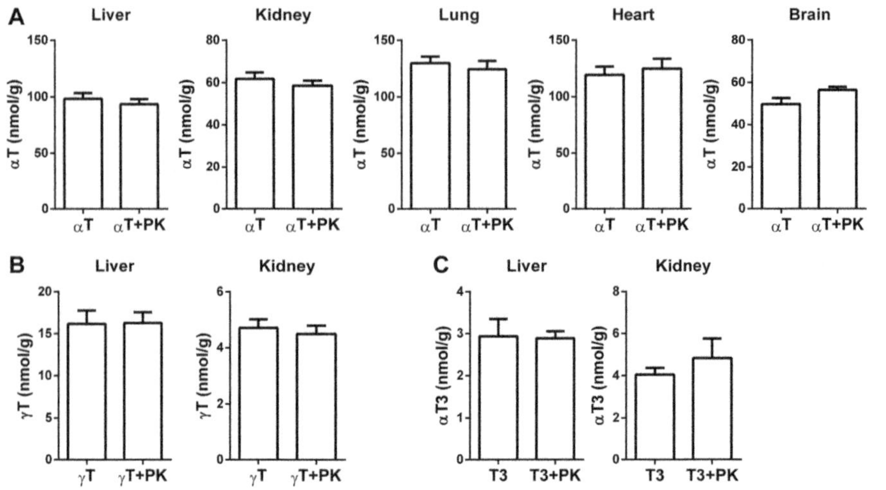

Figure 18.5 Effect of phylloquinone (PK) intake on vitamin E concentration in tissues. (A) α-tocopherol (αT) concentration in the liver, kidney, lung, heart, and brain of rats fed a diet containing 50 mg αT with or without 0.75 mg PK per kg diet for four months; (B) γ-tocopherol (γT) concentration in the liver and kidney of rats fed a diet containing 70 mg γT per kg with or without 1 mg PK per kg for two months; (C) α-tocotrienol (αT3) concentration in the liver and kidney of rats fed a diet containing 55 mg αT3 per kg with or without 1 mg PK per kg for two months. Values are mean + SEM, $n = 6$.

kidney (Figure 18.5C). Thus, PK intake seemed to have little effect on vitamin E concentration in tissues. Dietary intake of vitamin K is about 1 : 50 to 1 : 100 of the total vitamin E intake, and the vitamin K concentration in the body is about 1 : 1000 of the vitamin E concentration. Therefore, vitamin K is unlikely to interfere with the metabolism of vitamin E due to its low concentration.

18.5 Effect of Excess Intake of α-Tocopherol on Physiological Activity of Vitamin K

Because α-tocopherol intake lowered extrahepatic PK concentration, we examined the effect of α-tocopherol intake on the blood clotting activity of rats. Consumption of a vitamin K-free diet for three months lowered blood clotting activity such as prothrombin time, activated partial thromboplastin time, thrombin time, and hepaplastin test. However, consumption of an α-tocopherol-rich diet (500 mg *RRR*-α-tocopherol per kg diet) did not change these markers. The concentrations of PK and MK-4 in the liver where blood coagulation factors are synthesized increased due to consumption of the α-tocopherol-rich diet. Therefore, it was unlikely that reduction of PK concentration by excess intake of α-tocopherol lowered blood clotting activity.[7]

Vitamin K is also necessary for bone health by γ-glutamyl carboxylation of some bone matrix proteins including osteocalcin and matrix Gla protein. Vitamin K deficiency is known to correlate with fracture risk of the femur.[12,13] Moreover, Fujita *et al.* recently reported that excess intake of α-tocopherol reduced bone mass in rats and mice.[14] Therefore, we examined the effect of α-tocopherol intake on vitamin K concentration in bone and markers of bone metabolism in rats. The PK concentration in the bone of rats fed an α-tocopherol-rich diet (500 mg *RRR*-α-tocopherol per kg diet) for three months was less than 60% that of rats fed a vitamin E-free diet. However, consumption of the α-tocopherol-rich diet did not affect uncarboxylated-osteocalcin concentration in the serum or the ratio of bone volume to total volume of the femur. The consumption of the α-tocopherol-rich diet did not change the mRNA levels of genes related to bone resorption such as tartrate-resistant acid phosphatase, receptor activator of NF-κB ligand, cathepsin K, tumor necrosis factor-α, and c-Fos, and genes related to bone formation such as alkaline phosphatase, Runt-related transcription factor 2, activating transcription factor 4, osteocalcin, osteopontin, and osteonectin in the femur. These data suggested that reduction of PK concentration in bone due to excess α-tocopherol had little effect on bone metabolism.[7] Excess intake of α-tocopherol lowered the PK concentration in bone while MK-4 concentration did not decrease, suggesting an importance of MK-4 for bone metabolism.

18.6 Conclusion

The results of our study are summarized as follows. (i) α-Tocopherol intake lowers PK concentration in tissues other than the livers of rats fed PK but has little effect on concentration of MK-4 converted from PK in the body. (ii) α-Tocopherol intake does not affect the MK-4 concentration in rats fed MK-4 or MK-7 and MK-4 concentrations in rats fed MK-7. (iii) The PK-lowering effect of γ-tocopherol is weaker than that of α-tocopherol. (iv) Feeding excess α-tocopherol has little effect on blood coagulation and bone metabolism in rats.

The mechanism by which vitamin E intake lowers PK concentration in tissues except for the liver is unknown. Vitamin E is absorbed from the small intestine and transported to the liver by chylomicrons[15] and further carried to various tissues by VLDL.[16] In contrast, because PK is also carried by chylomicrons and VLDL in blood, most ingested PK is thought to be transported to various tissues *via* the liver.[17] It is also known that some of the vitamin E and K absorbed in the small intestine is absorbed by NPC1L1, the same transporter as the cholesterol transporter.[18] Therefore, considering the results of oral administration (Figure 18.3), large amounts of vitamin E may interfere with the transport of vitamin K when it is transported to each tissue from absorption in the small intestine. In addition, some ingested PK is converted to menadione (MD), a high-polarity form without a side chain, in the small intestine. MD is transported to each tissue without passing through the liver and converted into MK-4 by prenylation in each tissue.[19] Decrease of PK concentration in extrahepatic tissues due to vitamin E intake had little effect on the concentration of MK-4 synthesized from PK.[6] This indicates that much of the MK-4 in tissues is dependent on the amount of MD carried from the small intestine, rather than the amount of PK carried from the liver. In order to clarify this point, it is necessary to measure the MD concentration in future research.

In the present study, markers of blood coagulation and bone metabolism were measured as an indication of vitamin K status, but these did not change due to excess vitamin E intake. Other researchers have also reported that excess vitamin E had little effect on bone metabolism in older animals, ovariectomized animals, or animals on a high fat diet.[20–22] It seems that there is a low possibility that dietary vitamin E affects blood coagulation and bone metabolism.

Acknowledgements

This study was supported in part by JSPS KAKENHI (grant number 17H01970), Japan Society of Nutrition and Food Science, and Research Committee of Fat-Soluble Vitamins. We acknowledge the generous donation of vitamin E isoforms from Mitsubishi-Chemical Foods Corporation and menaquinone-3 from Eisai Co., Ltd.

References

1. M. G. Traber, *Nutr. Rev.*, 2008, **66**, 624.
2. M. K. Horwitt, *Am. J. Clin. Nutr.*, 1976, **29**, 569.
3. G. H. Wheldon, A. Bhatt, P. Keller and H. Hummler, *Int. J. Vitam. Nutr. Res.*, 1983, **53**, 287.
4. K. M. Abdo, G. Rao and C. A. Montgomery, *Food Chem. Toxicol.*, 1986, **24**, 1043.
5. P. G. Reeves, F. H. Nielsen and G. C. Fahey Jr, *J. Nutr.*, 1993, **123**, 1939.
6. F. Hanzawa, E. Sakuma, S. Nomura, T. Uchida, H. Oda and S. Ikeda, *Mol. Nutr. Food Res.*, 2014, **58**, 1601.
7. S. Ikeda, S. Nomura, F. Hanzawa, S. Takahashi, H. Oda, Y. Fujiwara and T. Uchida, *J. Nutr. Sci. Vitaminol.*, 2018, **64**, 243.
8. R. Yamamoto, M. Komai, K. Kojima, Y. Furukawa and S. Kimura, *J. Nutr. Sci. Vitaminol.*, 1997, **43**, 133.
9. T. Okano, Y. Shimomura, M. Yamane, Y. Suhara, M. Kamao, M. Sugiura and K. Nakagawa, *J. Biol. Chem.*, 2008, **283**, 11270.
10. M. Kamao, Y. Suhara and N. Tsugawa, *et al.*, *J. Nutr. Sci. Vitaminol.*, 2007, **53**, 464.
11. F. Hanzawa, S. Nomura, E. Sakuma, T. Uchida and S. Ikeda, *J. Nutr.*, 2013, **143**, 1067.

12. S. L. Booth, K. L. Tucker and H. Chen, *et al.*, *Am. J. Clin. Nutr.*, 2000, **71**, 1201.
13. N. Tsugawa, M. Shiraki, Y. Suhara, M. Kamao, R. Ozaki, K. Tanaka and T. Okano, *J. Bone Miner. Metab.*, 2008, **26**, 79.
14. K. Fujita, M. Iwasaki and H. Ochi, *et al.*, *Nat. Med.*, 2012, **18**, 589.
15. C. Abe, S. Ikeda, T. Uchida, K. Yamashita and T. Ichikawa, *J. Nutr.*, 2007, **137**, 345.
16. M. G. Traber and H. Arai, *Annu. Rev. Nutr.*, 1999, **19**, 343.
17. S. Lamon-Fava, J. A. Sadowski, K. W. Davidson, M. E. O'Brien, J. R. McNamara and E. J. Schaefer, *Am. J. Clin. Nutr.*, 1998, **67**, 1226.
18. Y. Yamanashi, T. Takada, R. Kurauchi, Y. Tanaka, T. Komine and H. Suzuki, *J. Atheroscler. Thromb.*, 2017, **24**, 347.
19. Y. Hirota, N. Tsugawa and K. Nakagawa, *et al.*, *J. Biol. Chem.*, 2013, **288**, 33071.
20. U. T. Iwaniec, R. T. Turner and B. J. Smith, *et al.*, *Free Radical Biol. Med.*, 2013, **65**, 1209.
21. S. Kasai, A. Ito, K. Shindo, T. Toyoshi and M. Bando, *PLoS One*, 2015, **10**, e0132059.
22. H. Ikegami, R. Kawawa, I. Ichi, T. Ishikawa, T. Koike, Y. Aoki and Y. Fujiwara, *J. Nutr.*, 2017, **147**, 1932.
23. S. Ikeda, F. Hanzawa, S. Takahashi, N. Suzuki, T. Sano, H. Oda and T. Uchida, *J. Nutr. Sci. Vitaminol.*, 2018, **64**, 391.

Subject Index